BASIC MATHEMATICS

BASIC MATHEMATICS

O. & J. PERRY

M

MACMILLAN EDUCATION

First published 1982

Published by
MACMILLAN EDUCATION LIMITED
Houndmills Basingstoke Hampshire
RG21 2XS and London
Associated companies in Delhi
Dublin Hong Kong Johannesburg
Lagos Melbourne New York
Singapore and Tokyo.

Printed in Hong Kong

This book is also available under the
title *Mastering Mathematics* published
by Macmillan Press.

CONTENTS

CONTENTS

CONTENTS

CONTENTS

NOTATION

$\{a, b, c, \ldots\}$	the set of $a, b, c, \ldots$
$:$	such that
$\in$	is an element of
$\notin$	is not an element of
$n(\ )$	the number of elements in the set of
ϕ	the empty (null) set
$\mathscr{E}$	the universal set
$\cup$	union
$\cap$	intersection
$\subset$	is a subset of
A'	the complement of the set A
N	the set of natural numbers
Z	the set of integers
R	the set of real numbers
PQ	operation Q followed by operation P
$f: x \rightarrow y$	the function of mapping the set X into the set Y
$f(x)$	the image of x under the function f
f^{-1}	the inverse of the function f
fg	the function f of the function g
—o—o—	open interval on the number line
—●—●—	closed interval on the number line
$\{x: -2 < x < 7\}$	the set of values of x such that . . .
$\Rightarrow$	implies that
$\Leftrightarrow$	implies and is implied by
$=$	is equal to
$\equiv$	is identically equal to
$\approx$	is approximately equal to
$\neq$	is not equal to
$<$	is less than
$\leqslant$	is less than or equal to
$>$	is greater than
$\geqslant$	is greater than or equal to
$\mid\ \mid$	the unsigned part of a signed number, that is the modulus
∞	infinity

PREFACE

In writing this book, we have assumed that you know at least the counting numbers and the addition and multiplication tables up to 9 by 9. The language of mathematics contains a number of specialised words, and these are printed in italics the first time they occur in the text. Some of them are new words, such as logarithm, calculus, cosine, which have been invented by mathematicians in the past; others are ordinary English words such as operation, which have a special meaning in mathematics. While it is not essential to your understanding of the subject, the use of the correct words and symbols makes communication between mathematicians easier, and it is worthwhile trying to learn the language.

The book covers a very wide field and so the introduction to each section is necessarily concise; you should work through and understand the examples in a section of the work before attempting the exercises that follow it. The use of an electronic calculator will save time and enable you to spend longer on studying the basic mathematical ideas and on practising with past examination papers. Remember that the use of calculators is not allowed in all examinations.

We hope you will enjoy working with this book and that it will bring you success.

Joyce Perry

Owen Perry.

CHAPTER 1

INTEGERS

1.1 NUMBERS

Numbers were first used for counting, and the roman numerals, which were based on finger counting, were used in Britain up to the seventeenth century. They were replaced by Hindu–Arabic symbols similar to the familiar 1, 2, 3, 4, 5, 6, 7, 8, 9, 0, used today, and these are called *digits* after the Latin word for 'finger'. In most countries the *denary system* of counting in tens is used, but other systems are considered in Chapter 6, including binary arithmetic.

The four fundamental processes of arithmetic are addition (+), subtraction (–), multiplication (×) and division (÷) and these are called arithmetic *operations*. Even at the end of the seventeenth century multiplication of numbers greater than 5 was considered a very difficult operation, while division could be performed only by professional mathematicians.

1.2 POSITIVE INTEGERS

An integer is a whole number and in mathematics the counting numbers are called *natural numbers* and also *positive integers*. In this book a knowledge of these numbers is assumed, and the addition and multiplication tables up to 9 by 9.

Raising to a power

The operation of multiplication is an extension of addition: it is repeated addition of the same number; 5 × 3 is 5 + 5 + 5 and 27 × 2 is 27 + 27. Repeated multiplication by the same number is the operation *raising to a power*; 2 × 2 × 2 is called 'two to the power three' and written 2^3; 10 × 10 × 10 × 10 is 10^4. This is *index notation*; 10 is called the base and 4 is the index; the theory and manipulation of numbers in index form will be found in Chapter 4.

1.3 PLACE VALUE

For numbers greater than 9 the digits are written in columns, starting with units at the right; each column is a different power of 10, and the counting number represented by a digit depends on which column it occupies. Noughts are inserted in the empty columns between other digits. For example, 25 means two tens and five units; 250 means two hundreds and five tens; 205 means two hundreds and five units.

Table 1.1

Millions 10^6	Hundred thousands 10^5	Ten thousands 10^4	Thousands 10^3	Hundreds 10^2	Tens 10	Units 1
				7	2	6
1	0	2	4	0	6	1
					3	5
				3	5	0
			3	5	0	0

In Table 1.1, 726 means $7 \times 10^2 + 2 \times 10 + 6 \times 1$, or in words seven hundred and twenty-six. 1 024 061 means $1 \times 10^6 + 2 \times 10^4 + 4 \times 10^3 + 6 \times 10 + 1$, or in words one million twenty-four thousand and sixty-one.

Multiplying and dividing by a power of ten

When a number is multiplied by 10, each digit is moved to the next column on the left, the digit nought being placed in the units column as a place holder. In Table 1.1

$$35 \times 10 = 350$$

and

$$35 \times 100 = 350 \times 10 = 3500$$

Moving each digit one place further to the right is equivalent to dividing the number by 10, so that $350 \div 10 = 35$.

Division is the *inverse operation* of multiplication. The symbol = means 'is equal to' or 'equals' and the arithmetic expressions on both sides of an equals sign must always represent the same number.

Example 1.1

Evaluate and write the answer in figures and in words (a) $186\,000 \div 100$, (b) 720×1000.

Solution

(a) $186\,000 \div 100 = 1860$ (b) $720 \times 1000 = 720\,000$

Answer: (a) 1860, one thousand eight hundred and sixty; (b) 720 000, seven hundred and twenty thousand.

Exercise 1.1

(1) Write as a power (a) 4 × 4 × 4, (b) 10 × 10 × 10 × 10, (c) 5 × 5.
(2) Express in words (a) 274, (b) 3063, (c) 10100.
(3) Write as a denary number (a) two thousand three hundred and six, (b) three hundred and five thousand four hundred and nine.
(4) Evaluate (a) 480 × 10, (b) 307 × 100, (c) 2682 × 100.
(5) Find the value of (a) 690 ÷ 10, (b) 30 000 ÷ 1000.
(6) If there are 100 hooks in a packet, how many such packets could be made up from 25 000 hooks?

1.4 ADDITION AND SUBTRACTION OF NATURAL NUMBERS

A number obtained by addition is called the *sum* and the result of a subtraction is called the *difference* of two numbers. Different powers of 10 cannot be combined directly by addition, and so it is very important to keep every digit in its correct column. The following examples show the method used, adding units to units and 10s to 10s, etc.

Example 1.2
Find the sum of 315, 2076 and 87.

Solution The units are added first, and when the sum of any column is more than 9 the 10s are added to the next column on the left

$$\begin{array}{r} 315 \\ 2076 \\ +\quad 87 \\ \hline 2478 \end{array}$$

Answer: the sum of the numbers is 2478.

Example 1.3
What is the difference of (a) 2253 and 1067, (b) 1645 and 873?

Solution

(a)

$$\begin{array}{r} 2253 \\ -\ 1067 \\ \hline 1186 \end{array} \qquad \begin{array}{r} 13-7=6 \\ 14-6=8 \\ 1-0=1 \\ 2-1=1 \end{array}$$

Starting with the units, 7 is more than 3 so a 10 is taken from the 5 in the next column and added to the 3 to make 13. The 5 becomes 4, and again a 10 is taken from the 2 in the next column, to make 14. Answer: the difference is 1186.

(b)

```
   1645        5 - 3 = 2
 -  873       14 - 7 = 7
   ----       15 - 8 = 7
    772
```

Answer: the difference is 772.

Exercise 1.2

(1) Find the sum of (a) 274 and 308, (b) 2983, 462 and 3084, (c) 282, 984 and 729.

(2) Evaluate (a) 374 – 146, (b) 208 – 97, (c) 2942 – 1685, (d) 2004 – 1076.

(3) From the sum of 273 and 4192 subtract 1066.

(4) What number must be added to 624 to give (a) 3721, (b) 4109, (c) 8210?

(5) The following table shows the reading (in litres) on each of three petrol pumps at the beginning of a day and at closing time. Calculate (a) the number of litres sold from each pump (b) the total amount sold that day.

	Pump *A*	Pump *B*	Pump *C*
Start	29 834	26 491	5831
End	30 482	27 384	6924

1.5 MULTIPLICATION AND DIVISION OF NATURAL NUMBERS

(a) Multiplication by a single digit

Starting with the units, each digit in turn is multiplied by the single multiplier and any 10s in the product are 'carried' to the next column. The result of a multiplication operation is called the *product* of the numbers.

Example 1.4

Find the product of (a) 523 × 4, (b) 2068 × 7.

Solution

(a)

```
    523
 ×    4
   ----
   2092
```

(b)

```
    2068
 ×     7
   -----
   14476
```

Answer: (a) 2092, (b) 14 476.

(b) Long multiplication

The number to be multiplied is written down with the multiplier below it. It is multiplied by each digit separately, and the units digit of the product is placed immediately beneath the multiplier as in Example 1.5. The separate products are then added together to give the final product.

Example 1.5

Find the exact value of (a) 523 × 46, (b) 2068 × 77.

Solution

(a)

```
    5 2 3
 ×    4 6
  3 1 3 8
2 0 9 2
2 4 0 5 8
```

(b)

```
      2 0 6 8
 ×        7 7
    1 4 4 7 6
  1 4 4 7 6
  1 5 9 2 3 6
```

Answer: (a) 24 058, (b) 159 236.

(c) Division

The result of dividing one number by another is called the *quotient* of the numbers, and for low numbers the quotient can be obtained from multiplication tables: 9 × 7 = 63 and therefore 63 ÷ 7 = 9. The general method is shown in the following Examples.

Example 1.6

Evaluate (a) 234 ÷ 3, (b) 347 ÷ 5.

Solution

(a)

```
       7 8
  3 ) 2 3 4      23 = 3 × 7 + 2
      2 1
        2 4      24 = 3 × 8
```

Starting at the left, 2 is less than 3 and so 23 is divided by 3. After subtracting 7 × 3 the next digit, 4, is brought down and 24 is divided by 3. Answer: 78.

(b)

```
       6 9
  5 ) 3 4 7      34 = 5 × 6 + 4
      3 0
        4 7      47 = 5 × 9 + 2
        4 5
          2
```

Answer: 69 remainder 2.

Example 1.7
Evaluate (a) 2880 ÷ 12, (b) 326 ÷ 17.

Solution

(a)

```
      2 4 0
12 ) 2 8 8 0      28 ÷ 12 = 2 remainder 4
     2 4          48 ÷ 12 = 4
     ---
       4 8
       4 8
       ---
         0
```

(b)

```
        1 9
17 ) 3 2 6        32 = 17 × 1 + 15
     1 7          156 = 17 × 9 + 3
     -----
     1 5 6
     1 5 3
     -----
         3
```

Answer: (a) 240, (b) 19 remainder 3.

Exercise 1.3

(1) Evaluate (a) 28 × 35, (b) 2073 × 64, (c) 493 ÷ 17, (d) 4004 ÷ 143.

(2) Express as a single number (a) 966 ÷ 23 × 7, (b) 1260 ÷ 28 × 9.

(3) What is the remainder when (a) 1088 is divided by 24, (b) 530 is divided by 17, (c) 7824 is divided by 15?

(4) A sub-assembly part needs 14 screws. (a) How many screws are needed for 125 of the parts? (b) What is the greatest number of the parts that could be made with 500 screws, and how many screws would be left over?

(5) In three storage trays there are 1765, 550, and 2780 buttons. (a) What is the total number of buttons in the trays? (b) How many packets of 15 buttons could be made up from them?

1.6 MIXED OPERATIONS

When a calculation involves more than one operation the order in which the numbers are combined is important and brackets ( ) are used to show which to do first. If there are no brackets, multiplication and division should be carried out before addition and subtraction. The reason for this is discussed later, in Chapter 7, but the correct answer to a calculation is obtained only if the procedure is right. The order is

(i) brackets
(ii) multiplication and division
(iii) addition and subtraction.

Example 1.8

(a) (i) $7 \times (4 + 2) = 7 \times 6 = 42$
(ii) $(7 \times 4) + 2 = 28 + 2 = 30$

(b) (i) $9 + (6 \div 3) = 9 + 2 = 11$
(ii) $(9 + 6) \div 3 = 15 \div 3 = 5$

(c) (i) $7 - 2 \times 3 = 7 - 6 = 1$
(ii) $(7 - 2) \times 3 = 5 \times 3 = 15$

Example 1.9

(a) $3 + 2 \times 7 - 4 = 3 + 14 - 4 = 13$
(b) $(3 + 2) \times 7 - 4 = 5 \times 7 - 4 = 31$
(c) $3 + 2 \times (7 - 4) = 3 + 2 \times 3 = 9$
(d) $(3 + 2) \times (7 - 4) = 5 \times 3 = 15$

Exercise 1.4

Find the value of the following.

(1) $5 + (2 \times 7)$. (2) $(4 \times 3) - 2 + 4$.
(3) $33 - (5 \times 4)$. (4) $(3 \times 7) - (2 \times 8)$.
(5) $13 + 4 - 2 \times 8$. (6) $2 + 3 + 6 \times 4$.
(7) $2 \times 14 - (3 + 9)$ (8) $3 \times 8 - 6 \times 2$.

1.7 MULTIPLES AND FACTORS

Definitions: When numbers are multiplied together the product is a *multiple* of each of the numbers, and each number is a *factor* of their product. For instance, $3 \times 5 = 15$ and so 15 is a multiple of 3 and a multiple of 5. 3 and 5 are both factors of 15.

Every number ending in nought is a multiple of 10, and also every even number has 2 as a factor.

(a) Prime numbers

A number which has no factors except itself and unity (1) is called a *prime* number; the primes up to 50 are 2, 3, 5, 7, 11, 13, 17, 19, 23, 29, 31, 37, 41, 43, 47.

Example 1.10

Express the following numbers as a product of prime factors

(a) $10 = 2 \times 5$ (b) $12 = 2 \times 2 \times 3 = 2^2 \times 3$
(c) $18 = 2 \times 3 \times 3 = 2 \times 3^2$ (d) $33 = 3 \times 11$
(e) $40 = 2 \times 2 \times 2 \times 5 = 2^3 \times 5$

(b) Highest common factor

The highest common factor (HCF) of a given set of numbers is the greatest number which is a factor of each of them.

To find the HCF, each number is written as a product of primes and the lowest power of any factor present in each number is noted.

Example 1.11

Find the HCF of 24, 28, 40.

Solution

$$24 = 2 \times 12 = 2^3 \times 3, \qquad 28 = 2 \times 14 = 2^2 \times 7, \qquad 40 = 2^3 \times 5$$

2 is the only number which is a factor of all three numbers and the lowest power present is 2^2. Answer: The HCF is <u>4</u>.

Example 1.12

What is the HCF of 42, 56, 140?

Solution

$$42 = 2 \times 3 \times 7, \quad 56 = 2 \times 2 \times 2 \times 7, \quad 140 = 2 \times 2 \times 5 \times 7$$

The common factors are 2 and 7, and therefore the HCF is 2×7. Answer: The HCF is <u>14</u>.

(c) Lowest common multiple

The lowest common multiple (LCM) of a given set of numbers is the smallest number that contains as a factor each of the given numbers. For example, the LCM of 2, 3 and 5 is 30, because 30 is the lowest number into which 2, 3 and 5 all divide exactly. Similarly, 24 is the LCM of 3, 8 and 12.

The general method of finding the LCM of given numbers is to write each of them as a product of prime factors. The highest power of each factor is noted and the product of these powers is the LCM.

Example 1.13

Find the LCM of (a) 12, 18, 48, (b) 14, 35, 49.

Solution

(a) $12 = 2 \times 2 \times 3 = 2^2 \times 3$

$18 = 2 \times 3 \times 3 = 2 \times 3^2$

$48 = 2 \times 2 \times 2 \times 2 \times 3 = 2^4 \times 3$

The LCM is $2^4 \times 3^2 = 16 \times 9 =$ <u>144</u>.

(b) $14 = 2 \times 7$
$35 = 5 \times 7$
$49 = 7 \times 7 = 7^2$

The LCM is $2 \times 5 \times 7^2 = \underline{490}$.

Exercise 1.5

Write each of the following as a product of prime factors

(1) 38. (2) 36. (3) 75. (4) 88.

Find the lowest common multiple of each of the following.

(5) 9, 18. (6) 4, 8, 12. (7) 5, 7. (8) 3, 7, 12.
(9) 6, 22, 121.

Find the highest common factor of each of the following.

(10) 27, 45, 54. (11) 20, 24, 28. (12) 84, 126, 196.

(13) Two cars are driven at constant speed on a test circuit. One makes a complete circuit in 84 seconds and the other in 60 seconds. If the time at which the two cars are level is noted, how many seconds later will they next be side by side?

1.8 NEGATIVE INTEGERS

Using only positive integers we cannot subtract a number from one which is smaller and to give a meaning to a difference such as 5 – 7 the number system has been extended to include zero and the negative integers. There is a negative integer corresponding to each positive integer and the symbols used are + for positive and – for negative written in front of the digits as in –3 and –15. Since the positive integers are the same as the natural numbers the + sign is usually dropped, so that 3 and +3 both represent the positive integer 3. Zero is represented by the digit 0 and is neither positive nor negative.

1.9 ORDER RELATIONS AND THE NUMBER LINE

Every integer can be represented by a point on a straight line called the *real number line* with the positive integers to the right of zero and the negative integers to the left

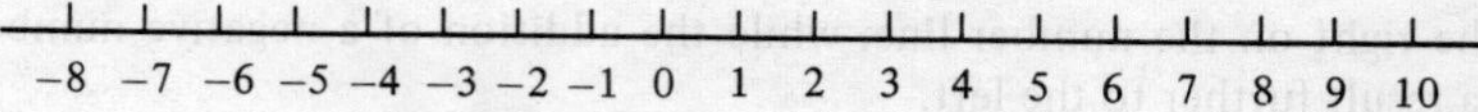

Any given integer is less than every integer further to the right and greater than every integer further to the left; numbers having the same position on the number line must be equal to each other.

The inequality symbols are $<$ for less than and $>$ for greater than. For example, $3 < 5$ and $5 > 3$. Similarly, $-7 < -2$ and $-2 > -7$.

Example 1.14
(a) Write the integers 4, –7, –2, 1 in ascending order.

Solution Ascending order means lowest first and –7 is the furthest left on the number lines. Answer: $-7 < -2 < 1 < 4$.
(b) Arrange in descending order –15, 3, –4, 0.

Solution 3 is furthest to the right on the number line, and is the greatest. Answer: $3 > 0 > -4 > -15$.

Negative integers are all less than zero, positive integers are greater than zero, and it follows that every positive number is greater than every negative number.

Exercise 1.6
Write the given numbers in ascending order using the symbol $<$.

(1) 2, –4, 3. (2) –1, –8, 0. (3) 15, –2, –1.

State whether the following relations are TRUE or FALSE.

(4) $7 < 9$. (5) $-1 < -4$. (6) $-3 > -5$.
(7) $0 < 4$. (7) $-74 < -100$.

1.10 ARITHMETIC OPERATIONS WITH DIRECTED NUMBERS

Numbers having a positive or negative sign are called *directed numbers* and it is unfortunate that the symbols + and – are the same as those for the operations addition and subtraction. However, there should be no difficulty in distinguishing them because an operation symbol must always come between two numbers.

For instance 4 + –5 means add negative 5 to 4 and –2 – 3 means subtract 3 from negative 2.

(a) Addition
The addition of a positive number to any given number has a result further to the right on the number line, while the addition of a negative number has a result further to the left.

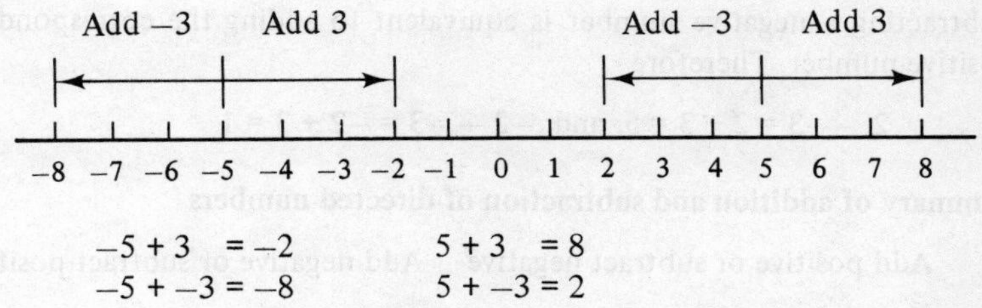

$$-5+3 = -2 \qquad 5+3 = 8$$
$$-5+-3 = -8 \qquad 5+-3 = 2$$

The addition of zero to any number leaves the number unchanged; $4+0=4$, $-2+0=-2$.

It can be seen by referring to the number line that the sum of any positive integer and the corresponding negative integer is zero; zero is the *identity* for addition of directed numbers

$$4+-4=0 \quad -2+2=0 \quad 7+-7=0 \quad -6+6=0$$

The general rule for addition of directed numbers is

(i) To add numbers with the same sign, take the sum of the unsigned numbers and the common sign. For instance

$$2+7=9 \qquad -2+-7=-9$$
$$13+5=18 \qquad -13+-5=-18$$

(ii) To add numbers with different signs, one positive and one negative, take the difference of the unsigned numbers and the sign of the one which is numerically greater. Thus

$-2+7=5$ and $2+-7=-5$

because

$7-2=5$ and $7>2$

Similarly

$-1+4=3$ and $1+-4=-3$
$-6+11=5$ and $6+-11=-5$

(b) Subtraction

Since subtraction is the inverse of addition, subtracting a positive number is equivalent to adding the corresponding negative number. For example

$$5-3 = 5+-3 = 2$$

and

$$-4-1=-4+-1=-5$$

Subtracting a negative number is equivalent to adding the corresponding positive number. Therefore

$$2 - -3 = 2 + 3 = 5 \text{ and } -2 - -3 = -2 + 3 = 1$$

Summary of addition and subtraction of directed numbers

Add positive or subtract negative	Add negative or subtract positive
Increase the number	Decrease the number
Move to the right on the number line	Move to the left on the number line

Example 1.15

(a) $17 + -8 - 3 + 15 = \underline{21}$ (b) $29 - -5 + 13 - 7 = \underline{40}$
(c) $-14 - 4 + -3 - 8 = \underline{-29}$ (d) $37 + -17 - -43 + 22 = \underline{85}$

Exercise 1.7
Evaluate the following.

(1) $14 + 5$. (2) $17 - 9$. (3) $-3 + 6$. (4) $8 + -8$.
(5) $-8 - 4$. (6) $4 + -9$. (7) $-2 + -6$. (8) $13 - -7$.
(9) $3 - -2 - 1$. (10) $-1 - -4 + 2$. (11) $27 + -2 + 3$.
(12) $2 - -10 + -4 - 7$. (13) $-15 + 33 - 12 + -11$.
(14) $47 - -9 + -7 - 8$. (15) $-10 - 15 + 10 - -5 + 5$.

(c) Multiplication and division
The rule for multiplying and dividing directed numbers can be stated in two parts.

(i) The product or quotient of numbers having the same sign is positive

$$4 \times 2 = 8 \text{ and } -4 \times -2 = 8$$
$$6 \div 3 = 2 \text{ and } -6 \div -3 = 2$$

(ii) The product or quotient of numbers having different signs is negative

$$4 \times -2 = -8 \text{ and } -4 \times 2 = -8$$
$$6 \div -2 = -3 \text{ and } -6 \div 2 = -3$$

Illustration on the number line
These rules may be justified by referring to the number line. For instance, considering multiplication as repeated addition and starting at zero

$$-4 \times 2 = 0 + -4 + -4 = -8$$

Similarly, by considering multiplication by a negative number as repeated subtraction, we have

$$-4 \times -2 = 0 - -4 - -4 = 0 + 4 + 4 = 8$$

Add −4 Add −4 Subtract −4 Subtract −4

−8 −7 −6 −5 −4 −3 −2 −1 0 1 2 3 4 5 6 7 8

$-4 \times 2 = 0 + -4 + -4 = -8$ $\quad$ $-4 \times -2 = 0 + 4 + 4 = 8$

The product of any number and zero is zero. Division by zero has no meaning.

Example 1.16

(a) $-21 \times -3 \div 7 = 63 \div 7 = \underline{9}$
(b) $112 \div -4 \div 2 = -28 \div 2 = \underline{-14}$
(c) $11 \times -3 \times -5 = -33 \times -5 = \underline{165}$
(d) $115 \div -1760 \times 0 \times 37 = \underline{0}$

Example 1.17
What is the HCF of −36 and 24?

Solution

$$-36 = -3 \times 12 \qquad 24 = 2 \times 12$$

and the highest common factor is $\underline{12}$. The negative sign makes no difference to the value of the HCF.

Exercise 1.8
Find the value of the following.

(1) 2×-3. (2) -4×6. (3) -2×-1.
(4) -9×0. (5) -16×-5. (6) $10 \div -2$.
(7) $-18 \div 3$. (8) $-12 \div -4$. (9) $-24 \div -3 \times -2$.
(10) $18 \div -9 \times -3$. (11) $-2 \times -12 \div 8 \times 8$.

CHAPTER 2

COMMON FRACTIONS

To most of us a fraction means 'part of a whole'. For instance, if a fruit flan is divided into six 'equal' portions, each one is one-sixth part of the whole flan, $1 \div 6$ or $1/6$, but in fact the portions can never be exactly equal. In mathematics fractions are exact, and they have a much wider meaning.

2.1 RATIONAL NUMBERS

When one integer is divided by another the resulting quotient is not necessarily an integer, and so the number system of integers is extended again to include the quotient of every pair of integers. They are called *rational numbers* and those that are not integers occupy positions between the integers on the real number line.

2.2 COMMON FRACTIONS

A rational number written with the two integers one above the other is called a *common fraction* (sometimes *vulgar* fraction)

$$\frac{1}{2}, \quad \frac{3}{5}, \quad \frac{7}{12}, \quad \frac{16}{4}, \quad \frac{-23}{5}$$

are all common fractions, but only 16/4 represents an integer.

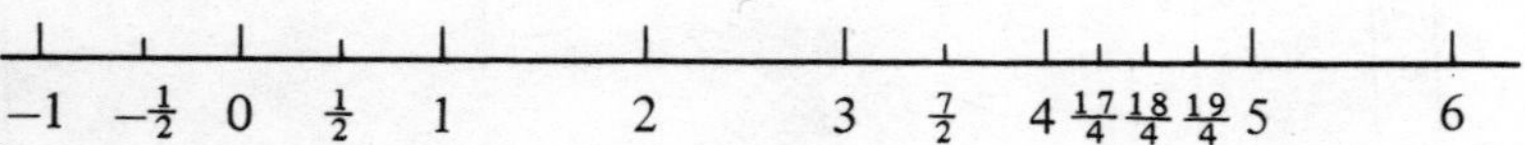

$16 \div 4 = 4$ and the fraction 16/4 is the integer 4. 20/4 is the integer 5 and the fractions 17/4, 18/4, 19/4, occupy places between the integers 4 and 5 on the number line.

(a) Equivalent fractions

Since the same quotient is obtained in many different ways, every rational number could be represented by many different fractions. For example

$$6 \div 3 = 2 \qquad 8 \div 4 = 2 \qquad 30 \div 15 = 2$$

and therefore

$$\frac{6}{3} = \frac{8}{4} = \frac{30}{15}$$

Every positive fraction in which the top integer is less than the one below has a value between zero and 1, and it is obvious from Fig. 2.1 that the fractions 1/2, 2/4, 3/6 represent the same number. There is a negative fraction corresponding to every positive fraction.

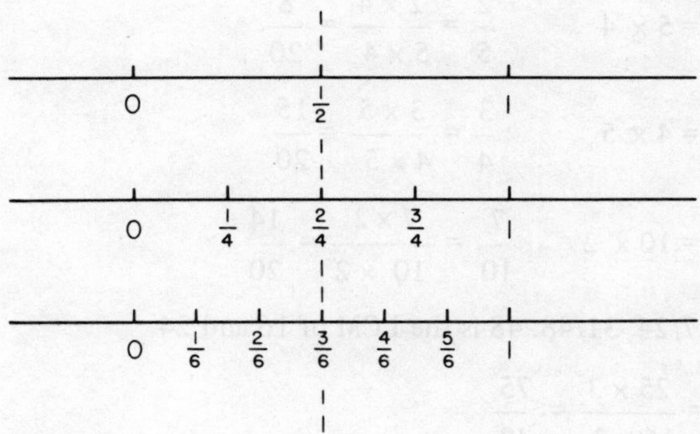

Fig 2.1

(b) Cancelling

The integer at the top of a fraction is called the *numerator* and that at the bottom is called the *denominator*.

A fraction is reduced to its lowest terms when the numerator and denominator have no common factor. The process of dividing by a common factor is called *cancelling*. For instance

$$\frac{21}{28} = \frac{3 \times 7}{4 \times 7} = \frac{3}{4} \qquad \text{(cancel by 7)}$$

and

$$\frac{15}{60} = \frac{3 \times 5}{2 \times 2 \times 3 \times 5} = \frac{1}{4} \qquad \text{(cancel by 3 and 5)}$$

Example 2.1
(a) $\frac{15}{165} = \frac{3}{33} = \frac{1}{11}$ (cancel by 5 and 3)
(All integers ending in 5 or 0 have 5 as a factor)
(b) $\frac{42}{56} = \frac{21}{28} = \frac{3}{4}$ (cancel by 2 and 7)
(All even numbers have 2 as a factor)
(c) $\frac{300}{500} = \frac{3}{5}$ (cancel by 100)

Example 2.2
Write the given fractions with the same denominator (a) 2/5, 3/4, 7/10, (b) 25/16, 17/24, 31/48.

Solution (a) 2/5, 3/4, 7/10. The common denominator must be a multiple of 5, 4, and 10 and the LCM, 20, is the most convenient since it is the smallest

$$20 = 5 \times 4 \qquad \frac{2}{5} = \frac{2 \times 4}{5 \times 4} = \frac{8}{20}$$

$$20 = 4 \times 5 \qquad \frac{3}{4} = \frac{3 \times 5}{4 \times 5} = \frac{15}{20}$$

$$20 = 10 \times 2 \qquad \frac{7}{10} = \frac{7 \times 2}{10 \times 2} = \frac{14}{20}$$

(b) 25/16, 17/24, 31/48. 48 is the LCM of 16 and 24

$$\frac{25}{16} = \frac{25 \times 3}{16 \times 3} = \frac{75}{48}$$

$$\frac{17}{24} = \frac{17 \times 2}{24 \times 2} = \frac{34}{28}$$

The numbers are 75/48, 34/48, 31/48.

2.3 ORDER RELATIONS FOR COMMON FRACTIONS

In order to compare fractions they must be given the same denominator, as in Example 2.2.

Example 2.3
Write the following fractions in ascending order: 4/9, 17/30, 23/36, 96/180.

Solution The LCM of the denominators is 180. 180 = 9 × 20, 180 = 30 × 6, 180 = 36 × 5

$$\frac{4}{9} = \frac{80}{180}, \quad \frac{17}{30} = \frac{102}{180}, \quad \frac{23}{36} = \frac{115}{180}$$

Comparing the numerators, $80 < 96 < 102 < 115$ therefore

$$4/9 < 96/180 < 17/30 < 23/36$$

Exercise 2.1

Reduce the following fractions to their lowest terms.

(1) $\frac{6}{16}$. (2) $\frac{16}{24}$. (3) $\frac{18}{48}$. (4) $\frac{60}{240}$. (5) $\frac{45}{75}$.
(6) $\frac{72}{84}$. (7) $\frac{44}{121}$. (8) $\frac{120}{140}$.

Express the following fractions with their lowest common denominator, and hence write the given fractions in ascending order.

(9) $\frac{3}{5}, \frac{4}{7}$. (10) $\frac{3}{8}, \frac{5}{12}$. (11) $\frac{1}{3}, \frac{2}{9}, \frac{4}{5}$. (12) $\frac{3}{8}, \frac{1}{3}, \frac{2}{5}$.
(13) $\frac{7}{10}, \frac{3}{4}, \frac{4}{5}$. (14) $\frac{5}{9}, \frac{1}{3}, \frac{5}{6}$.

2.4 IMPROPER FRACTIONS AND MIXED NUMBERS

An *improper fraction* has the numerator numerically greater than the denominator, regardless of sign; $-23/17$ and $115/22$ are improper fractions.

A *mixed number* has an integer part and a fractional part, and can be obtained from the corresponding improper fraction by division. Thus $7 \div 3 = 2$ remainder 1, and so the improper fraction $7/3$ is equal to the mixed number $2\frac{1}{3}$.

Example 2.4

Write as a mixed number (a) $37/4$, (b) $175/15$, (c) $-151/21$, (d) $135/12$.

Solution

(a) $37 \div 4 = 9$ remainder 1

therefore

$37/4 = 9\frac{1}{4}$

(b) $\frac{175}{15} = \frac{35}{3}$ (cancel by 5)

$35 \div 3 = 11$ remainder 2,

therefore

$35/3 = 11\frac{2}{3}$

(c) $-151 \div 21 = -7$ remainder -4

$-151/21 = -7\frac{4}{21}$

(d) $135 \div 12 = 11$ remainder 3

$135/12 = 11\frac{3}{12} = 11\frac{1}{4}$

In answer to a problem, fractions should be given in their lowest terms; cancelling at an earlier stage makes for easier arithmetic but does not affect the final answer.

By the reverse process, a mixed number can be changed to an improper fraction by multiplication. Thus

$$9\tfrac{1}{4} = \frac{9 \times 4 + 1}{4} = \frac{37}{4} \quad \text{and} \quad 11\tfrac{2}{3} = \frac{11 \times 3 + 2}{3} = \frac{35}{3}$$

Example 2.5
Change the following to improper fractions.

(a) $7\tfrac{1}{4}$. $\dfrac{7 \times 4 + 1}{4} = \dfrac{29}{4}$ (b) $33\tfrac{1}{3}$. $\dfrac{33 \times 3 + 1}{3} = \dfrac{100}{3}$

(c) $16\tfrac{4}{11}$. $\dfrac{16 \times 11 + 4}{11} = \dfrac{180}{11}$ (d) $10\tfrac{11}{15}$. $\dfrac{150 + 11}{15} = \dfrac{161}{15}$

Exercise 2.2
Express the following as mixed numbers.
(1) $\frac{5}{2}$. (2) $\frac{10}{3}$. (3) $\frac{18}{8}$. (4) $\frac{44}{14}$. (5) $\frac{86}{16}$.
(6) $\frac{44}{32}$.
Write the following as improper fractions.
(7) $2\frac{1}{4}$. (8) $4\frac{2}{3}$. (9) $3\frac{2}{9}$. (10) $6\frac{5}{7}$. (11) $5\frac{12}{13}$.
(12) $10\frac{7}{24}$.

2.5 OPERATIONS DEFINED ON COMMON FRACTIONS

(a) Addition and subtraction
Fractions having the same denominator may be added directly, so that for instance

$$\frac{3}{5} + \frac{1}{5} = \frac{4}{5} \quad \text{and} \quad \frac{2}{7} + \frac{3}{7} = \frac{5}{7}$$

Similarly for subtraction

$$\frac{6}{7} - \frac{2}{7} = \frac{4}{7} \quad \text{and} \quad \frac{9}{10} - \frac{5}{10} = \frac{4}{10}$$

Fractions to be added or subtracted must first be given a common denominator and in general the LCM of the denominators is used. The integer parts of mixed numbers are treated separately.

Example 2.6
Find the sum of the fractions $2\frac{1}{2}$, $3\frac{1}{7}$, $1\frac{2}{5}$.

Solution The LCM of 2, 7, and 5 is 70. 1/2 = 35/70, 1/7 = 10/70, 2/5 = 28/70.

$$2\tfrac{1}{2} + 3\tfrac{1}{7} + 1\tfrac{2}{5} = (2 + 3 + 1) + (\tfrac{1}{2} + \tfrac{1}{7} + \tfrac{2}{5})$$

$$= 6 + \frac{35 + 10 + 28}{70}$$

$$= 6 + \frac{73}{70} \quad \left(\frac{73}{70} = 1\tfrac{3}{70}\right)$$

$$= \underline{7\tfrac{3}{70}}$$

Example 2.7
What is the difference of $3\frac{1}{2}$ and 9/14?

Solution

$$3\tfrac{1}{2} - \frac{9}{14} = 3 + \frac{1}{2} - \frac{9}{14} = 3 + \frac{7 - 9}{14}$$

$$= 2 + \frac{14 + 7 - 9}{14} \quad \left(1 = \frac{14}{14}\right)$$

$$= 2 + \frac{12}{14}$$

$$= \underline{2\tfrac{6}{7}}$$

Example 2.8
Express as a single fraction $5\frac{1}{3} - 1\frac{11}{12} + 3\frac{2}{9}$.

Solution The LCM is 36

$$5\tfrac{1}{3} - 1\tfrac{11}{12} + 3\tfrac{2}{9} = 5 - 1 + 3 + \frac{1}{3} - \frac{11}{12} + \frac{2}{9}$$

$$= 7 + \frac{12 - 33 + 8}{36}$$

$$= 7 - \frac{13}{36} = 6\ \frac{36 - 13}{36}$$

$$= \underline{6\tfrac{23}{36}}$$

Example 2.9
In a jar containing pink, green and yellow sweets, one-third were pink and two-fifths were green. What fraction were yellow?

Solution All the sweets in the jar make up one whole number, and we have to find the fraction $1 - \frac{1}{3} - \frac{2}{5}$. Making the common denominator 15

$$\frac{15 - 5 - 6}{15} = \frac{4}{15}$$

and this is the fraction of the sweets that were yellow. Answer: $\underline{4/15}$.

Exercise 2.3

(1) $\frac{2}{5} + \frac{1}{3}$. (2) $\frac{4}{9} + \frac{5}{6}$. (3) $1\frac{1}{2} + \frac{4}{5}$. (4) $2\frac{3}{7} + 3\frac{5}{12}$.
(5) $1\frac{1}{3} + 3\frac{5}{14}$. (6) $2\frac{4}{7} + 1\frac{3}{8}$. (7) $\frac{3}{4} - \frac{1}{8}$. (8) $5\frac{2}{3} - 3\frac{4}{9}$.
(9) $3\frac{4}{7} - \frac{3}{8}$. (10) $5\frac{1}{5} - 2\frac{3}{10}$ (11) $2\frac{1}{2} - 1\frac{5}{8}$. (12) $4 - 1\frac{1}{3} - 2\frac{1}{2}$.
(13) $1 - \frac{1}{2} - \frac{1}{3}$. (14) $3\frac{1}{2} + \frac{1}{4} - 2\frac{1}{3}$. (15) $4\frac{14}{15} - 2\frac{2}{3} + 3\frac{5}{12}$.
(16) In a group of A-level students, one-third study arts subjects, two-fifths study science subjects, one-tenth study crafts, and the remainder study humanities. What fraction of the group study humanities?
(17) A technician works an 8-hour day. If $2\frac{2}{3}$ hours are charged to project A, $1\frac{1}{2}$ hours to project B and the remainder to project C, how many hours were charged to project C?

(b) Multiplication

The product of common fractions is obtained by multiplying the numerators together, multiplying the denominators together, and then cancelling if possible. Mixed numbers must first be converted to improper fractions.

Example 2.10

(a) $\frac{2}{5} \times \frac{5}{8} = \frac{2 \times 5}{5 \times 8} = \underline{\frac{1}{4}}$ (cancel by 5 and 2)

(b) $\frac{14}{15} \times \frac{5}{21} = \frac{14 \times 5}{15 \times 21} = \underline{\frac{2}{9}}$ (cancel by 7 and 5)

Example 2.11

Write as a single number $3\frac{1}{7} \times 4\frac{1}{2} \times 1\frac{5}{9}$.

Solution

$3\frac{1}{7} = 22/7, \quad 4\frac{1}{2} = 9/2, \quad 1\frac{5}{9} = 14/9$

$$3\tfrac{1}{7} \times 4\tfrac{1}{2} \times 1\tfrac{5}{9} = \frac{22 \times 9 \times 14}{7 \times 2 \times 9} \quad \text{(cancel by 2, 7 and 9)}$$

$$= \underline{22}$$

Example 2.12
If two-fifths of a £45 telephone account was the charge for rental how much was spent on calls?

Solution In arithmetic, 'of' means multiplied by. Then either

$\frac{2}{5} \times 45 = 18$

and so £18 was spent on rental charges.

£45 − £18 = £27

Or

$1 - \frac{2}{5} = \frac{3}{5}$

$\frac{3}{5} \times 45 = 27$

Answer: £27 was spent on calls.

(c) Reciprocals
Multiplying any number by the integer 1 leaves it unchanged and 1 is the identity element for multiplication

$$7 \times 1 = 7, \quad \frac{1}{2} \times 1 = \frac{1}{2}, \quad -\frac{3}{5} \times 1 = -\frac{3}{5}$$

The product of any number and its reciprocal is defined as 1, so that the reciprocal of a number is the inverse for multiplication. In general, the reciprocal of a common fraction has the numerator and denominator interchanged

$$\frac{3}{5} \times \frac{5}{3} = 1, \quad \frac{2}{7} \times \frac{7}{2} = 1$$

Example 2.13
Find the reciprocal of (a) $2\frac{1}{5}$, (b) $4\frac{3}{11}$, (c) 15/41.

Solution

(a) $2\frac{1}{5}$ = 11/5 the reciprocal is 5/11

(b) $4\frac{3}{11}$ = 47/11 the reciprocal is 11/47

(c) The reciprocal of 15/41 is 41/15 = $2\frac{11}{15}$

(d) Division of common fractions
Division is the inverse operation of multiplication, and dividing by a common fraction is equivalent to multiplying by its reciprocal.

Example 2.14

(a) $\frac{1}{2} \div \frac{3}{4} = \frac{1}{2} \times \frac{4}{3} = \frac{1 \times 4}{2 \times 3} = \frac{2}{3}$

(b) $\frac{7}{12} \div 1\frac{1}{2} = \frac{7}{12} \div \frac{3}{2} = \frac{7}{12} \times \frac{2}{3} = \frac{7 \times 2}{12 \times 3} = \frac{7}{18}$

(c) $6 \div \frac{1}{3} = 6 \times 3 = \underline{18}$

(d) $\frac{3}{5} \div 4 = \frac{3}{5} \times \frac{1}{4} = \underline{\frac{3}{20}}$

(e) $3\frac{1}{7} \div 4\frac{5}{11} = \frac{22}{7} \div \frac{49}{11} = \frac{22}{7} \times \frac{11}{49} = \underline{\frac{242}{343}}$

Example 2.15
The greatest distance across a circle can be found approximately by dividing the distance round the perimeter by $3\frac{1}{7}$. If the distance round the edge of a circular pond is 11 metres, calculate the distance across it.

Solution

$$11 \div 3\tfrac{1}{7} = 11 \div \frac{22}{7} = 11 \times \frac{7}{22} = \frac{7}{2} = 3\tfrac{1}{2}$$

Answer: $\underline{3\frac{1}{2}\text{ metres}}$.

Example 2.16
Evaluate $7\frac{1}{4} \times 1\frac{2}{5} \div 8\frac{2}{7}$.

Solution Changing to improper fractions

$7\frac{1}{4} = 29/4, \qquad 1\frac{2}{5} = 7/5, \qquad 8\frac{2}{7} = 58/7$

$$\frac{29}{4} \times \frac{7}{5} \div \frac{58}{7} = \frac{29}{4} \times \frac{7}{5} \times \frac{7}{58}$$

$$= \frac{49}{40} \quad (29 \times 2 = 58)$$

$$= \underline{1\tfrac{9}{40}}$$

Exercise 2.4

(1) $\frac{3}{4} \times \frac{2}{3}$. (2) $2 \times \frac{7}{8}$. (3) $1\frac{5}{8} \times \frac{2}{39}$. (4) $1\frac{3}{4} \times 1\frac{3}{4}$.
(5) $7\frac{4}{5} \times 1\frac{1}{4} \times 5\frac{5}{13}$. (6) $\frac{5}{9} \div 7\frac{1}{7} \times 1\frac{4}{5}$.

(7) What is one-fifth of $6\frac{2}{3}$? (8) Evaluate three-eighths of £96.

(9) A firm's total annual expenditure was £494 200 and $\frac{5}{14}$ of this was for wages. What was the annual wage bill?

(10) Three-fifths of the cost of a holiday tour of California was spent on the plane fare and car rental, four-fifteenths on hotels, and the rest on food and entertainment. If the total cost was £525, what was the amount spent on (a) hotel bills, (b) food and entertainment?

(e) Mixed operations with fractions

The precedence rule for operations applies to fractions in the same way as integers (Section 1.6); brackets are evaluated first, and multiplication and division take precedence over addition and subtraction.

Example 2.17

(a) $1\frac{1}{4} \times \frac{1}{2} + \frac{1}{3} = (\frac{5}{4} \times \frac{1}{2}) + \frac{1}{3} = \frac{5}{8} + \frac{1}{3}$

$$\frac{5}{8} + \frac{1}{3} = \frac{15 + 8}{24} = \frac{23}{24}$$

(b) $$\frac{1\frac{1}{4} + \frac{1}{2}}{2\frac{3}{8} - 1\frac{1}{4}} = \frac{1\frac{3}{4}}{1\frac{1}{8}} = \frac{7}{4} \div \frac{9}{8}$$

$$\frac{7}{4} \div \frac{9}{8} = \frac{7}{4} \times \frac{8}{9} = \frac{14}{9} = 1\frac{5}{9}$$

(c) $\frac{1}{3}$ of $5 - \frac{1}{5}$ of $3 = (\frac{1}{3} \times 5) - (\frac{1}{5} \times 3)$

$$= \frac{5}{3} - \frac{3}{5} = \frac{25 - 9}{15} = 1\frac{1}{15}$$

Exercise 2.5

Write down the reciprocals of the following fractions.

(1) $\frac{2}{3}$. (2) $\frac{1}{2}$. (3) $\frac{3}{4}$. (4) $2\frac{1}{4}$. (5) $3\frac{1}{7}$.

(6) $\frac{4}{53}$.

Calculate the following.

(7) $\frac{2}{3} \div \frac{4}{9}$. (8) $\frac{1}{4} \div \frac{1}{2}$. (9) $1\frac{1}{7} \div 1\frac{1}{3}$. (10) $3\frac{3}{4} \div 2\frac{1}{2}$.

(11) $4\frac{1}{3} \div 2\frac{4}{5} \div 3\frac{5}{7}$. (12) $\dfrac{2\frac{1}{2} + 3\frac{1}{3}}{1\frac{1}{2} \times \frac{1}{3}}$.

(13) $\dfrac{3\frac{1}{4} - 1\frac{3}{8}}{2\frac{1}{2} \div 1\frac{1}{3}}$. (14) $\dfrac{2\frac{1}{2} \div 9\frac{1}{6}}{3\frac{1}{4} + 1\frac{1}{12}} + 2$.

CHAPTER 3

DECIMAL FRACTIONS

In the denary system of powers of 10, fractions occupy columns to the right of the units column. They are called *decimal fractions* (or just decimals) and the fractional part of a number is separated from the integer part by a dot (.) called a decimal point.

Moving a digit one place further to the right is equivalent to dividing it by 10, so that the first column after the decimal point is headed 1/10 or tenths.

Thousands	Hundreds	Tens	Units	.	Tenths	Hundredths	Thousandths
10^3	10^2	10	1	.	1/10	$1/10^2$	$1/10^3$
		7	2	.	5		
		1	0	.	2	5	
			1	.	0	2	
			6	.	3	0	3

72.5, seventy-two point five, means 72 + 5/10, 10.25, ten point two five, means 10 + 2/10 + 5/100, 1.02, one point nought two, means 1 + 2/100, and 6.303 means 6 + 3/10 + 3/1000.

3.1 RELATING DECIMALS TO COMMON FRACTIONS

(a) Changing from decimal to common fractions

(i) For the numerator, the digits after the decimal point are written down, and any noughts at the beginning are deleted

(ii) The denominator is the power of 10 equal to the number of digits after the decimal point

(iii) The integer part of the decimal fraction remains unchanged, giving a mixed number.

Example 3.1

(a) $5.12 = 5 + \frac{12}{10^2} = 5\frac{12}{100} = \underline{5\frac{3}{25}}$

(b) $1.205 = 1 + \frac{205}{10^3} = 1\frac{205}{1000} = \underline{1\frac{41}{200}}$

(c) $0.03 = \frac{3}{10^2} = \underline{\frac{3}{100}}$

(d) $17.008 = 17 + \frac{8}{10^3} = 17\frac{8}{1000} = \underline{17\frac{1}{125}}$

(b) Changing from common to decimal fractions
Since every common fraction is the quotient of two integers, the equivalent decimal fraction is obtained by direct division. The decimal point is placed after the units digit of the numerator and as many noughts as required are added.

Example 3.2

(a) $\frac{3}{5} = 3 \div 5 = \underline{0.6}$ (b) $\frac{5}{8} = 5 \div 8 = 0.625$

$$5\overline{)3.0} \quad \underline{0.6}$$

$$\begin{array}{r} 0.625 \\ 8\overline{)5.000} \\ \underline{4\,8} \\ 20 \\ \underline{16} \\ 40 \\ \underline{40} \end{array}$$

(c) Recurring decimals
Sometimes in a decimal fraction the same digit is repeated indefinitely and is said to be recurring. This is indicated by a dot placed over the digit the first time it occurs. If a sequence of digits is repeated the first and last digits in the sequence are indicated by dots placed above them, as shown in Example 3.3.

Example 3.3

(a) $\dfrac{17}{3} = 17 \div 3$

$= 5.666\ldots$

$= \underline{5.\dot{6}}$

(b) $\dfrac{2}{7} = 2 \div 7$

$= 0.285\,714\,28\ldots$

$= \underline{0.\dot{2}85\,71\dot{4}}$

Every rational number can be expressed as either an exact or a recurring decimal fraction.

3.2 DECIMAL PLACES AND SIGNIFICANT FIGURES

Decimal places (dp) are the digits following the decimal point. Significant figures (sig. fig.) include all the digits in a number except noughts at the beginning and end. For example, 35 000 and 0.002 6 each have 2 significant figures and 30 500 and 2.06 each have 3 significant figures.

The multiplication or division of decimal numbers can result in a large number of digits, and it is usual to give answers 'corrected' or 'rounded' to an appropriate number of decimal places or significant figures.

Correcting or rounding a decimal number

Rounding a number consists of discarding any digits beyond the specified amount and adjusting the last digit retained *where necessary*. The simpler method which some people still use is to 'round up' when the first figure to be discarded is 5 or over 5. In this book the British Standard method is used.

The rule for adjustment as described in British Standard BS 1957 is

(i) If the first digit to be discarded is greater than 5, round up
(ii) If the first digit to be discarded is less than 5, round down
(iii) If the first digit to be discarded is 5 and there are other non-zero digits following it, round up
(iv) If the first digit to be discarded is 5 and there are no digits except noughts following it, round to an even number.

Example 3.4

Correct the following numbers to 2 sig. fig. or 1 dp

7.2607 rounds to 7.3 (rule i)
7.2407 rounds to 7.2 (rule ii)
7.2507 rounds to 7.3 (rule iii)
7.2500 rounds to 7.2 (rule iv)
7.1500 rounds to 7.2 (rule iv)
7.0500 rounds to 7.0 (rule iv) 0 counts even

Written correct to 3 sig. fig. or 2 dp

2.7283 is 2.73
2.7243 is 2.72
2.7253 is 2.73
2.7250 is 2.72 (2 is even, rule iv)
2.7350 is 2.74 (3 is odd, rule iv)

These rules apply to negative numbers as well as positive.

Rounding should be carried out as late in a calculation as is practicable in order to retain as much information as possible.

Exercise 3.1

Change the following to common fractions.
(1) 0.6. (2) 0.75. (3) 2.125. (4) 4.375.
Change the following to decimal fractions.
(5) 3/4. (6) 5/16. (7) 4/3 (8) 6/25.
Write, correct to (a) 2 dp, (b) 3 sig. fig., the following numbers.
(9) 0.507. (10) 0.00859. (11) 41.125. (12) 34.827.
(13) 1985.026. (14) 3895.7151.
Write the following correct to 3 significant figures.
(15) 47908. (16) 3004. (17) 5.00657 (18) 129.76.
Write the following common fractions as decimals correct to 3 dp.
(19) 7/40. (20) 22/7. (21) 25/21. (22) 79/9. (23) 13/14.

3.3 NUMBERS IN STANDARD INDEX FORM

In Section 1.3 it was shown that moving a digit to the next column on the right decreased the index by 1, and extending this to the columns after the decimal point leads to the idea of a negative index

$$10^4 \quad 10^3 \quad 10^2 \quad 10^1 \quad 10^0 . \quad 10^{-1} \quad 10^{-2} \quad 10^{-3}$$

so that $10^0 = 1, \quad 10^{-1} = \dfrac{1}{10}, \quad 10^{-2} = \dfrac{1}{100}$

Every decimal number, whether an integer or a fraction, can be written as the product of a number between 1 and 10 and a power of 10, and this is called *standard index form*.

Example 3.5

	Number	*Standard index form*
(a)	752 = 7.52 × 100	7.52×10^2
(b)	75 200 = 7.52 × 10 000	7.52×10^4
(c)	0.0752 = 7.52 ÷ 100	7.52×10^{-2}
(d)	0.7052 = 7.052 ÷ 10	7.052×10^{-1}

Numbers in standard form may be written correct to a given number of significant figures in the same way as single decimal numbers.

Example 3.6
Written in standard index form correct to 3 sig. fig.

(a) 41460 is 4.15×10^4 (c) 4201.6 is 4.20×10^3

(b) 32.785 is 3.28×10^1 (d) 0.003525 is 3.52×10^{-3}

A number with three significant figures must have at least three digits, so that in (c) the final nought was not discarded.

This method of writing decimal numbers is particularly useful in scientific work for numbers with many noughts, such as

The speed of light is 3×10^8 metres per second
The charge on an electron is 1.6×10^{-19} coulomb and the mass is 9.1×10^{-31} kilogram.

3.4 DECIMAL CURRENCY AND METRIC UNITS

Many countries, including the United States, use dollars and cents, francs and centimes are used in France and Switzerland, and in the United Kingdom, pounds sterling and pence are used, with 100 pennies making £1

£2.43 means 2 pounds and 43 pence

$115.05 means 115 dollars and 5 cents

and F3.95 means 3 francs and 95 centimes.

The metric system of measurements also uses a decimal notation. 1 metre (m) is equivalent to 100 centimetres (cm) or 1000 millimetres (mm) so that 150 cm = 150 ÷ 100 m = 1.5 m and 2.85 m = 2.85 × 1000 mm = 2850 mm.

3.5 ADDITION AND SUBTRACTION OF DECIMAL FRACTIONS

The method used is the same as that for integers (Section 1.4), but the decimal points must be placed correctly so that like powers of 10 are combined.

Example 3.7
(a) Add together 31.25, 6.081, 20.707.
(b) Evaluate 17.34 − 23.89 + 54.19.

Solution

(a)
```
   31.25
    6.081
 + 20.707
 --------
   58.038
```

(b)
```
   17.34
 + 54.19
 -------
   71.53
 − 23.89
 -------
   47.64
```

Answer: (a) 58.038, (b) 47.64.

Example 3.8
How much change should I receive from a £20 note after buying articles priced at £2.39, £15.70 and 35p?

Solution

```
     £             £
    2.39         20.00
   15.70       − 18.44
 +   .35       -------
 -------          1.56
   18.44
```

Answer: £1.56.

Addition of numbers in standard index form
Since only numbers with the same power can be added directly, before numbers with different powers are added the one with the lower power is written in terms of the higher power.

Example 3.9

(a) $3.15 \times 10^2 + 52.7 \times 10^2 = (3.15 + 52.7) \times 10^2 = 55.85 \times 10^2$

(b) $3.5 \times 10^3 + 1.7 \times 10^2 = (3.5 + 0.17) \times 10^3 = 3.67 \times 10^3$

Similarly for subtraction

(c) $17.43 \times 10^{-1} - 8.25 \times 10^{-1} = (17.43 - 8.25) \times 10^{-1} = 9.18 \times 10^{-1}$

(d) $4.6 \times 10^{-2} - 2.1 \times 10^{-4} = (4.6 - 0.021) \times 10^{-2} = 4.579 \times 10^{-2}$

because $2.1 \times 10^{-4} = 0.021 \times 10^{-2}$.

Exercise 3.2
Write the following in standard index form.

(1) 0.05. (2) 0.0049. (3) 62.75. (4) 5972. (5) 273051.

Evaluate the following.

(6) $4.52 + 6.03 + 0.45$. (7) $3.65 - 1.095$. (8) $25.1 - 5.08$.

(9) $12.83 + 0.54 - 6.09$. (10) $0.80 - 2.1 + 3.64$.

Express as a single number in standard index form each of the following.

(11) $3 \times 10^2 + 7 \times 10^2$. (12) $6 \times 10^{-4} + 5 \times 10^{-3}$

(13) $9.3 \times 10^2 + 5.2 \times 10^3$. (14) $3.6 \times 10^{-2} - 4.1 \times 10^{-3}$.

(15) $2.8 \times 10^4 - 3.1 \times 10^3$.

(16) The lengths of the sides of a paddock are 70.2 m, 98.3 m, 60.9 m and 111.7 m. What is the total distance around the paddock and how far short is it of 350 m?

3.6 MULTIPLICATION OF DECIMAL FRACTIONS

Two decimal numbers are multiplied as though they are integers, ignoring the decimal points at first, and the position of the decimal point in the product is determined separately. The number of decimal places in the product is equal to the total number of decimal places in the numbers being multiplied. The reason for this will be explained in Section 4.3.

Example 3.10
Evaluate the products (a) 1.2×0.04, (b) 0.15×0.08, (c) $2.5 \times 0.5 \times 0.01$, (d) 17.52×2.134.

Solution

(a) $12 \times 4 = 48$
Decimal places: $1 + 2 = 3$

Hence

$1.2 \times 0.04 = \underline{0.048}$

(b) $15 \times 8 = 120$
Decimal places: $2 + 2 = 4$

$0.15 \times 0.08 = \underline{0.012}$

(c) $25 \times 5 \times 1 = 125$
Decimal places: $1 + 1 + 2 = 4$

Hence

$2.5 \times 0.5 \times 0.01 = \underline{0.0125}$

(d) $1752 \times 2134 = 3738768$
Decimal places: $2 + 3 = 5$

The product is $\underline{37.38768}$.

3.7 DIVISION OF DECIMAL FRACTIONS

To divide two decimal fractions, both numbers are multiplied by the power of 10 required to make the divisor a whole number.

Example 3.11

(a) $36 \div 1.2 = 360 \div 12 = \underline{30}$

(b) $1000 \div 1.25 = 100\,000 \div 125 = \underline{800}$

(c) $0.02 \div 0.5 = 0.2 \div 5 = \underline{0.04}$

Example 3.12

(a) Evaluate $0.055 \div 7.2$ correct to 3 sig. fig.

(b) Divide 606.879 by 3.51 (i) exactly, (ii) correct to 2 sig. fig.

Solution

(a) $0.055 \div 7.2 = 0.55 \div 72$

```
       0.007638
   72) 0.550000
          504
           460
           432
            280
            216
             640
             576
```

Answer: $\underline{0.00764}$.

(b) $606.879 \div 3.51 = 60687.9 \div 351$

```
         172.9
   351) 60687.9
        351
        2558
        2457
         1017
          702
         3159
         3159
```

Answer: $\underline{\text{(i) } 172.9, \text{ (ii) } 170}$.

Example 3.13

Calculate the number of strips of wallpaper each 52 cm wide which would be needed to paper a wall 6.35 m long.

Solution The values must be in the same units, and 6.35 m = 635 cm. $635 \div 52 = 12.21$ and so 13 strips would be needed. Answer: $\underline{13}$.

Exercise 3.3

Calculate exactly the following.

(1) 22 x 0.05. (2) 0.003 x 1.4. (3) 0.21 ÷ 0.7. (4) 0.081 ÷ 0.09.
(5) 2.06 x 3.1. (6) 12.1 x 0.04. (7) 0.478 x 0.12.
(8) 13.02 ÷ 2.1. (9) 0.432 ÷ 0.12. (10) 457.6 ÷ 14.3.

Evaluate, correct to 2 decimal places, the following.

(11) 17.48 ÷ 32.5. (12) 0.06 ÷ 1.7. (13) 2.85 x 0.62 x 0.8.

(14) Find the total cost of 28 files at 65 p each, 4 packets of file paper at £2.42 a packet and 30 pens at 35 p each.

(15) Calculate the total weekly wage of 4 cashiers at £49.75 each, 2 cleaners at £32.38 each and 3 sales-staff at £65.70 each, plus the manager at £126.25.

(16) How many pieces or wire, each 180 mm long, could be cut from a 10 metre coil?

(17) Evaluate correct to the nearest penny (a) 2/3 of £49.39, (b) $1\frac{1}{8}$ of £9.48.

(18) Find the cost to the nearest penny of one toy if 78 of them cost £94.

CHAPTER 4

ROOTS, INDICES, FOUR-FIGURE TABLES, CALCULATORS

4.1 IRRATIONAL NUMBERS

In Chapter 3 we discussed exact and recurring decimal fractions but there are some decimal numbers that never end or repeat however many decimal places are calculated. These are called *irrational* numbers because they cannot be written as common fractions and they are corrected to a specified number of decimal places. An irrational number is a real number and on the number line it has a position between two rational numbers, although we cannot locate the position exactly.

4.2 ROOTS AND SURDS

(a) Roots

The arithmetic operation which is the inverse of 'raising to a power' is 'taking a root' and the root symbol $\sqrt{}$, placed in front of a number, was first used in the early sixteenth century

$2^2 = 4$ and $2 = \sqrt{4}$ (the square root of four)

$2^3 = 8$ and $2 = \sqrt[3]{8}$ (the cube root of eight)

$3^4 = 81$ and $3 = \sqrt[4]{81}$ (the fourth root of 81)

Comment: $(-2)^2 = 4$ and so the square root of 4 could be either +2 or −2. By convention the symbol $\sqrt{4}$ is taken to have a positive value.

(b) Surds

Surd is a special name given to a root which cannot be evaluated exactly, and the root of a prime number is a surd because a prime number is not an exact power of any number. All surds are irrational numbers since they are not exact decimals.

For example, $\sqrt{3}$ is a surd since there is no rational number which has 3 as its square. The square root of any prime number is a surd.

(i) Addition and subtraction of surds
Different surds cannot be added or subtracted directly, they must first be written in decimal form. Methods of converting roots and surds to decimal fractions are considered in Sections 4.9 and 4.12.

Multiples of the same surd may be combined. For instance

$$2\sqrt{3} + 5\sqrt{3} = 7\sqrt{3}$$

and

$$3\sqrt{2} - \sqrt{2} = 2\sqrt{2}$$

(ii) Multiplication and division of surds
Surds of the same degree may be combined

$$\sqrt{2} \times \sqrt{3} = \sqrt{(2 \times 3)} = \sqrt{6}$$

and

$$2\sqrt{2} \times 3\sqrt{2} = 6\sqrt{(2^2)} = 6 \times 2 = 12$$

Similarly for division

$$\sqrt{10} \div \sqrt{2} = \sqrt{(10 \div 2)} = \sqrt{5}$$

and

$$10\sqrt{3} \div 2\sqrt{3} = 5\sqrt{\left(\frac{3}{3}\right)} = 5 \quad (\sqrt{1} = 1)$$

Example 4.1
Evaluate the following expressions

(a) $\sqrt{50} \times \sqrt{98}$: $\sqrt{(50 \times 98)} = \sqrt{(100 \times 49)} = 10 \times 7 = \underline{70}$

(b) $\sqrt{72} \div \sqrt{2}$: $\sqrt{(72 \div 2)} = \sqrt{36} = \underline{6}$

(c) $2\sqrt{13} \times 5\sqrt{13}$: $2 \times 5 \times \sqrt{(13^2)} = 10 \times 13 = \underline{130}$

4.3 INDICES

Positive and negative powers of 10 were introduced in Chapters 1 and 3 of this book, but index notation and the rules for combining numbers in index form may be applied to any base and any index. These rules are stated here, together with simple numerical examples, and they are given in a general algebraic form in Section 7.2.

(1) To multiply powers of the same base, add the indices

$$2^3 \times 2^5 = 2^{3+5} = 2^8$$

(2) To divide powers of the same base number, subtract the indices

$$3^6 \div 3^2 = 3^{6-2} = 3^4$$

(3) To raise to a power a number in index form, multiply the indices

$$(5^3)^2 = 5^{3\times 2} = 5^6$$

(4) A number with a negative index is the reciprocal of the same base number with the corresponding positive index

$$(7^2)^{-3} = 7^{2\times -3} = 7^{-6} = \frac{1}{7^6}$$

Fractional indices

By the definition of a square root, $\sqrt{5} \times \sqrt{5} = 5$ and by the rule for multiplying powers $5^{1/2} \times 5^{1/2} = 5^{1/2+1/2} = 5^1 = 5$. Since $\sqrt{5} \times \sqrt{5} = 5$ and $5^{1/2} \times 5^{1/2} = 5$, $\sqrt{5}$ and $5^{1/2}$ must represent the same number, and a positive number raised to any power has a positive value. Similarly $\sqrt[3]{8} \times \sqrt[3]{8} \times \sqrt[3]{8} = \sqrt[3]{(8 \times 8 \times 8)} = 8$ and $8^{1/3} \times 8^{1/3} \times 8^{1/3} = 8^1 = 8$. Therefore $8^{1/3}$ and $^3\sqrt{8}$ must represent the same number.

Example 4.2

Written in index form

(a) $\sqrt[5]{6^2} = (6^2)^{1/5} = \underline{6^{2/5}}$

(b) $\sqrt[4]{5^3} = (5^3)^{1/4} = \underline{5^{3/4}}$

Written in surd form

(c) $80^{1/2} = \sqrt{(16 \times 5)} = \sqrt{16} \times \sqrt{5} = \underline{4\sqrt{5}}$

(d) $16^{-2/3} = \dfrac{1}{16^{2/3}} = \dfrac{1}{\sqrt[3]{16^2}} = \dfrac{1}{\sqrt[3]{(64 \times 4)}} = \underline{\dfrac{1}{4\sqrt[3]{4}}}$

It is usual to reduce the surd as far as possible.

Example 4.3

Evaluate the following expressions (a) $25^{1/2}$, (b) $(7^{1/4})^4$, (c) $2^2 \times 2^{-4}$, (d) $81^{-1/4}$, (e) $14^3 \div 7^3$, (f) $8 \div 8^{1/3}$.

Solution

(a) $25^{1/2} = \sqrt{25} = \underline{5}$

(b) $(7^{1/4})^4 = 7^{1/4 \times 4} = 7^1 = \underline{7}$

(c) $2^2 \times 2^{-4} = 2^{2+-4} = 2^{-2} = \dfrac{1}{2^2} = \underline{\dfrac{1}{4}}$

(d) $$81^{-1/4} = \frac{1}{81^{1/4}} = \frac{1}{\sqrt[4]{81}} = \underline{\frac{1}{3}}$$

(e) $$14^3 \div 7^3 = \left(\frac{14}{7}\right)^3 = 2^3 = \underline{8}$$

(f) $$8 \div 8^{1/3} = 8^{1-1/3} = 8^{2/3} = \sqrt[3]{8^2} = \underline{4}$$

Exercise 4.1

Evaluate the following without using tables or a calculator.

(1) $\sqrt{5} \times \sqrt{20}$. (2) $\sqrt{12} \times \sqrt{27}$. (3) $\sqrt{8} \times \sqrt{50}$.
(4) $\sqrt{48} \div \sqrt{12}$. (5) $\sqrt{108} \div \sqrt{27}$. (6) $9^{1/2}$.
(7) $8^{1/3}$. (8) $25^{1/2}$. (9) $27^{2/3}$. (10) $8^{-2/3}$.

Express the following surds in their lowest terms.

(11) $\sqrt{12}$. (12) $\sqrt{50}$. (13) $\sqrt{32}$. (14) $\sqrt{50} - \sqrt{18}$.

Express as a single power each of the following.

(15) $3^{1/4} \times 3^{3/4}$. (16) $2^{-1/2} \times 2^{7/2}$. (17) $5^{-1/2} \times 5^{-1/4}$.
(18) $3^4 \div 3^{1/2}$. (19) $7^{1/2} \div 7^{-3/2}$.

Write the following as single numbers in index form.

(20) $\dfrac{1}{2^5}$. (21) $\sqrt{5}$. (22) $\dfrac{1}{\sqrt{3}}$. (23) $\sqrt[3]{6^2}$.

(24) $\dfrac{1}{7^{-1/2}}$.

4.4 LIMITS OF ACCURACY AND MAXIMUM ERROR

In arithmetic a number has an exact value, but a measurement such as the length of a line or the mass of an object can never be exact, its value depends on the accuracy of the measuring instrument.

A length measured as 7.1 m must be between the limits 7.05 m and 7.15 m and the maximum error is 0.05 m; but a length of 7.10 m must be between 7.095 m and 7.105 m, so that the maximum error is 0.005 m.

Results of calculations should be corrected to at most one significant figure more than the least accurate value given. However, when four-figure tables are used the last digit is not completely reliable, and answers should be stated correct to three figures, even when all the given values have four figures.

Example 4.4

$$2.05 \times 12.15 = 24.9075 \text{ exactly}$$

The area of a strip of carpet 2.05 metres by 12.15 metres could be as low as 2.045 m × 12.145 m = 24.836525 m^2 or as high as 2.055 m × 12.155 m

= 24.978525 m^2. The answer should be corrected to at most four figures, since 2.05 has three figures, and would be 24.91 or 24.9 m^2.

4.5 AIDS TO COMPUTATION

Mathematical tables and slide rules were used extensively for calculations in science and engineering before the advent of the inexpensive electronic calculators which are available nowadays. Many Examining Boards allow the use of four-figure tables or calculators in at least part of their O-level examination papers, and students should read the syllabus and examination instructions to check whether or not their use is permitted in a particular paper or question.

4.6 OBTAINING VALUES FROM FOUR-FIGURE TABLES

Four-figure tables of squares, square roots, reciprocals, logarithms and antilogarithms all give values for four-digit numbers, usually for numbers between 1 and 10. The position of the decimal point is determined separately. The tables used in this book are those by Frank Castle (published by Macmillan) and part of a typical table is shown in Table 4.1.

To 'look up' a given four-digit number, the first two digits are in the left column of the table and locate the correct row. The third digit of the given number locates the correct column headed 0 to 9, and the fourth digit is in one of the mean differences columns.

Example 4.5
Look up in Table 4.1(a) 1.26, (b) 1.408, (c) 1.013.

Solution (a) 1.26. In row 1.2 and column 6 the value is 2311. Since the last digit of the given number is 0 there is nothing to add, and the required value is 0.2311.

(b) 1.408. The tabulated value in row 1.4 and column 0 is 0.3365. In the mean difference column 8 of that row the value is 55, and this is added to 3365 to give 3420. The required value is 0.3420.

(c) 1.013. In row 1.0 and column 1 the value is 0100, and the difference for 3 in that row is 29. 0100 + 29 = 0129. The required value is therefore 0.0129.

4.7 THE TABLE OF SQUARES

Squaring a number means raising it to the power 2, and since the values given in the table of squares are for numbers between 1 and 10, the given number is first written in standard index form as shown in Section 3.3.

Table 4.1

	0	1	2	3	4	5	6	7	8	9	mean differences 1	2	3	4	5	6	7	8	9
1.0	0.0000	0100	0198	0296	0392	0488	0583	0677	0770	0862	10	19	29	38	48	57	67	76	86
1.1	0.0953	1044	1133	1222	1310	1398	1484	1570	1655	1740	9	17	26	35	44	52	61	70	78
1.2	0.1823	1906	1989	2070	2151	2231	2311	2390	2469	2546	8	16	24	32	40	48	56	64	72
1.3	0.2624	2700	2776	2852	2927	3001	3075	3148	3221	3293	7	15	22	30	37	44	52	59	67
1.4	0.3365	3436	3507	3577	3646	3716	3784	3853	3920	3988	7	14	21	28	35	41	48	55	62
1.5	0.4055	4121	4187	4253	4318	4283	4447	4511	4574	4637	6	13	19	26	32	39	45	52	58
1.6	0.4700	4762	4824	4886	4947	5008	5068	5128	5188	5247	6	12	18	24	30	36	42	48	55
1.7	0.5306	5365	5423	5481	5539	5596	5653	5710	5766	5822	6	11	17	24	29	34	40	46	51
1.8	0.5878	5933	5988	6043	6098	6152	6206	6259	6313	6366	5	11	16	22	27	32	38	43	49
1.9	0.6419	6471	6523	6575	6627	6678	6729	6780	6831	6881	5	10	15	20	26	31	36	41	46

Example 4.6
Use four-figure tables of squares to find the value of (a) $0.037\,25^2$. (b) 7249^2.

Solution

(a) $0.037\,25 = 3.725 \times 10^{-2}$; $0.037\,25^2 = 3.725^2 \times 10^{-4}$ (1)

In row 3.7 and column 2 of the table of squares the value is 13.84 and in the mean differences column 5 of that row the value is 4. 1384 + 4 = 1388, therefore $3.725^2 = 13.88$. From (1) $0.03725^2 = 13.88 \times 10^{-4}$ or 0.00139.

(b) $7249 = 7.249 \times 10^3$; $7249^2 = 7.249^2 \times 10^6$ (2)

In row 7.2 column 4 the value is 52.42, and the mean difference for 9 is 13. 5242 + 13 = 5255 so the required value is 52.55. From (2) $7249^2 = 52.55 \times 10^6 = 5.26 \times 10^7$ to 3 sig. fig.

4.8 THE TABLE OF RECIPROCALS

Reciprocal tables are used in the same way as the tables of squares, except that the mean differences are subtracted. This is because a greater number has a smaller reciprocal.

Example 4.7
Use four-figure tables to find the reciprocal of (a) 320.7, (b) 0.008 125.

Solution (a) In standard form

$$320.7 = 3.207 \times 10^2 \quad \text{and} \quad \frac{1}{10^2} = 10^{-2}$$

Therefore

$$\frac{1}{320.7} = \frac{1}{3.207} \times 10^{-2}$$

From the table of reciprocals, the reciprocal of 3.20 is 0.3125. The mean difference is 7 which is subtracted to give 0.3118. The reciprocal of 320.7 is $0.3118 \times 10^{-2} = 0.00312$.

(b) $0.008\,125 = 8.125 \times 10^{-3}$

$$\frac{1}{0.008125} = \frac{1}{8.125} \times 10^3 \qquad \text{since} \quad \frac{1}{10^{-3}} = 10^3$$

From the tables the reciprocal of 8.125 is 0.1231 and so the reciprocal of 0.008 125 is $0.1231 \times 10^3 = 123$.

Example 4.8

Evaluate correct to 3 dp (a) $\frac{1}{2.015} - \frac{1}{13.7}$, (b) $\frac{1}{15^2} + \frac{1}{17^2}$

Solution

(a) $$\frac{1}{13.7} = \frac{1}{1.37} \times 10^{-1}$$

From reciprocal tables

$$\frac{1}{2.015} = 0.4963, \frac{1}{1.37} = 0.7299$$

$$0.4963 - 0.07299 = 0.42331 = \underline{0.423} \text{ to 3 dp}$$

(b) From the table of squares $15^2 = 225$, $17^2 = 289$.
From reciprocal tables

$$\frac{1}{225} = \frac{1}{2.25} \times 10^{-2} = 0.4444 \times 10^{-2}$$

and

$$\frac{1}{289} = \frac{1}{2.89} \times 10^{-2} = 0.3460 \times 10^{-2}$$

$$\frac{1}{15^2} + \frac{1}{17^2} = (0.4444 + 0.3460) \times 10^{-2} = 0.7904 \times 10^{-2}$$

$$= \underline{0.008} \text{ to 3 dp}$$

Exercise 4.2

Use four-figure tables to find the square of the following.

(1) 3.178. (2) 0.028. (3) 428.6. (4) 15×10^{-3}.

Find the reciprocal of the following.

(5) 4.82. (6) 62.9. (7) 0.52. (8) 0.024.

(9) 2.35×10^{-2}.

(10) Evaluate $\frac{1}{6.3} - \frac{1}{21.5}$. (11) Evaluate $\frac{1}{31.2} + \frac{1}{4.5}$.

(12) If the reciprocal of a number is 0.1736, what is the number?

4.9 SQUARE ROOT TABLES

To take the square root of a power of 10 the index is divided by 2 and so the given number is first written as the product of a number between 1 and 100 and an even power of 10. Thus

$$\sqrt{(3.5 \times 10^{-2})} = \sqrt{3.5} \times 10^{-1} \quad \text{and} \quad \sqrt{(17 \times 10^{2})} = \sqrt{17} \times 10^{1}$$

There are two different kinds of square root tables in common use. In one type there are separate tables for numbers between 1 and 10 and for numbers between 10 and 100. In the other kind two different values are tabulated for each number, with 1 to 10 at the top. The tables are used in the same way as the tables of squares, mean differences being added.

Example 4.9

Given that $\sqrt{1.2} = 1.095$, which of the following square roots can be stated without using tables (a) $\sqrt{12}$, (b) $\sqrt{120}$, (c) $\sqrt{0.12}$, (d) $\sqrt{0.012}$?

Solution

$$12 = 1.2 \times 10^{1} \qquad 120 = 1.2 \times 10^{2} \qquad 0.12 = 1.2 \times 10^{-1}$$
$$0.012 = 1.2 \times 10^{-2}$$

Only (b) and (d) can be found without using tables because they have even powers of 10

$$\sqrt{120} = 1.095 \times 10 = \underline{10.95}; \qquad \sqrt{0.012} = 1.095 \times 10^{-1} = \underline{0.1095}$$

Example 4.10

Using square root tables evaluate (a) $\sqrt{3.354}$, (b) $\sqrt{0.007\,29}$, (c) $\sqrt{316}$.

Solution

(a) $\sqrt{3.354} = 1.831 = \underline{1.83}$

(b) $\sqrt{0.007\,29} = \sqrt{(72.9 \times 10^{-4})} = \sqrt{72.9} \times 10^{-2}$

$= 8.538 \times 10^{-2} = \underline{0.0854}$

(c) $\sqrt{316} = \sqrt{(3.16 \times 10^{2})} = \sqrt{3.16} \times 10^{1}$

$= 1.778 \times 10 = \underline{17.8}$

Exercise 4.3

Evaluate the following.

(1) $\sqrt{4.8}$. (2) $\sqrt{72.92}$. (3) $\sqrt{0.581}$.

(4) $\sqrt{0.006085}$. (5) $\sqrt{(6.82 \times 10^{-3})}$. (6) $\sqrt{48.3} - \sqrt{9.65}$.

(7) $1/\sqrt{3.5} + 1/\sqrt{15.8}$.

(8) Without using tables state the value of (a) $\sqrt{0.04}$, (b) $\sqrt{0.16}$, (c) $\sqrt{0.0064}$.

(9) Given that $\sqrt{5.6} = 2.3664$ write down the square root of (a) 560, (b) 0.056, (c) 0.000 56.

4.10 LOGARITHMS AND ANTILOGARITHMS

When a number is written as a power of 10, the index is called the logarithm to base 10 or the common logarithm of the number. Since $100 = 10^2$, 2 is the logarithm of 100, $\log 100 = 2$.

Example 4.11

Number	*Power of 10*	*Common logarithm*
1000	10^3	$\log 1000 = 3$
0.01	10^{-2}	$\log 0.01 = -2$
10	10^1	$\log 10 = 1$
1	10^0	$\log 1 = 0$

It can be seen from Example 4.11 that any number between 1 and 10 must have a common logarithm between 0 and 1, and these are the values given in the four-figure tables of logarithms.

The logarithms of numbers between 1 and 10 are found from logarithm tables in the same way as from the tables of squares, the mean differences are added.

Example 4.12

$\log 2.0000 = 0.3010$ $\qquad$ $\log 5.0000 = 0.6990$

$\log 3.520 = 0.5465$ $\qquad$ $\log 7.841 = 0.8944$

Numbers which are not between 1 and 10 are first written in standard index form. When powers of 10 are multiplied the logarithms are added because a logarithm is an index. For example

$$352 = 3.52 \times 10^2$$

$$\log 352 = \log(3.52 \times 10^2) = \log 3.52 + \log 10^2$$

$$= 0.5465 + 2$$

$$= 2.5465$$

$$0.002 = 2 \times 10^{-3}$$

$$\log 0.002 = \log(2 \times 10^{-3}) = \log 2 + \log 10^{-3}$$

$$= 0.3010 + -3$$

The logarithm of the number between 1 and 10 obtained from the table is called the *mantissa* and the integer added to it is called the *characteristic*. The mantissa is always positive, and so the negative sign of a negative characteristic is placed above the integer, which is called a bar number.

$$\log 0.002 = 0.3010 + -3 = \bar{3}.3010$$

'bar three point three nought one nought'.

Example 4.13

Number	*Standard form*	*Mantissa*	*Characteristic*	*Logarithm*
2276	2.276×10^3	.3572	3	3.3572
370.3	3.703×10^2	.5685	2	2.5685
79.54	7.954×10^1	.9006	1	1.9006
0.175	1.750×10^{-1}	.2430	$\bar{1}$	$\bar{1}.2430$
0.0357	3.570×10^{-2}	.5527	$\bar{2}$	$\bar{2}.5527$

Antilogarithms

If a logarithm is known, the number which it represents can be found by 'looking up' the mantissa is antilogarithm tables. These are used in the same way as the other four-figure tables. The position of the decimal point in the answer is determined by the value of the characteristic.

Example 4.14

Logarithm	*Antilogarithm*	*Number*
0.4948	3.125×10^0	3.125
1.3973	2.497×10^1	24.97
3.9256	8.426×10^3	8426
$\bar{1}.5527$	3.571×10^{-1}	0.3571

Exercise 4.4

Find the common logarithm of the following.

(1) 4.86. (2) 38.16 (3) 72 900. (4) 0.561.
(5) 0.0080 243. (6) 0.000 2854. (7) 17×10^{-20}.

What is the antilogarithm of the following?

(8) 0.5794. (9) 2.8470. (10) $\bar{1}.2423$.
(11) $\bar{2}.9121$. (12) 6.0000.

4.11 MULTIPLICATION AND DIVISION USING LOGARITHMS

To multiply powers of 10, the indices are added, and therefore to multiply numbers the logarithms are added. When numbers are divided the logarithms are subtracted.

Before working the following exercises it is advisable to revise the addition and subtraction of directed numbers in Chapter 1.

$$\bar{1} + \bar{2} = \bar{3}, \qquad \bar{2} - \bar{1} = \bar{1}, \qquad \bar{1} - 1 = \bar{2}, \qquad \bar{1} - \bar{2} = 1$$

Example 4.15

(a) Use logarithm tables to evaluate $0.175 \times 3.124 \times 24.96$.

Solution

$0.175 = 1.75 \times 10^{-1}$

$24.96 = 2.496 \times 10^{1}$

Number	*Logarithm*
0.175	$\bar{1}.2430$
3.124	0.4948
24.96	1.3973
13.65	1.1351 Add

Answer: 13.7

(b) Find the value of $0.0357 \div 7.53$.

Solution

$0.0357 = 3.570 \times 10^{-2}$

Number	*Logarithm*
0.0357	$\bar{2}.5527$
7.53	0.8768
4.742×10^{-3}	$\bar{3}.6759$ Subtract

Answer: 4.74×10^{-3}.

Example 4.16

Evaluate $\dfrac{139.2 \times 7.634}{27.5 \times 0.0355}$.

Solution The logarithms of the numerator and denominator are calculated separately and then subtracted

$139.2 = 1.392 \times 10^{2}$

$27.5 = 2.75 \times 10^{1}$

$0.0355 = 3.550 \times 10^{-2}$

Number	*Logarithm*
27.5	1.4393
0.0355	$\bar{2}.5502$
	$\bar{1}.9895$ Add
139.2	2.1437
7.634	0.8827
	3.0264 Add
	$\bar{1}.9895$
1088	3.0369 Subtract

Answer: 1090.

Exercise 4.5

Write the following as single logarithms.

(1) $3.1020 + \bar{1}.9121$. (2) $\bar{1}.2314 + \bar{2}.3010$.

(3) $4.2735 - \bar{1}.9132$. (4) $\bar{2}.6834 - \bar{1}.3412$.

(5) $\bar{2}.4711 - \bar{4}.6990$.

Evaluate the following.

(6) 47.23×1.82. (7) $0.053\,41 \times 7.12$. (8) $0.241 \div 3.1$.

(9) $\dfrac{3.142 \times 0.825}{0.003\,57}$. (10) $\dfrac{8.48 \times 3.47}{340.6}$.

(11) A pile of paper 29 mm thick contains 556 sheets. Assuming that they are all the same size, calculate the thickness of one sheet.

(12) Calculate the value of Q when $Q = \frac{1}{2} \times 0.784 \times 13.91$.

(13) Find the total cost (to nearest 10 p) of 138 reels of film at £3.84 per reel and 76 packets of splicing joints at £1.15 per packet.

(14) A car travels 2188 km and uses 148 litres of petrol. What is the petrol consumption in kilometres per litre?

4.12 CALCULATION OF POWERS AND ROOTS USING LOGARITHMS

When a given number is raised to a positive power, the logarithm of the given number is multiplied by the index. For example

$$\log (5^3) = \log (5 \times 5 \times 5) = \log 5 + \log 5 + \log 5$$

$$= (\log 5) \times 3 \text{ or } 3 \log 5$$

Example 4.17
Evaluate (a) 17.45^3, (b) 1.024^5, (c) 0.0075^4.

Solution

(a) $\log 17.45 = 1.2418$

$\log 17.45^3 = 1.2418 \times 3$

$= 3.7254$

antilog $3.7254 = 5314$

$17.45^3 = \underline{5310}$

(b) $\log 1.024 = 0.0103$

$\log 1.024^5 = 0.0103 \times 5$

$= 0.0515$

antilog $0.0515 = 1.126$

$1.024^5 = \underline{1.13}$

(c) $\log 0.0075 = \bar{3}.8751$

$\log 0.0075^4 = \bar{3}.8751 \times 4$ $\qquad \bar{3} \times 4 = \overline{12}$

$= \bar{9}.5004$ $\qquad .8751 \times 4 = 3.5004$

antilog $\bar{9}.5004 = 3.165 \times 10^{-9}$

$0.0075^4 = \underline{3.17 \times 10^{-9}}$

Since extracting a root is equivalent to raising to a fractional power logarithm tables may also be used to evaluate roots. For instance, $\sqrt[3]{24} = 24^{1/3}$ and so

$$\log \sqrt[3]{24} = \log 24^{1/3} = \tfrac{1}{3} \times \log 24 = \log 24 \div 3$$

Example 4.18
Evaluate (a) $\sqrt[3]{1762}$, (b) $\sqrt{1.351}$, (c) $\sqrt[5]{27.41}$.

Solution

(a) $\log 1762 = 3.2460$

$\log \sqrt[3]{1762} = 3.2460 \div 3$

$= 1.0820$

antilog $1.0820 = 12.08$

$\sqrt[3]{1762} = \underline{12.1}$

(b) $\log 1.351 = 0.1306$

$\log \sqrt{1.351} = 0.1306 \div 2$

$= 0.0653$

antilog $0.0653 = 1.162$

$\sqrt{1.351} = \underline{1.16}$

(c) $\log 27.41 = 1.4380$

$\log \sqrt[5]{27.41} = 1.4380 \div 5$

$= 0.2876$

antilog $0.2876 = 1.939$

$\sqrt[5]{27.41} = \underline{1.94}$

Dividing a logarithm with a negative characteristic

In a bar number only the characteristic is negative and it must be made up to a complete multiple before dividing.

Example 4.19

(a) $\bar{3}.2180 \div 2 = (\bar{4} + 1.2180) \div 2 = \underline{\bar{2}.6090}$

(b) $\bar{2}.7283 \div 3 = (\bar{3} + 1.7283) \div 3 = \underline{\bar{1}.5761}$

(c) $\bar{4}.6256 \div 3 = (\bar{6} + 2.6256) \div 3 = \underline{\bar{2}.8752}$

(d) $\bar{3}.8965 \div 5 = (\bar{5} + 2.8965) \div 5 = \underline{\bar{1}.5793}$

Example 4.20

Evaluate (a) $\sqrt[3]{0.026\,65}$, (b) $\sqrt[4]{0.005\,862}$.

Solution

(a) $\log 0.026\,65 = \bar{2}.4257$

$\log \sqrt[3]{0.026\,65} = \bar{2}.4257 \div 3$

$= (\bar{3} + 1.4257) \div 3$

$= \bar{1}.4752$

antilog $\bar{1}.4752 = 0.2986$

$\sqrt[3]{0.026\,65} = \underline{0.299}$

(b) $\log 0.005\,862 = \bar{3}.7680$

$\log \sqrt[4]{0.005\,862} = \bar{3}.7680 \div 4$

$= (\bar{4} + 1.7680) \div 4$

$= \bar{1}.4420$

antilog $\bar{1}.4420 = 0.2767$

$\sqrt[4]{0.005\,862} = \underline{0.278}$

Exercise 4.6

Write as a single logarithm the following.

(1) $\bar{2}.3010 \times 3$. (2) $\bar{3}.4771 \times 6$. (3) $\bar{4}.6990 \times 3$.

(4) $4.8603 \div 3$. (5) $0.6360 \div 4$. (6) $\bar{3}.6615 \div 3$.
(7) $\bar{1}.4372 \div 2$. (8) $\bar{2}.6412 \div 2$. (9) $\bar{3}.4257 \div 5$.

Evaluate the following.

(10) 4.367^3. (11) 0.645^4. (12) $\sqrt[3]{1.059}$.
(13) $\sqrt[4]{6129}$. (14) $\sqrt[3]{0.05216}$. (15) $\sqrt[4]{0.00273}$.
(16) $\sqrt[5]{0.0465}$.

(17) $\dfrac{30 \times 2.056 \times 0.784^3}{0.032^{1/2}}$. (18) $\dfrac{21.16^4 - 16.21^4}{0.18}$.

(19) An investment of £580 in a building society for 7 years at 11.2% amounts to £580 $\times$ $(1.112)^7$. What is the amount?

(20) Calculate the energy, E joules, radiated per second from a body at a temperature of 492 K, given by $E = 0.8 \times 5.7 \times 10^{-8} \times 492^4$.

(21) Find the sum S of the first ten terms of the series $\frac{1}{9} + \frac{1}{3} + 1 + \ldots$ given that $S = (3^{10} - 1)/18$.

4.13 THE USE OF ELECTRONIC CALCULATORS

Electronic calculators are widely used in commercial and industrial situations as well as in scientific work, and they are becoming more common in schools and colleges. Most calculators have keys for evaluating powers and roots as well as the basic arithmetic operations, and many have additional function keys and a memory storage facility.

The calculations in the instruction manual provided with the calculator should be studied carefully and then practised for speed with accuracy. When the instrument is familiar, Exercises 4.3 to 4.6 should be worked, and the answers compared with those at the end of the book, which are given correct to 3 sig. fig.

The following general rules apply to most models:

(1) Sacrifice some of the speed of calculation for the sake of accuracy. Write down the answer and then repeat the calculation.

(2) Answers should be corrected to the number of figures warranted by the data used. (Limits of accuracy were discussed in Section 4.4.)

(3) Not all models take account of the precedence rules for operations. For example, if the keys $3 \times 5 + 2 \times 6$ are pressed in that order, some will give the correct answer 27, but others require the use of a memory store in place of the brackets $(3 \times 5) + (2 \times 6)$. Check whether $14 - 3 \times 4$ gives the correct answer 2 on your machine.

(4) If a battery is used in the calculator, switch off between calculations. If the display figures begin to fade, recharge or replace the batteries according to the instructions in the manual.

(5) Remember to charge the battery or buy a spare before an examination.

CHAPTER 5

PERCENTAGE, RATIO AND PROPORTION

5.1 PERCENTAGE

Per cent (symbol %) means per hundred, and a percentage may be considered as a common fraction with 100 as denominator

$$3\% = \frac{3}{100}, \quad 72\% = \frac{72}{100} \text{ or } \frac{18}{25}, \quad 2\tfrac{1}{2}\% = \frac{5}{200} \text{ or } \frac{1}{40}$$

A percentage may also be expressed as a decimal fraction, so that

$$7\% = 7 \div 100 = 0.07, \quad 2.5\% = 0.025, \quad 100\% = 1.00$$

To change a percentage to a fraction, divide by one hundred.

Example 5.1

(a) Write as common fractions 13%, $33\frac{1}{3}\%$, $7\frac{1}{2}\%$.

Solution

$$13\% = \frac{13}{100}, \quad 33\tfrac{1}{3}\% = \frac{100}{3} \div 100 = \frac{1}{3}, \quad 7\tfrac{1}{2}\% = \frac{15}{200} = \frac{3}{40}$$

(b) Write as decimal fractions 7.15%, 85.2%, 1355%.

Solution

$$7.15\% = 7.15 \div 100 = \underline{0.0715}$$

$$85.2\% = 85.2 \div 100 = \underline{0.852}$$

$$1355\% = 1355 \div 100 = \underline{13.55}$$

The reverse process, changing a fraction to a percentage, is equivalent to multiplying by 100

$\frac{1}{2} = (\frac{1}{2} \times 100)\% = 50\%$, $\quad \frac{1}{4} = (\frac{1}{4} \times 100)\% = 25\%$

In mixed numbers, each whole number is 100%, so that $2\frac{1}{2} = (200 + 50)\%$ and $1\frac{1}{4} = 125\%$.

Example 5.2
Write as percentages (a) $\frac{3}{5}$, $1\frac{5}{12}$, $3\frac{7}{8}$, (b) 0.015, 3.047.

Solution

(a) $\frac{3}{5} = (\frac{3}{5} \times 100)\% = \underline{60\%}$

$1\frac{5}{12} = (100 + \frac{5}{12} \times 100)\% = \underline{141\frac{2}{3}\%}$

$3\frac{7}{8} = (300 + \frac{7}{8} \times 100)\% = \underline{387\frac{1}{2}\%}$

(b) $0.015 = \underline{1.5\%}$, $\quad 3.047 = \underline{304.7\%}$

Example 5.3
(a) Calculate 14% of $3\frac{1}{7}$.

Solution

$$\frac{14}{100} \times \frac{22}{7} = \underline{\frac{11}{25}}$$

(b) What length in millimetres is 0.2% of 7 metres?

Solution

0.2% = 0.002, 7 m = 7000 mm

0.002 × 7000 mm = $\underline{14 \text{ mm}}$

(c) If the owner of a store offers a discount of 5% on all goods, what was the reduction in price of an article marked at £7.20?

Solution

5% = 0.05, 0.05 × 7.20 = 0.36

The reduction in price was $\underline{\text{£0.36 or 36 p.}}$

Exercise 5.1
Express the following fractions as percentages.
(1) $\frac{1}{4}$. (2) $\frac{3}{8}$. (3) $\frac{4}{5}$. (4) $2\frac{1}{3}$. (5) $4\frac{1}{8}$. (6) 0.28.
(7) 12.7.
Write the following percentages as common fractions.
(8) 50%. (9) 28%. (10) $66\frac{2}{3}\%$. (11) $78\frac{1}{2}\%$.

Evaluate the following.

(12) 15% of £98.20. (13) 35% of 520 mm.

(14) 12% of 2800 bricks. (15) $2\frac{1}{4}$% of $17\frac{1}{9}$.

(16) If 48 employees at a factory were absent and this was 4% of the people employed, calculate the total number employed at the factory.

(17) Calculate the percentage of rejects correct to one decimal place, if 70 out of 1230 machine parts are sub-standard.

(18) If I spend £85.50 a month on household expenses, what percentage is this of a monthly salary of £380?

(19) An agent received £235.75 which was 2% commission on the price of an article sold. What was the sale price?

(20) A tennis racquet priced at £32 was reduced by 10% in a sale. If a sales tax was charged at 15% of the selling price, how much was paid for the racquet?

(21) 70% of the students in a college eat sweets during the weekend, but of these only 20% eat chocolate. What percentage of the college students eat chocolate?

(22) In a certain firm, the basic pay of a salesman is £52 per week, and in addition he receives $1\frac{1}{2}$% commission on his sales over £3500 and 3% on sales over £6000. How much pay did he receive in a week when he sold £8700 worth of goods?

5.2 RATIO AND PROPORTION

A ratio is a fraction. For instance, if in a class there are 12 girls and 15 boys, the *ratio* of girls to boys may be written 12/15 or 12 : 15. A ratio may be cancelled in the same way as a common fraction, and both values must be in the same units.

Example 5.4

(a) What is the ratio of 20 cm to 15 m in the form 1 : *n*?

Solution 15 m = 1500 cm. The required ratio is 20 : 1500 or 1 : 75.

(b) Find the ratio of 24 p to £36.

Solution 24 p : 3600 p = 1 : 150.

Proportion is another name for fraction. If there are 12 girls in a class of 27, the *proportion* of girls in the class is 12/27 and the proportion of boys is 15/27.

Example 5.5

What proportion of the output of light bulbs were satisfactory if 150 were rejected from a batch of 30 000?

Solution The proportion rejected was

$$\frac{150}{30\,000} = \frac{1}{200} \text{ or } 0.5\%$$

The proportion that were satisfactory was 199/200 or 99.5%.

5.3 CHANGING A QUANTITY IN A GIVEN RATIO

To change a quantity in a given ratio it is multiplied by the equivalent fraction. If a price increases in the ratio 3 : 2 it is multiplied by 3/2.

Example 5.6

(a) Increase £4.50 in the ratio 7 : 5.

Solution

£4.50 × (7/5) = £6.30

(b) Decrease 60 kilometres in the ratio 5 : 8.

Solution

60 km × (5/8) = 37.5 km

Example 5.7

If 4 men can load 5 tonnes of goods in 2 hours how long would it take 6 men working at the same rate to load 8 tonnes?

Solution The time, 2 hours, is increased in the ratio 8 : 5 for the weight and decreased in the ratio 4 : 6 for the number of men, because more men would take less time

2 hours × (8/5) × (4/6) = (32/15) hours = 2 hours 8 minutes

Example 5.8

Pauline pays £14.52 tax which is 22% of her weekly salary. Calculate her weekly salary.

Solution The whole salary is 100% and £14.52 is 22%. £14.52 is increased in the ratio 100 : 22

£14.52 × (100/22) = £66

Exercise 5.2

Express in the ratio 1 : *n* the following.

(1) 3 : 1.5. (2) 0.5 : 4.2. (3) $3\frac{1}{2}$: 0.7 (4) 4 p : £1.

(5) 5 mm to 3.5 cm. (6) 5 seconds to 4 minutes.
(7) Increase £16.25 in the ratio 9 : 5.
(8) Decrease 5.85 km in the ratio 1 : 6.5.
(9) After a rise of $7\frac{1}{2}$% a person's salary was £5807.15. How much did he earn before the increase?
(10) In a canteen it costs 97.5 p to make 30 cups of tea. What is the cost of making 156 cups?
(11) A tablet-making machine uses 12 g of powder every 5 s. How many complete minutes would it take to empty a hopper containing 15 kg of the powder?
(12) If two painters take 6 h to paint 54 m of railing on a promenade, calculate the time taken by 5 people to paint 450 m at the same rate.

5.4 CHANGE OF UNITS

In scientific work it is necessary to express measurements in different units, and in commerce, too, money is changed from one currency to another. These are further examples of changing a quantity in a given ratio.

(a) SI units

The following table shows the metric units of length and mass in common use. The basic SI unit of length is the metre and of mass is the kilogram, and these, together with the recommended decimal multiples and sub-multiples are marked by asterisks.

Prefix	kilo	hecto	deca		deci	centi	milli	micro
Symbol	k	h	da		d	c	m	μ
Multiple	10^3	10^2	10	1	10^{-1}	10^{-2}	10^{-3}	10^{-6}
Length	km*			m*		cm*	mm*	μm*
Mass	kg*			g*			mg*	μg*

1 metric tonne = 1000 kg

The common unit for fluid measure is the litre (l).

Example 5.9

(a) Express a speed of 50 centimetres per second in kilometres per hour.

Solution 1 km = 100 000 cm 1 h = 3600 s. 50 is increased in the ratio 3600 : 1 for the time and decreased in the ratio 1 : 100 000 for the distance.

$$50 \text{ cm/s} = 50 \times \frac{3600}{1} \times \frac{1}{100\,000} = \underline{1.8 \text{ km/h}}$$

(b) Calculate the mass in kg of 800 m of uniform wire if the mass per mm is 1.5 mg.

Solution 800 m = 800 000 mm 1 kg = 1 000 000 mg

$$\text{mass of 800 m} = 800\,000 \times 1.5 \text{ mg}$$

$$= \frac{800\,000 \times 1.5}{1\,000\,000} \text{ kg}$$

$$= \underline{1.2 \text{ kg}}$$

Example 5.10

On a certain day £1 sterling was worth $2.21, 4.37 guilders, or 1850 lire. Calculate (a) the number of lire exchanged for 10 guilders, (b) the value in guilders of $265.20.

Solution

(a) 1850 lire is increased in the ratio 10 : 4.37

$$\frac{1850 \times 10}{4.37} = \underline{4233} \quad \text{(to nearest lira)}$$

(b) 4.37 guilders is increased in the ratio 265.20 : 2.21

$$4.37 \times \frac{265.20}{2.21} = \underline{524.40 \text{ guilders}}$$

(b) Scales and maps

Maps and diagrams which are drawn to scale are reduced in a fixed ratio. If a survey map is drawn to a scale of 1 : 25 000, then all distances are decreased in the ratio 1 : 25 000, and 1 mm on the map represents 25 000 mm on the ground.

Example 5.11

On a map drawn to a scale of 1 in 50 000 (a) what distance is represented by a line of length 24 mm, (b) what length on the map represents a 7.5 km length of motorway?

Solution

(a) 1 : 50 000 = 24 mm : 50 000 × 24 mm or $\underline{1200 \text{ m}}$

(b) 7.5 km = 7 500 000 mm

$$7.5 \text{ km reduced in the ratio } 1 : 50\,000 = 7\,500\,000 \div 50\,000 \text{ mm}$$

$$= \underline{150 \text{ mm}}$$

Example 5.12
If an architect's plans for a new college building were drawn to a scale of 1 : 750, calculate (a) the size on the plan of a lecture theatre 30 m by 18 m, (b) the length of a corridor shown on the plan as 54 mm.

Solution

(a) $$30 \text{ m reduced in the ratio } 1 : 750 = \frac{30}{750} \text{ m} = \frac{30\,000}{750} = 40 \text{ mm}$$

$$18 \text{ m reduced in the same ratio} = \frac{18\,000}{750} \text{ mm} = 24 \text{ mm}$$

Answer: 40 mm by 24 mm.

(b) $$54 \text{ mm increased in the ratio } 750 : 1 = 54 \times 750 \text{ mm}$$
$$= 40.5 \text{ m}$$

Answer: 40.5 m.

Exercise 5.3

(1) Express (a) 90 km/h in m/s, (b) 45 m/minute in km/h.
(2) A line 32 mm long joins two towns on a map of scale 1 : 100 000. (a) How far apart are the towns, (b) what length on the map represents a distance of 1.5 km?
(3) A living room is 7.2 m long and 4.5 m wide. What size would it be on a plan drawn to a scale of 1 : 25?
(4) In a scale model of a boat, a davit 3 m high is made 5 mm. If the boat is 136 m long, what is the length of the model to the nearest mm?
(5) What scale is used if a line 84 mm long represents a fence 63 m long?
(6) On a certain day £1 could buy \$2.25, 105.5 escudos, or 9.25 francs. Calculate (a) the number of dollars exchanged for £38, (b) the amount received in pounds sterling for 875 francs, (c) the value in escudos of \$125.

5.5 PROFIT AND LOSS

Increasing or decreasing a price by a given percentage is equivalent to changing it in a given ratio. For example, increasing a price by 5% changes it in the ratio 105 : 100. Decreasing by 5% changes it in the ratio 95 : 100.

Profit and loss are usually calculated as a percentage of the cost price, and discounts as a percentage of the marked or catalogue price.

Example 5.13
If 15% is added to the marked price of goods in a shop, calculate the new price of an article marked (a) £1.40, (b) £9.00, (c) £132.

Solution The price is increased in the ratio 115 : 100

(a) $£1.40 \times \dfrac{115}{100} = \underline{£1.61}$

(b) $£9 \times 1.15 = \underline{£10.35}$

(c) $£132 \times 1.15 = \underline{£151.80}$

Example 5.14
If a dealer sold for £2880 a car he bought for £2400, what was his percentage profit?

Solution The actual profit was £2880 – £2400 = £480. This is 480/2400 of the cost price or 1/5 × 100%. The profit was therefore $\underline{20\%}$.

Example 5.15
A discount store offers a reduction of 10% on all prices. Calculate the original price of (a) a cassette player sold for £82.80, (b) a music centre sold for £348.21.

Solution The price was reduced from 100% to 90% and so the original price was 100/90 of the selling price

(a) $£82.80 \times \dfrac{100}{90} = \underline{£92}$

(b) $£348.21 \times \dfrac{100}{90} = \underline{£386.90}$

5.6 SIMPLE INTEREST

When a sum of money is invested at 5% per annum simple interest, every £100 invested for a full year earns £5 interest. If the rate of interest is 8%, each £100 earns £8 in a year.

The sum invested is called the *principal*, and the principal plus the interest is called the *amount*.

£100 invested for 1 year at R% earns £R

£P invested for 1 year at R% earns £$P/100 \times R$

£P invested for T years at R% earns £$P/100 \times T \times R$

£R is increased in the ratio P : 100 and T : 1

Simple interest, I, is calculated using an algebraic formula

$$I = \frac{P \times T \times R}{100}$$

Formulae are considered in Chapter 7.

Example 5.16
Find the amount of £1250 invested for 3 years at $7\frac{1}{2}\%$ per annum simple interest.

Solution

$$P = £1250, \qquad T = 3, \qquad R = 7\tfrac{1}{2}$$

$$I = \frac{P \times T \times R}{100} = \frac{£1250 \times 3 \times 15}{100 \times 2}$$

$$= £281.25$$

Amount $= P + I = \underline{£1531.25}$.

Example 5.17
David borrowed £800 from a finance company to buy a car, and agreed to pay £33 each month for 36 months. Calculate the rate per annum of simple interest.

Solution

$$£33 \times 36 = £1188, \qquad £1188 - £800 = £388$$

He paid £388 interest on £800 in 3 years

$$I = \frac{P \times T \times R}{100} \Rightarrow 388 = \frac{800 \times 3 \times R}{100} = 24 \times R$$

$$\text{Rate} = 388 \div 24 = \underline{16\tfrac{1}{6}\%}$$

Exercise 5.4

(1) The owner of a boutique makes a profit of 80% on all clothes sold. At what price must she sell a garment which cost (a) £6.50, (b) £9.80?

(2) If Angela sold for £414 a car that cost £450, calculate her percentage loss.

(3) A jacket is sold in a shop for £26.22 at a profit of 15%. What price did the shopkeeper pay for the jacket, and how many could he buy for £250?

(4) Calculate the original cost of shares sold at 96 pence each at a profit of 28%.

(5) A manufacturer makes a brush for 90 p and sells it to a wholesaler at a profit of 10%. The wholesaler in turn sells it to a retailer at a profit of 5%. Find the cost to the retailer of 20 of the brushes.

(6) Find the simple interest on (a) £240 for 3 years at 5% per annum, (b) £725 for 4 years at $3\frac{1}{2}$% per annum, (c) £820 for 3 months at 20% per annum.

(7) What sum of money must be invested at 13% simple interest per annum in order to obtain interest of £5 per week?

(8) Calculate the rate per cent per annum if £13.50 interest is paid on an investment of £50 after 2 years. Find the time taken for £450 to amount to £632.25 at this rate of interest.

5.7 DIVIDING A QUANTITY IN A GIVEN RATIO

When a line of length 100 mm is divided in the ratio 2 : 3, the proportional parts are 2/(2 + 3) and 3/(2 + 3) or 2/5 and 3/5 of the 100 mm. The shorter part is 40 mm and the longer part 60 mm. If the line is divided in the ratio 2 : 3 : 5, the proportional parts are 2/(2 + 3 + 5), 3/(2 + 3 + 5), 5/(2 + 3 + 5), that is 2/10, 3/10, 5/10 of the 100 mm. (See Fig. 5.1.)

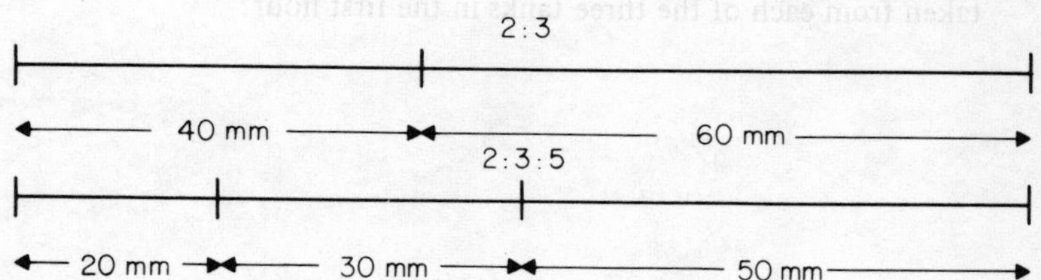

Fig 5.1

Example 5.18

A grocer bought three grades of coffee beans at £1.20, £1.50, and £2.00 per kg. After roasting and grinding the beans he made up bags containing 0.2 kg of a mixture of 5 parts of the cheapest, 3 parts of the next and 2 parts of the dearest coffee. What price did he ask for each bag if he made a profit of 200% on the cost of the beans?

Solution The mixture was in the ratio 5 : 3 : 2 and the proportional parts are 5/10, 3/10, 2/10 of 0.2 kg. Each bag contained 0.1 kg of the cheapest, 0.06 of the next, and 0.04 kg of the dearest coffee, and the cost was

$$0.1 \times £1.20 + 0.06 \times £1.50 + 0.04 \times £2.00 = 12\text{ p} + 9\text{ p} + 8\text{ p}$$
$$= 29\text{ p}$$

A profit of 200% of 29 p = 58 p. Therefore each bag of coffee was sold for 58 + 29 = 87 pence.

Exercise 5.5

(1) In a restaurant the waiter, chef and receptionist shared tips in the ratio 5 : 4 : 3. If the tips in one week were £75, how much did each of them receive?

(2) A bronze consists of copper and tin in the ratio 17 : 3. What mass of each metal would be required to make a casting of mass 35 kg?

(3) Three partners invested £4500, £3000 and £1500 in a business and shared the profit in the ratio of their investment. If the difference between the largest and smallest was £1800 in one year, calculate the total profit and the amount each partner received.

(4) An ingot of mass 1148 kg contains copper, tin and lead in the ratio 9 : 4 : 1. Calculate the value of the ingot if the price per tonne of the metals is copper: £1250, lead: £680, tin: £8600.

(5) When a tanker is discharging oil the oil is pumped out simultaneously from the forward, midship, and aft tanks in the ratio 6 : 4 : 2. If a total of 9000 tonnes is pumped out in the first 45 minutes, how much oil is taken from each of the three tanks in the first hour?

CHAPTER 6

OTHER NUMBER BASES

The denary system of counting used in Chapter 1 has ten digits and the columns are powers of ten. Any positive integer other than one could be used as the base of a counting system, for instance the octal system, base eight, uses the eight digits 1, 2, 3, 4, 5, 6, 7, 0 and the columns are powers of eight.

Apart from ten, the only system used extensively is the binary system, base two, which has the two digits 1 and 0. These are associated with the switch-on/switch-off mode of an electrical circuit, and also with the pulse/no-pulse mode of waves. For this reason binary numbers form the basis of computers, some telephonic communication systems, and the digital recording and reproduction of music. For example, in computerised public libraries each book and each borrower's ticket has a different binary code number consisting of a series of black and white lines. The librarian moves a light pen across the code number and the information is transferred to a computer.

6.1 COUNTING IN BASES OTHER THAN TEN

For all bases except ten the base number is written as a subscript after the number, as in 101_2 and 45_8. The subscript is always written in base ten so that the meaning is clear and the column headings are also written as denary numbers.

Table 6.1, of numbers written in every base from two to nine, shows how they compare.

Example 6.1

(a) $5 = 101_2 = 11_4$ (b) $7 = 21_3 = 12_5$

(c) $15 = 17_8 = 23_6$ (d) $20 = 40_5 = 110_4$

Table 6.1

Denary	Other bases							
10	2	3	4	5	6	7	8	9
2	10	2	2	2	2	2	2	2
3	11	10	3	3	3	3	3	3
4	100	11	10	4	4	4	4	4
5	101	12	11	10	5	5	5	5
6	110	20	12	11	10	6	6	6
7	111	21	13	12	11	10	7	7
8	1 000	22	20	13	12	11	10	8
9	1 001	100	21	14	13	12	11	10
10	1 010	101	22	20	14	13	12	11
15	1 111	120	33	30	23	21	17	16
20	10 100	202	110	40	32	26	24	22

6.2 CHANGING BASES

(a) Changing to base ten from other bases

The column method is used.

Example 6.2

Write the following binary numbers in base ten (a) 111, (b) 10 101, (c) 111 010.

Solution

Power of base number	2^5 32	2^4 16	2^3 8	2^2 4	2^1 2	2^0 1
Binary number				1	1	1
		1	0	1	0	1
	1	1	1	0	1	0

(a) $111_2 = 4 + 2 + 1 = \underline{7}$

(b) $10\,101_2 = 16 + 4 + 1 = \underline{21}$

(c) $111\,010_2 = 32 + 16 + 8 + 2 = \underline{58}$

Example 6.3

Write as denary numbers (a) 234_5, (b) 234_8.

Solution

(a)

5^2	5^1	5^0
25	5	1
2	3	4

$$234_5 = 25 \times 2 + 5 \times 3 + 4$$
$$= 50 + 15 + 4$$
$$= \underline{69}$$

(b)

8^2	8^1	8^0
64	8	1
2	3	4

$$234_8 = 64 \times 2 + 8 \times 3 + 4$$
$$= 128 + 24 + 4$$
$$= \underline{156}$$

(b) Changing from base ten to other bases

The simplest method is called *successive division*. The given denary number is divided repeatedly by the appropriate base number and the remainder is noted at each stage.

Example 6.4

Write the denary number 176 in (a) base eight, (b) base five, (c) base two.

Solution

(a)

	remainder
8)176	
8) 22	0
8) 2	6
0	2

$\underline{176 = 260_8}$

(b)

5)176	
5) 35	1
5) 7	0
5) 1	2
0	1

$\underline{176 = 1201_5}$

(c)

2)176	
2) 88	0
2) 44	0
2) 22	0
2) 11	0
2) 5	1
2) 2	1
2) 1	0
0	1

$\underline{176 = 10\,110\,000_2}$

(c) Changing to bases other than ten

In general, the simplest method of changing from one base to another is to convert to base ten as an intermediate stage.

Example 6.5

Write 305_8 in (a) base six, (b) base four.

Solution As a denary number, $305_8 = 64 \times 3 + 5 = 197$

(a)

$$\begin{array}{r|l} 6)\underline{197} & \\ 6)\underline{\ 32} & 5 \\ 6)\underline{\ \ 5} & 2 \\ 0 & 5 \end{array} \uparrow$$

By successive division

$197 = 525_6$

Therefore $\underline{305_8 = 525_6}$.

(b)

$$\begin{array}{r|l} 4)\underline{197} & \\ 4)\underline{\ 49} & 1 \\ 4)\underline{\ 12} & 1 \\ 4)\underline{\ \ 3} & 0 \\ 0 & 3 \end{array} \uparrow$$

By successive division

$197 = 3011_4$

Therefore $\underline{305_8 = 3011_4}$.

Doublets and triads

There is a shorter method than that shown in Example 6.5 which can be used when one of the bases is a power of the other, such as eight and two, four and two, nine and three.

Since $8 = 2^3$, each digit in the octal system can be written as a three-digit binary number called a *triad.*

0 = 000, 1 = 001, 2 = 010, 3 = 011, 4 = 100,

5 = 101, 6 = 110, 7 = 111

Example 6.6

(a) Write 724_8 as a binary number.

Solution Each separate octal digit is written as a triad of binary digits

7 = 111 2 = 010 4 = 100

Therefore $\underline{724_8 = 111\,010\,100_2}$.

(b) Write $110\,111_2$ in base eight.

Solution Each triad of binary digits is replaced by a single octal digit

$110_2 = 6$ $111_2 = 7$

Therefore $\underline{110\,111_2 = 67_8}$.

When changing denary numbers to the binary system, time can be saved by changing first to base eight by successive division and then writing down the binary equivalent in triads. From Example 6.4(a), $176 = 260_8$ and hence in Example 6.4(c), 176 could be written as a binary number $10\,110\,000_2$ using triads. The extra 0 on the left is omitted.

By a similar method, each digit in base nine can be written as a two-digit number in base three, called a *doublet*. To write a denary number in base three, change to base nine first by successive division and then write down the base three number in doublets. For example, $186_9 = 1\,22\,20_3$ and $305_9 = 10\,00\,12_3$.

Exercise 6.1

Express the following numbers in the denary scale.

(1) 101_2. (2) $10\,110_2$. (3) 1011_2. (4) 243_5.

(5) 212_3. (6) 46_7.

Change the given denary number to the base shown in the square brackets.

(7) 9 [2]. (8) 15 [2]. (9) 35 [2].

(10) 26 [5]. (11) 133 [9]. (12) 201 [6].

(13) 476 [8].

Find the number represented by X in the following.

(14) $X_2 = 200_3$. (15) $X_4 = 61_7$. (16) $15_X = 13_9$.

(17) $666_8 = X_2$. (18) $10\,101_2 = 111_X$.

(19) State which of the following binary numbers are even (a) 1111, (b) 11 011, (c) 1 100 110.

(20) Write the following numbers in *descending* order of magnitude $1011_2, 141_6, 28_9, 2113_4$.

6.3 PRIME NUMBERS AND FACTORS

The position of a number on the number line remains the same whatever base is used for counting, and so a prime number is prime in every base. If one number is a factor of another in base ten it remains a factor when the base is changed.

Example 6.7

Which of these numbers are prime $23_4, 17_9, 21_6$?

Solution Converting to base ten

$$23_4 = 4 \times 2 + 3 = 11$$

$$17_9 = 9 + 7 = 16$$

$$21_6 = 6 \times 2 + 1 = 13$$

Therefore $\underline{23_4 \text{ and } 21_6}$ are prime numbers.

6.4 ARITHMETIC OPERATIONS IN OTHER NUMBER BASES

(a) Multiplication by a power of the base number
In any base, multiplying a number by 10 takes each digit one place further to the left, and a 0 is inserted in the units column as a place holder. Dividing a number by 10 takes each digit to the next column on the right.

Example 6.8
(a) $1001_2 \times 100_2 = 100\,100_2$
(b) $375_8 \times 10_8 = 3750_8$
(c) $10\,110_2 \div 10_2 = 1011_2$
(d) $243\,000_5 \div 100_5 = 2430_5$

(b) Addition and subtraction
The ordinary processes of arithmetic are the same in all number bases, 'one carried' being one of the particular base: 2 in binary arithmetic, 3 in ternary, 8 in octal, and so on. The following examples illustrate the method.

Example 6.9
Calculate the sum of the binary numbers (a) 1101 and 111, (b) 1001, 101 and 11.

Solution

(a)
```
  1101
 + 111
 10100
```

$1 + 1 = 10$, carry one

$0 + 1 + 1 = 10$, carry one

$1 + 1 + 1 = 11$, carry one

$1 + 1 = 10$

Answer: 10 100.

(b)
```
  1001
   101
 +  11
 10001
```

$1 + 1 + 1 = 11$, carry one

$0 + 0 + 1 + 1 = 10$, carry one

$0 + 1 + 1 = 10$, carry one

$1 + 1 = 10$

Answer: 10 001.

Example 6.10
Evaluate (a) $1001_2 - 11_2$, (b) $475_8 - 236_8$.

Solution

(a)

$$\begin{array}{r} 1001 \\ -\quad 11 \\ \hline 110 \end{array}$$

Answer: $\underline{110_2}$.

(b)

$$\begin{array}{r} 475 \\ -236 \\ \hline 237 \end{array}$$

6 from 8 = 2, 2 + 5 = 7
3 from 6 = 3
2 from 4 = 2

Answer: $\underline{237_8}$.

(c) Multiplication

Example 6.11

(a) In base five, 4 × 3 = 22, 3 × 3 = 14, 3 × 2 = 11.
Evaluate $314_5 \times 23_5$.

Solution

$$\begin{array}{r} 314 \\ 23 \\ \hline 2002 \\ 1133 \\ \hline 13332 \end{array}$$

Answer: $\underline{13\,332_5}$.

(b) Multiply 1101 by 101 in base two and in base three.

Solution

In base two

$$\begin{array}{r} 1101 \\ 101 \\ \hline 1101 \\ 1101 \\ \hline 1000001 \end{array}$$

In base three

$$\begin{array}{r} 1101 \\ 101 \\ \hline 1101 \\ 1101 \\ \hline 111201 \end{array}$$

Answer: $\underline{1\,000\,001_2, 111\,201_3}$.

(d) Division

The method is the same in principle as for denary arithmetic, but for long division it is usually easier to convert both numbers to base ten, and then change the quotient back to the original base.

Exercise 6.2

(1) Which of the following numbers are primes (a) 1101_2, (b) 56_7, (c) 231_4, (d) 304_5, (e) 88_9?

Evaluate the following.

(2) $11_2 + 111_2$. (3) $101_2 + 111_2$. (4) $24_5 + 33_5$.

(5) $22_3 + 21_3$. (6) $101_2 - 11_2$. (7) $1011_2 - 111_2$.

(8) $67_8 - 46_8$. (9) $251_7 + 352_7 + 256_7$.

(10) $312_4 + 213_4 - 33_4$. (11) $101_2 \times 10_2$.

(12) $213_4 \times 100_4$. (13) $21_3 \times 2_3$. (14) $101_2 \times 11_2$.

(15) $54_6 \times 5_6$. (16) $73_8 \times 4_8$. (17) $77_8 \times 11_8$.

Write each of the following as a product of prime factors in the given base.

(18) $11\,110_2$. (19) 330_4. (20) 100_6.

PROGRESS TEST 1

A

(1) Evaluate (a) 378 + 925, (b) 2034 − 769, (c) 2529 ÷ 9, (d) 928 ÷ 16, (e) (−3 × 2) − (9 × −1). **(2)** Write as a single fraction (a) $2\frac{1}{6} + 7\frac{1}{8}$, (b) $3\frac{1}{5} - 1\frac{1}{2}$, (c) $4\frac{1}{4} + 1\frac{1}{3} - 2\frac{5}{6}$, (d) $2\frac{1}{8} \times 1\frac{1}{3}$, (e) $3\frac{1}{5} \div \frac{8}{15} \times \frac{1}{2}$, (f) $-2\frac{1}{3} \times -\frac{1}{7}$, (g) $-\frac{2}{9} \div -\frac{4}{9}$. **(3)** Evaluate without a calculator (a) 0.026 × 200, (b) 7.3 × 0.08, (c) 0.205 ÷ 0.03075, (d) −4.5 × 0.3, (e) $\sqrt{44} \times \sqrt{99}$, (f) $\sqrt{48} \div \sqrt{27}$, (g) $2.5 \times 10^{-3} + 0.2 \times 10^{-4}$, (h) 15% of £2.70, (i) $2\frac{1}{5}$% of £520. **(4)** Write as a single logarithm (a) $\overline{2}.7310 \div 3$, (b) $\overline{1}.8214 \times 3$, (c) $\overline{1}.4281 - \overline{3}.6140$. **(5)** Evaluate correct to (i) 2 dp, (ii) 2 sig. fig. (a) 0.03 ÷ 5, (b) 101.61 × 0.3, (c) $\frac{2}{3} \times \frac{4}{5}$, (d) $\frac{1}{3} + (\frac{3}{4} \div \frac{1}{6})$. **(6)** Divide £183.75 in the ratio 2 : 5 : 8. **(7)** Find the rate per cent simple interest if £680 invested for 4 years earned £204. **(8)** If five-sixths of a number is 1060, what is the number? **(9)** How many pieces of wire each 15 mm long can be cut from 1.1 m and what length is left over? **(10)** A coat was sold for £48 at a profit of 20%. (a) What was the cost price? At what price should the coat be sold to make (b) 50% profit, (c) 5% loss? **(11)** Express as a binary number (a) 16, (b) 27, (c) $101_2 + 111_2$, (d) $111_2 - 101_2$.

B

Evaluate without using a calculator **(1)** (a) 246 − 177 + 35, (b) 28.1 − 8.9, (c) $2\frac{1}{2} \times \frac{3}{4} \div 1\frac{1}{7}$, (d) $5\frac{1}{6} - 2\frac{4}{7}$, (e) 0.04 ÷ 0.005, (f) $3\frac{1}{2}$% of 49 kg. **(2)** (a) $2.6 \times 10^{-4} - 3.1 \times 10^{-3}$, (b) $27^{1/3}$, (c) $(8/27)^{-2/3}$, (d) $(0.4)^{-2}$, (e) $(-0.8)^2$, (f) $\sqrt{96} \times \sqrt{6}$, (g) $\sqrt{(0.64)}$, (h) $\sqrt{(0.0001)} \times (0.1)^2$, (i) $\sqrt{6000}$ given that $\sqrt{60}$ is 7.75, (j) $1011_2 \times 110_2$. **(3)** Express 500 mm/s in km/h. **(4)** A used car was sold for £287.50 including 15% tax. Calculate the price before tax was added. **(5)** What is the greatest integer that divides into 183 and 173 leaving a remainder 3? **(6)** Find the ratio of $(4\frac{1}{8} - 2\frac{1}{3})$ to $1\frac{1}{6}$. **(7)** After a decrease of 5%, the number of people in a work force was 76; what was the original number? **(8)** Express 28.7×10^{-4} in standard index form. **(9)** Given that £1 buys \$2.28 (a) how many dollars can be bought for £4.25, (b) how many pounds for \$17.10? **(10)** Add 15% of £25 to 5% of £60. **(11)** What distance is represented by 180 mm on a map with scale 1 : 125 000? **(12)** Write 346_{10} as a number in (a) base two, (b) base eight, (c) base five.

CHAPTER 7

ALGEBRAIC EXPRESSIONS

The name *algebra* is derived from an Arabic word, and most of the elementary algebra was known by the fourth century A.D. In algebra the processes of arithmetic are described in a general but compact form by using letter symbols to represent many different numbers and the ability to manipulate algebraic expressions and to solve equations is important in all branches of science as well as in mathematics.

7.1 ALGEBRAIC NOTATION

An *algebraic expression* is an arithmetic expression containing letters instead of some numbers, such as $2x$ metres, $t + 3$ seconds, or $4pq$ kilograms. Multiplication signs between the factors are usually omitted, so that $4pq$ means $4 \times p \times q$

$4pq$ has three factors: $4,\ p,\ q$

$7ab(p + q)$ has four factors: $7,\ a,\ b,\ (p + q)$

The expression $3pq + 2a$ has two *terms* and $ax^2 - bx + c$ has three *terms*.

When an algebraic term contains both letters and figures, the number is written first, and is called the *coefficient*. For example, in the expression $5m^2 + 12mn - 7n^2$, the coefficient of m^2 is 5, the coefficient of mn is 12, and the coefficient of n^2 is 7.

Example 7.1

Walking at a steady speed of 4 km/h, what distance would I walk in

(a) $2t$ hours, (b) $t/2$ hours, (c) N minutes (d) $2N + 3$ minutes?

Solution

(a) The distance is $2t \times 4$ km $=$ $\underline{8t \text{ km}}$

(b) The distance is $t/2 \times 4$ km $= t \times 2$ km $= \underline{2t \text{ km}}$

(c) N minutes $= N/60$ hours

$$\frac{N}{60} \times 4 \text{ km} = \underline{\frac{N}{15} \text{ km}}$$

(d) In $2N + 3$ minutes the distance is

$$\frac{2N + 3}{60} \times 4 \text{ km} = \underline{\frac{2N + 3}{15} \text{ km}}$$

Example 7.2

If oranges are 3 for 20 pence, what is the cost of (a) $3k$ oranges, (b) $k + 2$ oranges? How many oranges could be bought for (c) m pence (d) £n?

Solution (a) 1 orange costs 20/3 pence, $3k$ oranges cost

$$\frac{20}{3} \times 3k = \underline{20k \text{ pence}}$$

(b) $k + 2$ oranges cost

$$\frac{20}{3} \times (k + 2) \text{ pence} = \underline{\frac{20(k + 2)}{3} \text{ pence}}$$

(c) The number is changed in the ratio $m : 20$

$$3 \times \frac{m}{20} = \underline{\frac{3m}{20} \text{ oranges}}$$

(d) £n = $100n$ pence. The number of oranges is

$$3 \times \frac{100n}{20} = \underline{15n}$$

Example 7.3

If the charge for postage is X pence per kilogram, calculate the total cost of posting N parcels weighing $Y/10$ kg each and M parcels weighing $Z/12$ kg each.

Solution The total weight is

$$\frac{Y}{10} \times N + \frac{Z}{12} \times M \text{ kg}$$

or

$$\frac{NY}{10} + \frac{MZ}{12} \text{ kg}$$

Each kilogram costs X pence. The total cost is therefore

$$X\left(\frac{NY}{10} + \frac{MZ}{12}\right) \text{ pence}$$

or

$$\frac{NYZ}{10} + \frac{MZX}{12} \text{ pence}$$

Exercise 7.1

(1) Write an expression for the number of (a) centimetres in x metres, (b) kilograms in y grams, (c) pence in £$10k$, (d) millimetres in $(d + 3)$ metres.

(2) If Amanda is x years old now, what age will she be in (a) z years time, (b) $x + 3$ years from now? How old was she (c) 2 years ago, (d) y years ago?

(3) Write down as an expression in x the result of performing each of the following operations on the number x. (a) multiply by 2, (b) subtract 4, (c) divide by 10, (d) decrease by 8%.

(4) A box contained X bolts. If Y bolts were removed and then Z were replaced in the box, how many bolts were in the box?

(5) Lucinda bought w tins of peaches at x pence a tin, how much change did she have from £y? How many tins could she have bought for £z?

(6) If the cost per kilogram of apples is a pence and tomatoes t pence, (a) find the total cost of x kilograms of apples and y kilograms of tomatoes. (b) If the price of applies is increased by 15%, what is the new price per kilo?

(7) The average speed of a train is t km/h and of a plane is p km/h. Calculate the time taken by the train to travel s kilometres. What is the total distance travelled by a passenger in x hours by train and then y hours by plane?

7.2 INDICES IN ALGEBRA

Numbers in index form were introduced in Section 1.2 and the rules for combining them were stated in Section 4.3. Just as $2 \times 2 \times 2 = 2^3$ and $3^{-2} = 1/3^2$, so in algebra $a \times a \times a = a^3$ and $b^{-2} = 1/b^2$. The laws of indices are stated here in algebraic form, and they are valid for both integral and fractional values.

(1) $\underline{a^m \times a^n = a^{m+n}}$

To multiply powers of the same base add the indices

$$a^2 \times a^3 = a \times a \times a \times a \times a = a^5$$

(2) $\underline{a^m \div a^n = a^{m-n}}$

When dividing powers of the same base the indices are subtracted. For the special case when the indices are equal

$$\underline{a^m \div a^m = a^0 = 1}$$

For a negative index

$$a^{-m} = a^{0-m} = a^0 \div a^m = \frac{1}{a^m}$$

$$\underline{a^{-m} = \frac{1}{a^m}}$$

For example

$$a^4 \div a^2 = \frac{a \times a \times a \times a}{a \times a} = a^2$$

$$a^4 \div a^4 = \frac{a \times a \times a \times a}{a \times a \times a \times a} = 1$$

$$a^2 \div a^4 = \frac{a \times a}{a \times a \times a \times a} = \frac{1}{a^2}$$

(3) $\underline{(a^m)^n = a^{mn}}$

Thus

$$(a^2)^3 = a^2 \times a^2 \times a^2 = a^6$$

and

$$(a^{-3})^2 = a^{-3} \times a^{-3} = a^{-6}$$

(4) $\underline{\sqrt[n]{a^m} = (\sqrt[n]{a})^m = a^{m/n}}$

For example

$$\sqrt[3]{a^2} = a^{2/3} \quad \text{and} \quad (\sqrt[4]{a})^7 = a^{7/4}$$

Example 7.4

(a) $a^{1.2} \times a^{1.8} = a^{1.2+1.8} = \underline{a^3}$

(b) $p^{2/3} \div p^{1/3} = p^{2/3-1/3} = \underline{p^{1/3} \text{ or } \sqrt[3]{p}}$

(c) $(k^{1/5})^{-10} = k^{-10/5} = \underline{k^{-2} \text{ or } \dfrac{1}{k^2}}$

(d) $(t^{-2})^{1/4} = t^{-2/4} = \underline{t^{-1/2} \text{ or } \dfrac{1}{\sqrt{t}}}$

7.3 MULTIPLICATION AND DIVISION OF ALGEBRAIC TERMS

The numbers are combined first, and then each letter in turn, using the laws of indices.

Example 7.5

(a) $2ab \times 5a^2b^3 = 2 \times 5 \times a^{1+2} \times b^{1+3} = \underline{10a^3b^4}$

(b) $8p^4q^2 \div 4p^2q = 8 \div 4 \times p^{4-2} \times q^{2-1} = \underline{2p^2q}$

(c) $4x^2yz \times 3xy^2z^2 \div 6x^2y^2z^2 = 2x^{2+1-2}y^{1+2-2}z^{1+2-2} = \underline{2xyz}$

(d) $8x^2yz^{-2} \times 3xy^{-2}z^2 = 24x^{2+1}y^{1-2}z^{-2+2} = \underline{24x^3y^{-1}}$ $\quad (z^0 = 1)$

Exercise 7.2

Simplify the following.

(1) $a^3 \times a$. (2) $2t^2 \times 4t^3$. (3) $y^{1/3} \times y^{2/3}$.
(4) $b^{2/3} \times b^{4/3}$. (5) $x^4 \div x^3$. (6) $s^{3/2} \div s^{1/2}$.
(7) $a^2b \times ab^2$. (8) $5z^2y \times 3zy^3x$. (9) $6x^3y^4 \div 2x^3y$.
(10) $15a^3b^2c \div 3bc$. (11) $(m^2)^3$. (12) $(4x^3)^2$.
(13) $\left(\dfrac{y^2}{3}\right)^3$. (14) $3x^2w^2 \div 6xw^3$. (15) $2xyz^2 \div 8x^2y^2$.

(16) The area of a rectangle is the product of the lengths of two adjacent sides. Find an expression for the area of a rectangle with sides $3a^2$ and $4ab^2$ mm.

(17) The dimensions in mm of a block of wood are length $2x$, width x, height x^2. Given that the volume is the product of the three dimensions, express the volume in terms of x.

7.4 SUBSTITUTION

In engineering and scientific work and in the drawing of graphs it is often necessary to find the numerical value of an algebraic expression by assigning particular values to the letters. This process is called *substitution*. For instance, when $a = 2$ and $b = 3$

$$2a^2b = 2 \times 4 \times 3 = 24$$

and

$$ab - b^2 = 6 - 9 = -3$$

The work on directed numbers in Section 1.10 should be revised thoroughly before Exercise 7.3 is attempted.

Example 7.6

Evaluate the following expressions when $a = 2, b = -1, x = -3, y = 4$.
(a) $ax^2 + bx + xy$, (b) $ax^2 - 2abxy + by^2$, (c) $x^a + y^b$,
(d) $(b^2 - 4ax)^{1/2}$.

Solution

(a) $$\begin{aligned} ax^2 + bx + xy &= 2 \times (-3)^2 + (-1) \times (-3) + (-3) \times 4 \\ &= 2 \times 9 + 3 + -12 \\ &= \underline{9} \end{aligned}$$

(b) $$\begin{aligned} ax^2 - 2abxy + by^2 &= [2 \times (-3)^2] - (2 \times 2 \times -1 \times -3 \times 4) \\ &\quad + (-1 \times 4^2) \\ &= 18 - 48 - 16 \\ &= \underline{-46} \end{aligned}$$

(c) $$x^a + y^b = (-3)^2 + 4^{-1} = 9 + \tfrac{1}{4} = \underline{9\tfrac{1}{4}}$$

(d) $$\begin{aligned} (b^2 - 4ax)^{1/2} &= (1 - 4 \times 2 \times -3)^{1/2} \\ &= (1 - -24)^{1/2} \\ &= \sqrt{25} \quad \text{or} \quad \underline{5} \end{aligned}$$

Example 7.7

Given that $T = 2\pi\sqrt{(L/G)}$, calculate the value of T when $\pi = 22/7$, $L = 49$, $G = 100$.

Solution

$$T = 2 \times \frac{22}{7} \times \sqrt{\frac{49}{100}} = 2 \times \frac{22}{7} \times \frac{7}{10} = \underline{4.4}$$

Exercise 7.3

Evaluate the following expressions given that $a = -1, b = 2, c = 0, x = -2, y = 3, z = -3$.

(1) a^4. (2) ab^2. (3) a^2b^3c. (4) $3a^2x^2$. (5) $2xy^2$.

(6) $(x^2)^2$. (7) $(2bx^2)^{1/2}$. (8) $\dfrac{xz}{a^2}$. (9) $\dfrac{ab^2}{x^3}$. (10) $2a + b$.

(11) $3y - 2a^3$. (12) $2b^2 - 4b$. (13) $3x^2yz - ab^2$.

(14) $3x^2y^2z + 2a^4b^6c$. (15) $x^b - b^x$. (16) $y^c + b^a$.

(17) $(2a + 3b^2)^2$. (18) $y^4 - 2ab - z^2$.

(19) Find the value of $(1/u) + (1/v)$ when $u = 3/4$ and $v = -3/16$.

(20) If $s = ut - \frac{1}{2}at^2$ calculate the value of s when
(a) $u = 30,\ t = 4,\ a = 8$, (b) $u = 5,\ t = 0.2,\ a = 22$.

7.5 ADDITION AND SUBTRACTION OF ALGEBRAIC TERMS

Algebraic terms can be added or subtracted only when they contain exactly the same combination of letters and powers. For example

$$a + 2a = 3a$$

$$5x - 3x = 2x$$

and

$$3ab^2 - ab^2 = 2ab^2$$

but there is *no* single expression for

$$b + 2a, \qquad 5x - 2y \qquad \text{or} \qquad 3ab^2 - a^2b$$

Terms which have the same combination of letters are called *like terms* and algebraic expressions are simplifed by collecting like terms.

Example 7.8

(a) $5x - 2y + 3x + y = \underline{8x - y}$

(b) $3x^2y + 2xy - 2x^2y = \underline{x^2y + 2xy}$

(c) $5mn^2 - 2mn - 3mn^2 + 6mn = \underline{2mn^2 + 4mn}$

7.6 REMOVING BRACKETS USING THE DISTRIBUTIVE LAW

In arithmetic numbers in brackets are combined first, but in algebra a bracket usually contains unlike terms which cannot be combined, and brackets are removed using the distributive law. Written in algebraic form this is $a(b + c) = ab + ac$.

(a) Multiplying a bracket by a number

Each term in the bracket is multiplied by the number.

Example 7.9

(a) $7(x^2y - xy) = 7x^2y - 7xy$

(b) $-3(mn - 2n + 3k) = -3mn - -6n + -9k = -3mn + 6n - 9k$

(b) Multiplying a bracket by an algebraic expression

Example 7.10

(a) $3a(a^2b - 2ab) = 3a \times a^2b - 3a \times 2ab$
$= \underline{3a^3b - 6a^2b}$

(b) $-m^2n(mn - 2n + 3k) = -m^3n^2 - -2m^2n^2 + -3km^2n$
$= \underline{-m^3n^2 + 2m^2n^2 - 3km^2n}$

When a bracket is multiplied by a negative quantity, removing the bracket changes the sign of each term inside it.

(c) Multiplying a bracket by another bracket

Each term of one bracket is multiplied by each term of the other.

$$(a + b)(c + d) = a(c + d) + b(c + d)$$
$$= ac + ad + bc + bd$$

Example 7.11

(a) $(3x + 2)(x + 4) = 3x(x + 4) + 2(x + 4)$
$= 3x^2 + 12x + 2x + 8$
$= \underline{3x^2 + 14x + 8}$

(b) $(a - 2b)(3a + b) = a(3a + b) - 2b(3a + b)$
$= 3a^2 + ab - 6ab - 2b^2$
$= \underline{3a^2 - 5ab - 2b^2}$

Example 7.12

Simplify the following expressions

(a) $2p(p - 3q) - 3q(p + q)$
$2p \times p - 2p \times 3q - 3q \times p - 3q \times q$
$= 2p^2 - 6pq - 3pq - 3q^2$
$= \underline{2p^2 - 9pq - 3q^2}$

(b) $3(x^2 - xy) + (5x - 2)(x - 2y)$

$$3(x^2 - xy) + 5x(x - 2y) - 2(x - 2y)$$
$$= 3x^2 - 3xy + 5x^2 - 10xy - 2x + 4y$$
$$= \underline{8x^2 - 13xy - 2x + 4y}$$

Exercise 7.4

Simplify the following.

(1) $5x^2 - 2x^2$. (2) $3x + y + 2x$.
(3) $a + 2b^2 - 4a + b^2$. (4) $t^2 - 4t + 6t - 5$.
(5) $3x^2z - 6x^2z - 2zx^2$. (6) $3m - 6mn + 2n + 8mn$.
(7) Add $3x - 6y + z$ to $6x + 4y - z$.
(8) Add $4x^2 - 6x + 3$ to $3y^2 - 4x + 1$.
(9) Subtract $m^2 - 5n - 2$ from $3m^2 - n + 6$.
(10) Subtract $6x^2y + 9x^2 - 8$ from $5x^2y - 2x^2 + 4$.

Simplify the following.

(11) $s^2 + 4(s^2 + 2t)$. (12) $3x - 4(2y - 5x)$.
(13) $3a^3 + 2(4 + a^3)$. (14) $2(3u + 4v) - 4(2v - 3u)$.
(15) $5x^2 - 2x(x + x^2)$. (16) $p(p - q) + 2q(q - p)$.

Express the following without brackets.

(17) $(2x - 3)(x + 6)$. (18) $(4y - 3)(2y - 7)$.
(19) $(xy + 4x)(3xy - 2x)$. (20) $(2a - b)(2a + b)$.
(21) $(3x + 2)^2$.
(22) The lengths in metres of the four sides of a building are $3x$, $6x - 5$, $9x$, and $4x + 1$. (a) Calculate the total length around the building. (b) By how much does this differ from $30x$ metres?

7.7 ALGEBRAIC FRACTIONS

(a) Multiplication and division

Algebraic fractions may be multiplied, divided and cancelled by a common factor in the same way as numerical fractions. Two or more terms enclosed in a bracket are treated as a single factor.

Example 7.13

Write as a single fraction in its lowest terms

(a) $\dfrac{3x}{y} \times \dfrac{x^2}{2y} \div \dfrac{2x}{3y^2}$ (b) $\dfrac{3(a + b)}{2a} \times \dfrac{5a}{4b} \div \dfrac{5(a + b)}{b}$

Solution

(a) Division is treated as multiplication by the reciprocal

$$\frac{3x \times x^2 \times 3y^2}{y \times 2y \times 2x} = \frac{9x^3y^2}{4xy^2} = \frac{9x^2}{4} \quad \text{(cancel by } xy^2\text{)}$$

(b) $$\frac{3(a + b) \times 5a \times b}{2a \times 4b \times 5(a + b)} = \frac{3}{8} \qquad [\text{cancel by } 5ab(a + b)]$$

(b) Addition and subtraction

Before they are combined by addition or subtraction, algebraic fractions must have the same denominator, and this is usually the LCM of the denominators of the separate terms.

To find the LCM, the LCM of the coefficients is written first, and then the highest power present of each factor in turn.

Example 7.14

(a) The LCM of $2xy, x^2$ and y^2 is $2x^2y^2$

(b) The LCM of $2(a + b), (a + b)^2$ and $3b$ is $6b(a + b)^2$

(c) The LCM of $3x$, $2y$ and $4xy^2$ is $12xy^2$

(d) The LCM of $(x + 1), 2(x - 1)$ and 6 is $6(x + 1)(x - 1)$

Example 7.15

(a) $$\frac{2}{y^2} + \frac{1}{x^2} = \frac{2x^2 + y^2}{x^2y^2}$$

(b) $$\frac{b}{a} + \frac{2}{3a} = \frac{3b + 2}{3a}$$

(c) $$\frac{3}{(x + 1)} - \frac{2}{(x - 1)} = \frac{3(x - 1) - 2(x + 1)}{(x + 1)(x - 1)}$$

$$= \frac{3x - 3 - 2x - 2}{(x + 1)(x - 1)}$$

$$= \frac{x - 5}{(x + 1)(x - 1)}$$

(d) $\dfrac{b}{(a+b)} - \dfrac{ab}{(a+b)^2} = \dfrac{b(a+b) - ab}{(a+b)^2}$

$$= \underline{\dfrac{b^2}{(a+b)^2}}$$

Exercise 7.5

Simplify the following.

(1) $\dfrac{x^2}{y^3} \times \dfrac{y}{x}$. (2) $\dfrac{3x}{ab} \times \dfrac{4a^2b}{x^3}$. (3) $\dfrac{x^3y^3z}{a^2b^2} \times \dfrac{a^4b^2}{xyz}$

(4) $\dfrac{2x}{3y^2} \div \dfrac{x^2}{6y^4}$. (5) $\dfrac{2a^2b^3}{3z^2} \div \dfrac{ab^2}{2z}$. (6) $\dfrac{2x^3(a-b)}{3x(a-b)^2}$.

(7) $\dfrac{x}{4} + \dfrac{2x}{3}$. (8) $\dfrac{x}{2} - \dfrac{x}{3}$. (9) $x + \dfrac{1}{x}$. (10) $\dfrac{1}{x} + \dfrac{5}{2x}$.

(11) $\dfrac{1}{b} - \dfrac{1}{5b}$. (12) $\dfrac{1}{y} + \dfrac{1}{2y} + \dfrac{2}{3y}$. (13) $\dfrac{3-x}{4} + \dfrac{2-x}{3}$.

(14) $\dfrac{a-1}{2} - \dfrac{2a-1}{5}$. (15) $\dfrac{x}{x-1} - \dfrac{1}{x-1}$.

(16) $\dfrac{4}{x+1} - \dfrac{2}{2x-1}$. (17) $\dfrac{1}{v} + \dfrac{1}{2v} - \dfrac{2(v+2)}{4v}$.

(18) $\dfrac{3}{x+z} - \dfrac{2(x-z)}{(x+z)^2} + \dfrac{x-z}{x+z}$.

7.8 FACTORISING ALGEBRAIC EXPRESSIONS

To factorise an expression means to write it as a product of factors, and this is the reverse of removing brackets using the distributive law. For instance, since $a(b+c) = ab + ac$, the expression $ab + ac$ can be factorised as $a(b+c)$.

(a) Taking out a common factor

The HCF of the terms is found first, and then each term is divided by it to give the other factor.

Example 7.16

Write as a product of factors (i) $3ab - 6a^2b + 9\,ab^2$, (ii) $4x^2 + 2x^2y - 6x^3y^2$, (iii) $2(a-b) - 3c(a-b)$.

Solution (i) $3ab - 6a^2b + 9ab^2$. The HCF is $3ab$ and each term is divided by this as shown in Section 7.3

$$\frac{3ab}{3ab} = 1, \qquad \frac{6a^2b}{3ab} = 2a, \qquad \frac{9ab^2}{3ab} = 3b$$

and so the factors are $3ab$ and $1 - 2a + 3b$

$$3ab - 6ab^2 + 9ab^2 = \underline{3ab(1 - 2a + 3b)}$$

(ii) $4x^2 + 2x^2y - 6x^3y^2$
Here $2x^2$ is the HCF and the factors are $\underline{2x^2(2 + y - 3xy^2)}$.

(iii) $$2(a - b) - 3c(a - b)$$

$(a - b)$ is the common factor, and dividing by it gives $2 - 3c$ as the other factor. The expression factorises as $\underline{(a - b)(2 - 3c)}$.

After factorising an expression it is advisable to check by multiplying the brackets that the correct factors have been found.

(b) Grouping

For this method the expression to be factorised usually has four terms. If there is a common factor it is taken out first, and then the terms are grouped in pairs so that each pair has a common factor.

Example 7.17

Factorise the expressions (i) $abx + 2cx + aby + 2cy$, (ii) $6ap - 3qc + 3pc - 6aq$, (iii) $3st - 3t^2 - s + t$.

Solution (i) $abx + 2cx + aby + 2cy$. The first two terms have x as a factor and the last two terms have y as a factor, and the expression becomes

$$x(ab + 2c) + y(ab + 2c)$$

Taking out the factor $ab + 2c$ gives the result

$$abx + 2cx + aby + 2cy = \underline{(ab + 2c)(x + y)}$$

(ii) $6ap - 3qc + 3pc - 6aq$. Dividing first by the common factor 3 gives

$$3(2ap - qc + pc - 2aq)$$

Since $2ap$ and qc have no common factor the terms are rearranged as

$$\begin{aligned} &3(2ap - 2aq + pc - qc) \\ &= 3[2a(p - q) + c(p - q)] \\ &= \underline{3(p - q)(2a + c)} \end{aligned}$$

(iii) $3st - 3t^2 - s + t$

$= 3t(s - t) - (s - t)$ (Note the effect of the bracket on the sign of t)

$= \underline{(s - t)(3t - 1)}$

Exercise 7.6
Factorise the following.

(1) $2x^2 - 3x^3$. (2) $3pq^2 - p^2q^2 + 2pqx$.
(3) $4(x - y) + x(x - y)$. (4) $3t(2t + 1) - 6x(2t + 1)$.
(5) $6xz^2 - 8x + 9z^2 - 12$. (6) $2z^2 + 3z + 2yz + 3y$.
(7) $12a^3 + 8a^2b - 3ab - 2b^2$. (8) $3x^2 + 2xy - 6x - 4y$.
(9) $2mn + 2m^2 - 2pn - 2pm$. (10) $5a^2 - 15ab + 5ac - 15bc$.

(c) Factorising quadratic expressions
An expression with two terms is a binomial, and a quadratic expression is the product of two binomial factors. The first term in each bracket is a factor of the first term of the quadratic expression; the second term in each bracket is a factor of the last term of the expression.

Multiplying brackets is quite easy, but to find which factors were multiplied when the product is given is rather more difficult. There are a few rules which are helpful, and it is important to check by multiplying out the brackets that the original quadratic expression is obtained.

(i) *The difference of two squares*
The factors of $a^2 - b^2$ are $(a + b)(a - b)$ for all values of a and b.

Example 7.18
(a) $x^2 - 9 = x^2 - 3^2 = \underline{(x + 3)(x - 3)}$

(b) $4x^2 - 25 = (2x)^2 - 5^2 = \underline{(2x + 5)(2x - 5)}$

(c) $9p^2 - 16q^2 = (3p)^2 - (4q)^2 = \underline{(3p + 4q)(3p - 4q)}$

Since this rule applies to the difference of any two perfect squares it can be used in arithmetic as well.

Example 7.19
(a) $101^2 - 99^2 = (101 + 99)(101 - 99) = 200 \times 2 = \underline{400}$

(b) $11.3^2 - 8.7^2 = (11.3 + 8.7)(11.3 - 8.7) = 20 \times 2.6 = \underline{52}$

(ii) *Perfect squares*

Quadratic expressions which are the product of two identical factors are called perfect squares. For all values of a and b

$$a^2 + 2ab + b^2 = (a + b)^2$$

and

$$a^2 - 2ab + b^2 = (a - b)^2$$

Example 7.20

(a) $x^2 + 2x + 1$ has factors $\underline{(x + 1)^2}$

(b) $4x^2 - 12x + 9$ factorises as $\underline{(2x - 3)^2}$

(c) The factors of $25a^2 + 20ab + 4b^2$ are $\underline{(5a + 2b)^2}$

Check

$$(5a + 2b)(5a + 2b) = 5a(5a + 2b) + 2b(5a + 2b)$$
$$= 25a^2 + 10ab + 10ab + 4b^2$$

which is the correct expression.

Example 7.21

What number must be added to $9x^2 - 30x + 20$ to make it a perfect square?

Solution Suppose the square is $(a - b)^2$, then $a = 3x$ and $2ab = 30x$ which makes $b = 5$

$$(3x - 5)^2 = 9x^2 - 30x + 25$$

The number to be added is therefore $\underline{5}$.

(iii) *Factorising expressions of the type* $ax^2 + bx + c$ *when the coefficients* a, b, *and* c *are integers*.

Each of the binomial factors has a term in x and a number, and the general procedure is to try likely pairs until the correct combination is found. The following rules are a useful guide, and the facility of recognising likely combinations of factors improves with practice.

First divide throught by any common factor. If the constant term c is positive, the sign in each bracket is the same as the sign of the coefficient of x in the quadratic to be factorised. If a has the value 1, the numbers in the two brackets multiply to give c and add to give b.

Example 7.22

(i) Factorise $x^2 - 10x + 21$.

Solution Here a is 1, b is -10, and c is 21. The factors of 21 which add up to -10 are -7 and -3, and so the required factors are $x-7$ and $x-3$

$$x^2 - 10x + 21 = \underline{(x-7)(x-3)}$$

(ii) Express $x^2 + 4x - 5$ as a product of two factors.

Solution Here a is 1, b is 4 and c is -5. The only integers that add to 4 and multiply to -5 are 5 and -1, and so the factors are $\underline{(x+5)(x-1)}$.

When neither a nor c is a prime number or unity, each pair of factors of a is tried with each pair of factors of c in turn.

Example 7.23
(a) $3x^2 + 10x + 8$.

Solution Both brackets have a positive sign, the only factors of 3 are 3 and 1 and the possible factors of 8 are 4 and 2 or 8 and 1

$3x$	8	1	4	2
x	1	8	2	4

Multiplying diagonally

$$3x + 8x = 11x$$
$$24x + x = 25x$$
$$6x + 4x = 10x \quad \text{which is correct}$$

The factors are $\underline{(3x+4)(x+2)}$.

(b) $6x^2 - 20x - 16$

Solution Taking out the common factor 2 gives $2(3x^2 - 10x - 8)$. The factors of -8 must have opposite signs, and various pairs are tried in a systematic way as in the previous example

$3x$	8	-8	4	-4	2
x	-1	1	-2	2	-4

$8x - 3x \quad 3x - 8x \quad 4x - 6x \quad 6x - 4x \quad 2x - 12x = -10x$

The correct combination is $3x + 2$ and $x - 4$.

$$6x^2 - 20x - 16 \text{ factorises as } \underline{2(3x+2)(x-4)}$$

Example 7.24

Factorise the expression $4x^2 - 27x + 18$.

Solution Both factors of 18 must be negative since the coefficient of x is negative, and the lower numbers are tried first

$$2x \quad -6 \quad -9 \quad -18$$

$$2x \quad -3 \quad -2 \quad -1$$

$$-6x - 12x \qquad -4x - 18x \qquad -2x - 36x$$

None of these is correct so we must try $4x$ and x

$$4x \quad -6 \quad -3$$

$$x \quad -3 \quad -6$$

$$-12x - 6x \qquad -24x - 3x = -27x$$

The factors are $\underline{(4x - 3)(x - 6)}$.

Exercise 7.7

Factorise the following.

(1) $x^2 + 9x + 18$. (2) $x^2 + 5x - 14$. (3) $y^2 - 9y + 20$.
(4) $x^2 - 2x - 15$. (5) $2t^2 + 9t - 5$. (6) $3x^2 - 7x - 20$.
(7) $5y^2 - 21y + 4$. (8) $6 + 10x - 4x^2$. (9) $4x^2 - 9y^2$.
(10) $9b^2 - 1$. (11) $3xy^2 - 27x$. (12) $4x^2 + 12x + 9$.
(13) Calculate the exact value of $12.4985^2 - 10.4985^2$.
(14) Express $x^2 - 6x - t$ as a product of two factors given that t is a positive prime number.
(15) What number must be added to each expression to make it a perfect square (a) $4x^2 + 12x + 7$, (b) $x^2 - 14x + 56$?

CHAPTER 8

ALGEBRAIC EQUATIONS AND INEQUALITIES

8.1 THE USE OF EQUALITY AND INEQUALITY SYMBOLS

Placing an equals sign between two algebraic expressions implies that they have the same numerical value

$$2x = 10, \qquad 3x - 2 = x + 4, \qquad x^2 - 5x - 6 = 0$$

are all *equations* in one unknown, x, which is called the *variable*. In order to maintain the equality, whatever change is made to the expression on one side of an equation must also be made on the other side.

The inequality symbols are $<$ and $>$ which we have used already to show order relations

$2x < 10$ means $2x$ is less than 10

$3x > 4$ means $3x$ is greater than 4

The combined symbols $\geqslant$ and $\leqslant$ are used for 'greater than or equal to' and 'less than or equal to'.

8.2 THE SOLUTION OF LINEAR EQUATIONS AND INEQUALITIES

(a) Equations

A linear equation in one variable x contains only numbers and terms in x, there are no terms in x^2 or other powers. A *solution* is a numerical value of the variable which makes both sides of the equation the same number; it *satisfies* the equation. For example, if $3x = 6$, the solution is $x = 2$, since $3 \times 2 = 6$. In general a linear equation in one variable has exactly one solution. The exceptions to this rule are of two kinds.

(i) Equations of the type $x + 5 = x + 3$, which has no solution because there is no value of x which makes both sides the same number.

(ii) Equations such as $2x + x = 3x$, which is satisfied by every value of x.

An equation satisfied by all values of the variable is called an *identity*; the symbol used is $\equiv$ so that $2x + x \equiv 3x$.

To solve an equation it is manipulated to leave only numerals on one side of the equals sign, as shown in Examples 8.1 to 8.3.

Example 8.1

Solve the equations (a) $2x - 5 = 1$, (b) $\frac{y}{3} = -5$.

(a) $2x - 5 = 1$

Add 5

$$2x - 5 + 5 = 1 + 5$$

$$2x = 6$$

Divide by 2

$$\frac{2x}{2} = \frac{6}{2}$$

$$x = 3$$

The solution is $\underline{x = 3}$.

(b) $\frac{y}{3} = -5$

Multiply by 3

$$\frac{y}{3} \times 3 = -5 \times 3$$

$$y = -15$$

The solution is $\underline{y = -15}$.

Example 8.2
Solve the equations (a) $2(t - 3) = t + 7$, (b) $2(p + 2) - 5(2p - 4) = 10$.

Solution
(a) $2(t - 3) = t + 7$
Remove the brackets

$$2t - 6 = t + 7$$

Add 6

$$2t - 6 + 6 = t + 7 + 6$$

$$2t = t + 13$$

Subtract t

$$2t - t = t - t + 13$$

$$\underline{t = 13}$$

(b) $2(p + 2) - 5(2p - 4) = 10$.

Remove the brackets

$$2p + 4 - 10p + 20 = 10$$

Collect terms

$$2p - 10p + 24 = 10$$

Subtract 24

$$2p - 10p = 10 - 24$$

$$-8p = -14$$

Divide by −8

$$p = \frac{-14}{-8} = 1\tfrac{3}{4}$$

$$\underline{p = 1\tfrac{3}{4}}$$

Example 8.3

Solve the equations (a) $\dfrac{3x}{5} = x + 2$, (b) $\dfrac{4}{x} - \dfrac{3}{2x} = 5$.

(a) $$\frac{3x}{5} = x + 2$$

Multiply by 5

$$\frac{3x}{5} \times 5 = 5(x + 2)$$

Remove bracket

$$3x = 5x + 10$$

Subtract $5x$

$$3x - 5x = 5x - 5x + 10$$

$$-2x = 10$$

Divide by −2

$$x = \frac{10}{-2} = -5$$

The solution is $\underline{x = -5}$.

(b) $$\frac{4}{x} - \frac{3}{2x} = 5$$

Combine the fractions

$$\frac{8-3}{2x} = 5$$

Multiply by $2x$

$$\frac{5}{2x} \times 2x = 5 \times 2x$$

$$5 = 10x$$

Divide by 10

$$\frac{5}{10} = x$$

$$x = \frac{1}{2}$$

The solution is $x = 1/2$.

For the rest of the book, we shall omit some of the intermediate steps when solving equations in worked examples.

Exercise 8.1

Solve the following equations.

(1) $4x = -12$. (2) $-2t = -9$. (3) $y + 6 = 8$.
(4) $3s + 5 = -4$. (5) $3x + 2 + x = 7 - x$.
(6) $3p - 2p - 5 = 4p - 6$. (7) $13x - 6 = 20 + 3x$.
(8) $2x + 3(x + 5) = 0$. (9) $3 - 2(2y + 3) = 11$.
(10) $12/x = 3$. (11) $x + 2x/5 = -21$. (12) $0.4m = 0.02$.
(13) $0.2(2x + 0.1) = 0.3(x - 0.1)$. (14) $\frac{t}{4} - \frac{t+3}{6} = 1\frac{1}{6}$.
(15) $\frac{3x-2}{5} + \frac{4x}{10} = -1\frac{2}{5}$. (16) $\frac{3}{4x} - \frac{2(x+2)}{7x} = 1$.

(b) Inequalities (or inequations)

Inequalities are solved in the same way as equations, except that multiplying by a negative number changes the direction of the inequality. It was shown in Section 1.9 that $3 < 5$ but $-3 > -5$. In the same way, if $-x > 5$ then $x < -5$.

Taking the reciprocal of both sides also changes the direction of an inequality. Thus if $1/x < 1/5$, then $x > 5$.

(i) Conditional statements When the truth of one algebraic statement depends on the truth of another they are called conditional statements. If . . . then . . . may be replaced by the symbol ⇒ meaning 'implies that'

$$-x > 5 \Rightarrow x < -5$$

The double arrow ⇔ means that the converse is also true, and stands for 'if and only if'

$$3 < 5 \Leftrightarrow -3 > -5$$

(ii) Representing inequalities on the number line The solution of a linear equation is a single number which can be represented by a point on the number line, but an inequality is satisfied by a range of values of the variable and the solution can be represented by an interval, as in Fig. 8.1.

(a)

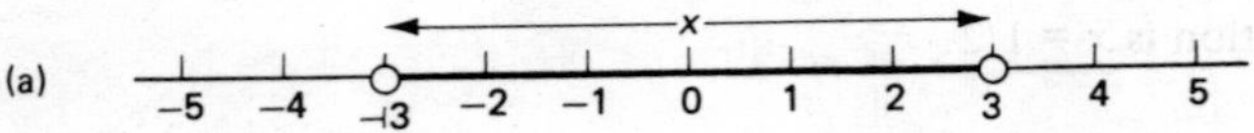

(b)

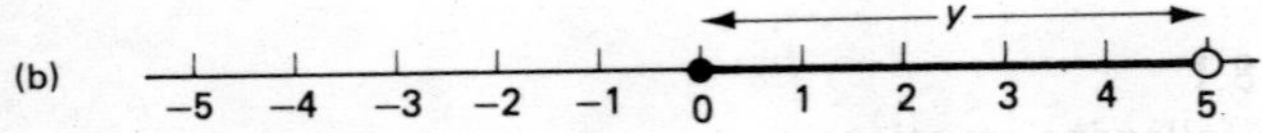

(c)

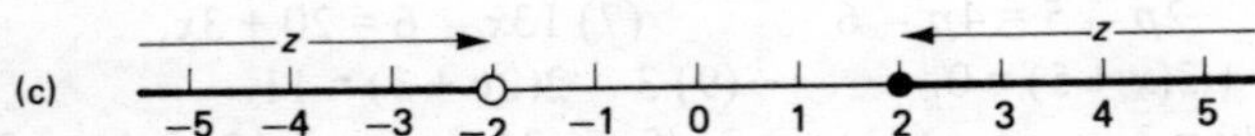

Fig 8.1

(a) $-3 < x < 3$ means that the variable x can take values between -3 and $+3$. The limits are not included and so the interval is open and marked ○.

(b) $0 \leqslant y < 5$ means that y is a number greater than or equal to 0 but less than 5. The interval is closed at the lower limit 0, marked ●.

(c) $z < -2, z \geqslant 2$ means that there are no values of z between -2 and $+2$, the interval is open at -2 but closed at $+2$.

Example 8.4
Solve the inequalities

(a) $2x - 4 < x + 1$ (b) $\frac{2t}{3} > t - 7$

$2x < x + 5$ $2t > 3t - 21$

$2x - x < 5$ $2t - 3t > -21$

$\underline{x < 5}$ $-t > -21$

$\underline{t < 21}$

Example 8.5
(a) For what range of values of x is $3(2x + 1)$ less than $8(x - 1)$?

Solution

$$3(2x + 1) < 8(x - 1)$$

$$6x + 3 < 8x - 8$$

$$6x - 8x < -8 - 3$$

$$-2x < -11$$

$$-x < -5\tfrac{1}{2}$$

$$\underline{x > 5\tfrac{1}{2}}$$

(b) Solve the inequality $3(2x + 1) - 5(x - 1) \geqslant 7x$.

Solution Removing the brackets first

$$6x + 3 - 5x + 5 \geqslant 7x$$

$$x + 8 \geqslant 7x$$

$$x - 7x \geqslant -8$$

$$-6x \geqslant -8$$

$$-x \geqslant -1\tfrac{1}{3}$$

$$\underline{x \leqslant 1\tfrac{1}{3}}$$

Exercise 8.2
Find the range of values of x satisfying the following inequalities.

(1) $3x < 15$. (2) $x - 7 \geqslant 4$. (3) $2x - 5 < x - 3$.
(4) $2x - 5x < 3$. (5) $4(x - 1) \leqslant 2x + 1$. (6) $1/x > 1/6$.
(7) $1/x < -4$. (8) $x/3 \geqslant 1\tfrac{1}{2}$. (9) $2 - 3(x - 1) > 11$.

(10) For a certain manufacturing programme the cost, x pence, of each component must satisfy both the inequalities $5x - 6 \leqslant 294$ and $4x - 15 \geqslant 173$. Find the range of values of the cost of a component.

8.3 FORMING AN EQUATION FROM GIVEN INFORMATION

Example 8.6
Sheldon is twice as old as Alan, who is 5 years older than Claire. If their total age is 27 years, how old is Claire?

Solution To write this as an algebraic equation let us call Claire's age x years. Alan is 5 years older, that is, $x + 5$ years. Sheldon is therefore $2(x + 5)$ or $2x + 10$ years, and so the total age is $x + x + 5 + 2x + 10$ years or $4x + 15$ years. But the total age is 27 years, and this leads to the equation in x

$$4x + 15 = 27$$

$$4x = 12$$

$$x = 3$$

Claire is 3 years old.

Example 8.7
A salesman is paid at a standard hourly rate of £1.40 for a 40-hour week and double pay for overtime. How many hours overtime did he work in a week when he was paid £70?

Solution Suppose the number of hours overtime is N. Then

$$40 \times £1.40 + N \times £2.80 = £70$$

$$56 + N \times 2.8 = 70$$

$$N \times 2.8 = 14$$

$$N = 14 \div 2.8 = 5$$

He worked 5 hours overtime.

Exercise 8.3

(1) I thought of a number, multiplied it by $3\frac{1}{2}$, added 16 and the answer was 30. What was the number I thought of?

(2) A father and daughter together took £150 each week from the profits of their shop as their wages, the father having twice as much as the daughter. How much did they each earn per week?

(3) The total length of the three sides of a triangle is 45.5 cm and two of the sides are respectively $2\frac{1}{2}$ times and $3\frac{1}{2}$ times the third side. By calling the length of the shortest side x cm form an equation in x and calculate the length of each side of the triangle.

(4) In one week Sarah worked 36 hours at the standard rate, 9 hours overtime at $1\frac{1}{2}$ times the standard rate and 3 hours at treble pay. If the total pay was £87.75, calculate the standard hourly rate.

(5) A consignment of magazines was made up into 45 parcels each weighing either 5 kg or 1 kg. Calculate the number of 5 kg parcels in the consignment if the total weight was 153 kg.

(6) Of the eighteen coins in my pocket, I had some pennies, twice as many tenpenny coins and the rest were fivepenny coins. If the total value amounted to 108 pence how many of the coins were fivepennies?

(7) An airliner left an airport and travelled at an average speed of 800 km/h. A second plane which took off half an hour later travelled on a parallel course at 960 km/h. Calculate the distance from the airport at which it passed the first plane.

8.4 THE SOLUTION OF SIMULTANEOUS LINEAR EQUATIONS

An equation such as $2x + 3y = 4$ is a linear equation in the two variables x and y. There is not a unique solution to this equation because any change in the value of one variable leads to a change in the other: the equation is satisfied by many pairs of values. When two equations are satisfied by the same pair of values they are called *simultaneous equations*.

The algebraic method of solving simultaneous equations which is used in this Chapter is called the elimination method. A second method using matrices is described in Section 10.4 and the graphical solution of equations is considered in Chapter 15.

Example 8.8
Solve the simultaneous equations $2x + 3y = 4$, $6x - y = 2$.

Solution The coefficient of one of the variables is made the same in both equations but with the opposite sign. That variable is then 'eliminated' by addition, leaving a simple equation in the other variable

$$2x + 3y = 4 \qquad (1)$$

$$6x - y = 2 \qquad (2)$$

Multiply equation (2) by 3

$$18x - 3y = 6 \qquad (3)$$

$$2x + 3y = 4 \qquad (1)$$

Add equations (1) and (3) term by term to eliminate y

$$20x = 10$$

$$x = \tfrac{1}{2}$$

This value for x is now substituted in one of the original equations to obtain a simple equation in y. Substitute $x = 1/2$ in (1)

$$1 + 3y = 4$$

$$3y = 4 - 1$$

$$y = 1$$

As a check both values should be substituted in the other equation

$$6 \times \tfrac{1}{2} - 1 = 3 - 1 = 2$$

which is correct. The solution is $\underline{x = 1/2, y = 1}$.

Example 8.9

Solve the equations $3s - 2t = 2s - 4t = 8$.

Solution

$$3s - 2t = 8 \qquad (1)$$

$$2s - 4t = 8 \qquad (2)$$

Multiply equation (1) by -2

$$-6s + 4t = -16 \qquad (3)$$

$$2s - 4t \;= 8 \qquad (1)$$

Add equations (1) and (3)

$$-4s = -8$$

$$s = 2$$

Substitute $s = 2$ in equation (1)

$$3 \times 2 - 2t = 8$$

$$-2t = 8 - 6 = 2$$

$$t = -1$$

The solution is $\underline{s = 2, t = -1}$.

The checking will be omitted in worked examples to save space.

Example 8.10
Find the values of the variables u and v which satisfy both the equations $3u - 4v + 11 = 0, 3v = 4u + 5\frac{1}{3}$.

Solution

$$3u - 4v = -11 \qquad (1)$$

$$4u - 3v = -5\tfrac{1}{3} \qquad (2)$$

Multiply equation (1) by 3 and equation (2) by –4 to eliminate v

$$9u - 12v = -33 \qquad (3)$$

$$-16u + 12v = 21\tfrac{1}{3} \qquad (4)$$

Add equations (3) and (4) term by term

$$-7u = -11\tfrac{2}{3}$$

$$u = -11\tfrac{2}{3} \div -7 = \frac{5}{3}$$

$$u = 1\tfrac{2}{3}$$

Substitute $u = 5/3$ in equation (1)

$$3 \times \frac{5}{3} - 4v = -11$$

$$5 - 4v = -11$$

$$-4v = -16$$

$$v = 4$$

The solution is $\underline{u = 1\tfrac{2}{3}, v = 4}$.

8.5 PROBLEMS LEADING TO SIMULTANEOUS LINEAR EQUATIONS

Example 8.11
Two variables T and t are connected by the equation $T = at + b$. If $T = 194$ when $t = 90$ and $T = 158$ when $t = 70$, find the value of the constants a and b and hence calculate the value of T when $t = 60$.

Solution Substituting the first pair of values

$$194 = 90a + b \qquad (1)$$

and from the second pair

$$158 = 70a + b \qquad (2)$$

$$-158 = -70a - b \qquad (3)$$

Add equations (1) and (3) to eliminate b

$$36 = 20a$$

$$a = \frac{36}{20} = \frac{9}{5}$$

Substitute this value in equation (1)

$$194 = 90 \times \frac{9}{5} + b$$

$$194 = 162 + b$$

$$b = 32$$

The equation connecting T and t is therefore $T = \frac{9}{5}t + 32$. When $t = 60$, $T = \frac{9}{5} \times 60 + 32 = 140$.
Answer: $a = 1\frac{4}{5}, b = 32, T = 140.$

Example 8.12
A telephone bill is made up of a fixed sum for rental charges and an additional fixed sum for each call made. If the total charge was £7.95 when 65 calls were made and £6.90 when only 30 calls were made, calculate the rental charge and the cost of each call.

Solution Suppose the rental is R pence and the cost of each call is C pence. From the first pair of values

$$795 = R + 65C \qquad (1)$$

and from the second pair

$$690 = R + 30C$$

or

$$-690 = -R + -30C \qquad (2)$$

Adding equations (1) and (2)

$$105 = 35C \Rightarrow C = 3$$

From equation (2)

$$R = 690 - 30C = 600$$

The rental charge is therefore £6 and the cost of each call 3 pence.

Exercise 8.4

Solve the following simultaneous equations.

(1) $x - 3y = 5, 2x + 3y = 1.$

(2) $2x + y = -4, 3x - 2y = -13.$

(3) $4s - 3t + 4 = 0, 6s + 5t - 13 = 0.$

(4) $4x - 3z = 10, 6x + 6z = 1.$

(5) $3a - 2b = -8.6, 5a + 7b = 27.$

(6) If $3x - 4y = 10 = x - 3y$, what is the value of $x - y$?

(7) The cost of postage on 8 small parcels and 3 large parcels is £3.70 and for the same price 1 large parcel and 15 of the same small parcels can be posted. Calculate the postage on each size of parcel.

(8) A teacher noticed that the sum of the marks given to Karen and Debbie in a mathematics test was 51 and the difference was 15. If Karen's mark was the higher, what were the marks?

(9) If $V = u + at$ and the value of V is 33 when $t = 2$, and 4 when $t = 3\frac{1}{2}$, find the constants u and a and hence the value of V when $t = 5\frac{1}{2}$.

(10) Find two numbers such that their difference is one-third of their sum and half their difference is 6.

8.6 THE SOLUTION OF QUADRATIC EQUATIONS AND INEQUALITIES

Quadratic equations in one variable are of the form $ax^2 + bx + c = 0$ where the coefficients a, b and c are constants, usually integers. In general there are two solutions which are called the *roots of the equation*.

(a) By factorisation

(i) Quadratic equations

When the product of two binomial factors is zero, at least one of them must be zero since $A \times 0 = 0$ for any value of A. If the expression on the left of a quadratic equation can be factorised as in Section 7.8, the equation can be solved by equating each factor to zero.

Example 8.13

Solve the equations (a) $4x^2 - 25 = 0$, (b) $2x^2 + 7x = 0$, (c) $2x^2 - 8x + 4 = 0$, (d) $2x^2 + 11x - 21 = 0$.

Solution

(a) $4x^2 - 25 = 0$

$x^2 = \frac{25}{4}$

$x = \pm\sqrt{\frac{25}{4}}$

$= \pm\frac{5}{2}$

$x = +2\frac{1}{2}$ or $-2\frac{1}{2}$.

(b) $2x^2 + 7x = 0$

$x(2x + 7) = 0$

Either $x = 0$ or $2x + 7 = 0$

$2x + 7 = 0 \Rightarrow x = -3\frac{1}{2}$

$x = 0$ or $-3\frac{1}{2}$.

(c) $2x^2 - 8x + 8 = 0$

$2(x^2 - 4x + 4) = 0$

$2(x - 2)^2 = 0$

$x - 2 = 0$ is the only solution, and the roots are said to be coincident. $x = 2$.

(d) $2x^2 + 11x - 21 = 0$

$(2x - 3)(x + 7) = 0$

Either $2x - 3 = 0$

$x = \frac{3}{2}$

or $x + 7 = 0$

$x = -7$

$x = 1\frac{1}{2}$ or -7.

(ii) Quadratic inequalities

By the rule for multiplying directed numbers given in Section 1.10 if the product of two factors is positive they have the same sign and if the product is negative they have opposite signs.

Example 8.14

Find the range of values of x for which (a) $x^2 - x - 6 \leqslant 0$, (b) $x^2 + 5x + 4 \geqslant 0$ and mark the solution as an interval on the number line.

Solution (a) $x^2 - x - 6 = (x + 2)(x - 3)$. When $(x + 2)(x - 3) = 0$, either

$x + 2 = 0$ or $x - 3 = 0$

and

$x = -2$ or $x = 3$

When $(x + 2)(x - 3) \leqslant 0$, one factor is positive and the other is negative since the product is negative. $x + 2$ is always greater than $x - 3$, and so $x + 2$ is the positive factor: the solution is therefore $x + 2 \geqslant 0, x - 3 \leqslant 0$.

Answer: $\underline{x \geqslant -2, x \leqslant 3}$.

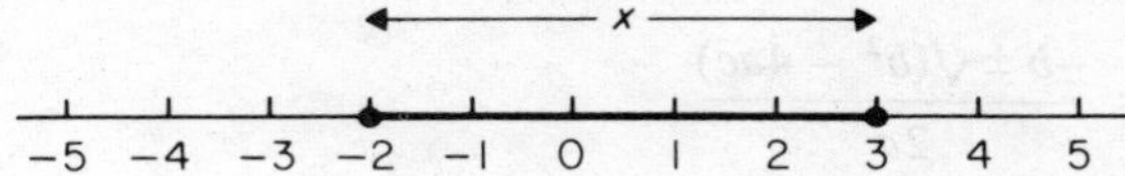

Fig 8.2

(b) $$x^2 + 5x + 4 = (x + 1)(x + 4)$$

$$(x + 1)(x + 4) = 0 \Rightarrow x + 1 = 0 \qquad \text{or } x + 4 = 0$$

$$x = -1 \qquad \text{or } x = -4$$

If $(x + 1)(x + 4) > 0$ both factors must have the same sign. Since $x + 1$ is less than $x + 4$, both factors are positive when $x + 1$ is positive

$$x + 1 > 0 \Rightarrow x > -1$$

Both factors are negative when $x + 4$ is negative

$$x + 4 < 0 \Rightarrow x < -4$$

The solution is $\underline{x \leqslant -4, x \geqslant -1}$.

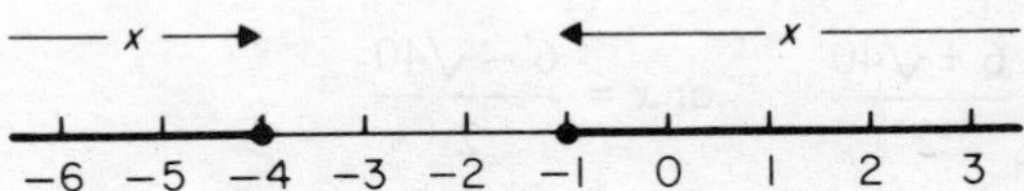

Fig 8.3

Exercise 8.5

Solve the following equations.

(1) $x^2 = 25$. (2) $16y^2 - 6 = 0$. (3) $y^2 = 9y$.

(4) $x^2 + 2x - 35 = 0$. (5) $2x^2 + x - 3 = 0$.

(6) $5x^2 + 34x = 7$. (7) $9y^2 - 30y + 25 = 0$.

(8) $6z^2 - 17z + 12 = 0$. (9) $4x^2 + 7x = 2$.

(10) $12t^2 - 12 = 10t$. (11) $7x - 3x^2 = 2$.

Solve the following inequalities.

(12) $9x^2 \geqslant 25$. (13) $x^2 - 10x + 24 \leqslant 0$.

(14) $2y^2 + 5y - 12 \leqslant 0$. (15) $6t^2 + 11t \geqslant 10$.

(b) By formula

A formula is a special type of equation in which one variable alone appears on one side, and the manipulation of formulae is considered later, in Section 8.9.

There is a formula for the solution of quadratic equations which can be used whether or not the expression factorises easily. If $ax^2 + bx + c = 0$, then

$$x = \frac{-b \pm \sqrt{(b^2 - 4ac)}}{2a}$$

and the coefficients in a given equation are substituted for a, b and c.

Example 8.15
Solve the equations (a) $x^2 - 6x - 1 = 0$, (b) $3x^2 + x - 1 = 0$ giving the solution correct to 2 dp.

Solution

(a) $x^2 - 6x - 1 = 0$

The fact that 2 dp was specified is an indication that the formula should be used instead of trying to find factors. By comparison with the standard equation $ax^2 + bx + c = 0$, we see that $a = 1, b = -6, c = -1$

$$x = \frac{-b \pm \sqrt{(b^2 - 4ac)}}{2a} = \frac{+6 \pm \sqrt{(36 + 4)}}{2}$$

$$x = \frac{6 + \sqrt{40}}{2} \qquad \text{or } x = \frac{6 - \sqrt{40}}{2}$$

$$x = \frac{6 + 6.324}{2} \qquad \text{or } x = \frac{6 - 6.324}{2}$$

$$x = 6.162 \qquad \text{or } x = -0.162$$

The solution is $\underline{x = 6.16 \text{ or } -0.16}$.

(b) $3x^2 + x - 1 = 0$ $\qquad a = 3, b = 1, c = -1$

$$x = \frac{-b \pm \sqrt{(b^2 - 4ac)}}{2a} = \frac{-1 \pm \sqrt{(1 - -12)}}{6}$$

$$x = \frac{-1 + \sqrt{13}}{6} \qquad \text{or } x = \frac{-1 - \sqrt{13}}{6}$$

$$x = \frac{-1 + 3.606}{6} \qquad \text{or } x = \frac{-1 - 3.606}{6}$$

$$x = 0.434 \qquad \text{or } x = -0.768$$

The solution is $\underline{x = 0.43 \text{ or } -0.77}$.

Example 8.16

Solve the equation $\dfrac{1}{2x-1}+\dfrac{2}{x}=5.$

Solution The fractions are combined first over a common denominator

$$\frac{x+2(2x-1)}{x(2x-1)}=5$$

$$x+4x-2=5x(2x-1)=10x^2-5x$$

This leads to the quadratic equation

$$10x^2-10x+2=0$$

or

$$5x^2-5x+1=0$$

Applying the formula with $a=5, b=-5, c=1$, gives

$$x=\frac{5\pm\sqrt{(25-20)}}{10}$$

$$x=\frac{5+\sqrt{5}}{10} \qquad \text{or } x=\frac{5-\sqrt{5}}{10}$$

$$x=0.7236 \text{ or } x=0.2764$$

The solution is $\underline{x=0.724 \text{ or } 0.276}$.

8.7 PROBLEMS INVOLVING QUADRATIC EQUATIONS

Example 8.17

If the difference of two numbers is 11 and their product is 80, what are the numbers?

Solution Suppose the greater number is N, then the other is $N-11$ and the second condition is

$$N(N-11)=80$$

$$N^2-11N=80$$

$$N^2-11N-80=0$$

By factorisation

$$(N+5)(N-16)=0$$

Therefore

$$N = -5 \text{ or } N = 16$$

The two numbers are $\underline{-5 \text{ and } -16 \text{ or } 16 \text{ and } 5}$.

Example 8.18
The time taken to travel a distance of 225 km was reduced by 20 minutes when the average speed was increased by 20 km/h. Calculate the slower speed.

Solution Call the slower speed x km/h, so that the faster speed is $(x + 20)$ km/h. Then the times taken for the journey of 225 km are $225/x$ and $225/(x + 20)$ hours and the difference in these times is 20 minutes or 1/3 hour. Therefore

$$\frac{225}{x} - \frac{225}{x + 20} = \frac{1}{3}$$

$$\frac{225(x + 20) - 225x}{x(x + 20)} = \frac{1}{3}$$

$$\frac{4500}{x(x + 20)} = \frac{1}{3}$$

$$4500 \times 3 = x(x + 20)$$

This leads to the quadratic equation

$$x^2 + 20x - 13\,500 = 0$$

and solving by the formula gives

$$x = \frac{-20 \pm \sqrt{(400 + 54\,000)}}{2}$$

Since we are finding a speed, only the positive value is required

$$x = \frac{-20 + \sqrt{54\,400}}{2} = \frac{213.2}{2} = 106.6$$

The speed was $\underline{107 \text{ km/h}}$.

Exercise 8.6
Solve the following equations giving your answers correct to 3 sig. fig.

(1) $x^2 = 18$. (2) $x^2 - 8x + 3 = 0$. (3) $3x^2 + 5x + 1 = 0$.
(4) $2x^2 - 6x - 3 = 0$. (5) $6x^2 - 5x = 4$.

(6) $2x + \frac{3}{x} = 7.$ (7) $\frac{3}{x} = 3x - 1.$

(8) $\frac{3}{x-1} - \frac{1}{x} = 1.$ (9) $\frac{1}{x} - \frac{3(x-1)}{5} = 12.$

(10) $\frac{1}{x+1} + \frac{1}{x+2} = 14.$ (11) $\frac{1}{2x-1} - \frac{3}{x+1} = 8.$

(12) Solve $x^2 - 12x + c = 0$ given that c is a negative prime number and x is an integer.

(13) If the sum of a number and its reciprocal is 2.9, what are the possible values of the number?

(14) Find two consecutive odd number such that the sum of their squares is 290.

8.8 THE SOLUTION OF SIMULTANEOUS LINEAR AND QUADRATIC EQUATIONS

The general method is to express one of the variables in terms of the other using the linear equation, and then to substitute the expression in the quadratic equation. The resulting quadratic equation in one variable is then solved by factors or formula, and the solutions are then substituted in the original linear equation to obtain the values of the second variable. This is known as the *substitution method* of solving simultaneous equations and it can be applied to two linear equations.

Example 8.19

Solve the simultaneous equations $x + y = 10$, $x^2 + y^2 = 58$.

Solution

$$x + y = 10 \quad (1)$$

$$x^2 + y^2 = 58 \quad (2)$$

From equation (1) $y = 10 - x$. Substitute for y in equation (2)

$$x^2 + (10 - x)^2 = 58$$

$$x^2 + 100 - 20x + x^2 = 58$$

$$2x^2 - 20x + 42 = 0$$

$$x^2 - 10x + 21 = 0$$

By factorisation

$$(x - 3)(x - 7) = 0$$

and substituting in equation (1) the solution is $\underline{x = 3, y = 7}$ or $\underline{x = 7, y = 3}$.

In this case the two solutions were equal because the original equations were symmetrical.

Example 8.20
Solve the simultaneous equations

$$2x^2 - x = y + 6 \qquad (1)$$

$$3x + y - 6 = 0 \qquad (2)$$

Solution From equation (2) $y = 6 - 3x$. Substitute for y in equation (1)

$$2x^2 - x = 6 - 3x + 6$$

Rearrange the terms

$$2x^2 + 2x - 12 = 0$$

$$x^2 + x - 6 = 0$$

Factorise

$$(x + 3)(x - 2) = 0$$

The solution is $x = -3$ or $x = 2$.

Substitute for x in equation (2)

$$x = -3 \Rightarrow y = 15$$

$$x = 2 \Rightarrow y = 0$$

The complete solution is $\underline{x = -3, y = 15 \text{ or } x = 2, y = 0}$.

Exercise 8.7
Solve the following equations.
(1) $x - y = 4, x^2 - y^2 = 8$. (2) $xy = -28, x + 2y = 10$.
(3) $2x - 3y = 4, x^2 + y^2 = 29$. (4) $2x + 3y = 2, 1/x + 1/y = 5$.
(5) $3x + 2y = 5, 2x^2 + xy + y^2 = 14$.
(6) If the sum of two numbers is 8 and the difference of their squares is also 8 what are the numbers?

8.9 TRANSPOSITION OF FORMULAE

The subject of a formula is the single variable on one side of the equation; transposing a formula means rearranging it to make a different variable the subject. For example, the formula for calculating simple interest was given in Section 5.6 as

$$I = \frac{PTR}{100}$$

Transposed for P it becomes

$$P = \frac{100I}{RT} \text{ and for } T, \qquad T = \frac{100I}{PR}$$

There are vast numbers of different formulae used in physics, chemistry and engineering as well as in mathematics, but a general method of approach is given here which can be applied to most of the formulae met at this level of mathematics.

(i) If there is a surd, raise both sides of the equation to the appropriate power. $F = \sqrt{(I/MH)} \Rightarrow F^2 = I/MH$; $\quad d = \sqrt[3]{(a^2 - b^2)} \Rightarrow d^3 = a^2 - b^2$

(ii) If there are fractions, remove them by multiplying each term by the common denominator. $1/u + 1/v = 1/f \Rightarrow fv + fu = uv$

(iii) Remove any brackets and bring all terms containing the required subject to the left side of the equation

(iv) If there is more than one term, take out the subject as a common factor

(v) Divide by the other factor, and take a root if necessary.

Example 8.21

(a) Make x the subject of

$$v = w\sqrt{(a^2 - x^2)}$$

$$v^2 = w^2(a^2 - x^2)$$

$$v^2 = w^2a^2 - w^2x^2$$

$$w^2x^2 = w^2a^2 - v^2$$

$$x^2 = \frac{w^2a^2 - v^2}{w^2}$$

$$x = \pm\sqrt{\left(\frac{w^2a^2 - v^2}{w^2}\right)}$$

(b) Transpose for s the formula

$$t = \frac{3s + 2}{s - 3}$$

$$t(s - 3) = 3s + 2$$

$$ts - 3t = 3s + 2$$

$$ts - 3s = 3t + 2$$

$$s(t - 3) = 3t + 2$$

$$s = \frac{3t + 2}{t - 3}$$

Example 8.22
(a) If $x = (ay + b)/(cy + b)$, express b in terms of the other variables.

$$x = \frac{ay + b}{cy + b}$$

$$x(cy + b) = ay + b$$

$$xcy + xb = ay + b$$

$$xb - b = ay - xcy$$

$$b(x - 1) = y(a - xc)$$

$$b = \frac{y(a - xc)}{x - 1}$$

(b) Express x in terms of y and z when $1/x + 2/y = 1/z$. The common denominator is xyz

$$yz + 2xz = xy$$

$$2xz - xy = -yz$$

$$x(2z - y) = -yz$$

$$x = \frac{-yz}{(2z - y)}$$

or

$$x = \frac{yz}{(y - 2z)}$$

Exercise 8.8

Make the letter in the bracket the subject of the given formula.

(1) $v = u + at$ (t). (2) $m = u - kt$ (k).

(3) $V = \sqrt{(2E/M)}$ (E). (4) $x = b + ay^2$ (y).

(5) $1/u + 1/v = 1/f$ (v). (6) $y = (a + x)/(b - x)$ (x).

(7) $z = xt^2/(u - t^2)$ (t). (8) $I = E/(R + r)$ (R).

(9) If $A = Kr^2 + 2Krh$, calculate the value of h when $r = 4$, $K = 3\frac{1}{7}$ and $A = 132$.

(10) Calculate the value of m from the formula $E = \frac{1}{2}mv^2$ when $v = 4 \times 10^3$ and $E = 2 \times 10^{-12}$.

CHAPTER 9

SETS

Sets are the basis of modern algebra, and set theory and notation are used in other branches of mathematics, particularly in probability and topology.

9.1 SET LANGUAGE AND NOTATION

(a) Elements of a set

A set is a collection of different objects called the elements of the set; it is usually represented by a capital letter and the elements are enclosed in a bracket $\{\ \}$.

For example, if B denotes the set of people in a certain team, we could write

$$B = \{\text{members of this team}\}$$

or $\quad B = \{\text{Angela, Caroline, Sarah, Alan, David, Paul, Sheldon}\}$

The set R of letters in the word random could be written

$$R = \{\text{letters in random}\} \quad \text{or} \quad \{\text{r, a, n, d, o, m}\}$$

The symbol $\in$ is used to mean 'is an element of'. Angela and Paul are members of the team but Julia is not, and this is written as

$$\text{Angela} \in B, \qquad \text{Paul} \in B, \qquad \text{Julia} \notin B$$

Similarly

$$\text{r} \in R, \qquad \text{n} \in R, \qquad \text{but x} \notin R$$

(b) Number of elements in a set

(i) *Finite sets*

The set B has seven elements since there are seven people in the team and this is written as $n(B) = 7$. The word random has six letters, and so $n(R) =$

6. If P represents the set of letters in the word 'panama', then $P = \{p, a, n, m\}$ and $n(P) = 4$. The letter a is only included once because a set must contain different elements.

(ii) Infinite sets
These include sets such as $N = \{\text{natural numbers}\}$ and $Z = \{\text{integers}\}$.

(c) Universal sets
A universal set is denoted by $\mathcal{E}$, and depends on the particular sets being considered. For the letters in various words the universal set would contain all the letters of the alphabet, so that

$$\mathcal{E} = \{\text{letters of the alphabet}\} \quad \text{and} \quad n(\mathcal{E}) = 26$$

For members of a team the universal set might be members of a sports club, or students at a college.

(d) Complement of a set
The complement of a set A is written A' and is the set containing all the elements of the universal set that are not elements of A. For example, if $A = \{\text{vowels}\}$, then $A' = \{\text{consonants}\}$ and $u \in A$ but $u \notin A'$: $p \in A'$ but $p \notin A$

$$n(A) + n(A') = N(\mathcal{E})$$

(e) Empty set
A set with no elements is called the empty set and denoted by the symbol $\emptyset$. One example of an empty set is the set of even prime numbers greater than 2.

9.2 SUBSETS

(a) Proper and improper subsets
If all the elements of a set B are elements of a set A, then B is a subset of A and A contains B as a subset. The symbols used are $\subset$ meaning 'is a subset of' and $\supset$ meaning 'contains as a subset'

$$B \subset A \Leftrightarrow A \supset B$$

Thus if $A = \{2, 4, 6, 8\}$, $B = \{2, 4\}$, and $C = \{1, 2, 3\}$ then

$B \subset A$ because 2 and 4 are both elements of A

$C \not\subset A$ because 1 and 3 are not elements of A

Alternatively we could write $A \supset B$ but $A \not\supset C$.

The whole set and the empty set are subsets of every set and they are called *improper* subsets. All other subsets are *proper* subsets.

(b) Number of subsets

In general any set with n elements has $2^n - 2$ proper subsets.
For example, the set $\{a, b\}$ has 2 proper subsets $\{a\}, \{b\}$.
The proper subsets of $\{x, y, z\}$ are $\{x, y\}, \{x, z\}, \{y, z\}, \{x\}, \{y\}, \{z\}$.

Example 9.1
List all the proper subsets of the set {red, blue, green, yellow}. How many of them contain green and how many contain neither green nor blue?

Solution There are $2^4 - 2$, that is 14 proper subsets: {red, blue, green}, {red, blue, yellow}, {red, green, yellow}, {blue, green, yellow}, {red, blue}, {red, green}, {red, yellow}, {blue, green}, {blue, yellow}, {green, yellow}, {red}, {blue}, {green}, {yellow}.
Answer: 7 of the sets contain the element green, 3 of them contain neither green nor blue.

(c) Equal sets

Two sets are equal only if they contain exactly the same elements. For example, if $A = \{a, b, c, d\}$ and $B = \{b, d, c, a\}$ A and B are equal sets because they contain the same elements, the order is not important. Each of them is an improper subset of the other. Hence

$$A = B \Leftrightarrow A \subset B \text{ and } B \subset A$$

9.3 VENN DIAGRAMS

Venn diagrams are used to illustrate sets, and they are helpful in the solution of problems involving sets. Fig. 9.1 shows two typical Venn diagrams; each set is represented by a closed figure, usually a rectangle or circle, and the listed elements or the number of elements may be written inside each region.

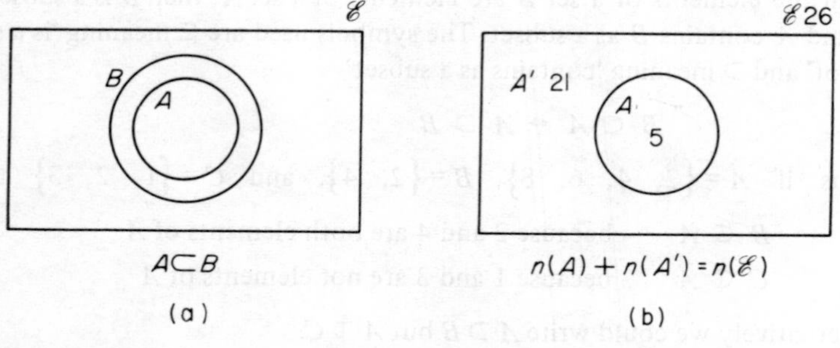

Fig 9.1

Example 9.2
Express the following statements in set notation and illustrate each by a Venn diagram.
(a) Not all who swim can dive, but all those who dive can also swim.

Solution The statement means that there are more swimmers than divers. If

$$S = \{\text{swimmers}\} \quad \text{and} \quad D = \{\text{divers}\}$$

written in set notation it becomes

$$D \subset S \quad \text{but} \quad S \not\subset D$$

A possible Venn diagram is shown in Fig. 9.2(a).
(b) Some integers are not natural numbers but all integers are real numbers.

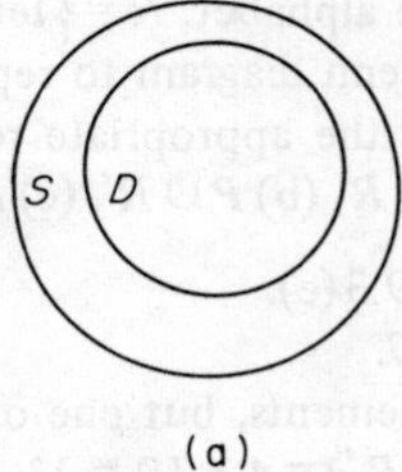

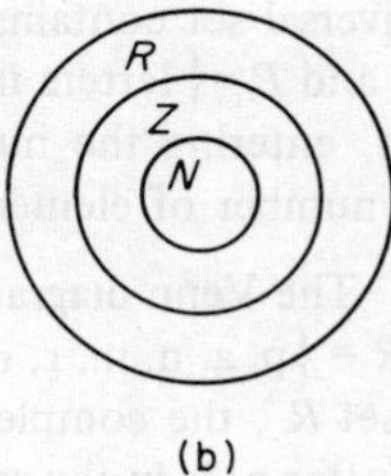

Fig 9.2

Solution

$N = \{\text{natural numbers}\}$, $Z = \{\text{integers}\}$, $R = \{\text{real numbers}\}$

In set notation the statements are equivalent to

$$Z \not\subset N \quad \text{but} \quad Z \subset R$$

and the Venn diagram is shown in Fig. 9.2(b).

9.4 METHODS OF COMBINING SETS

Since sets are not numbers the ordinary processes of addition and multiplication cannot be used; instead there are two special operations called *union* and *intersection* of sets.

(a) Union of sets

The symbol for union is $\cup$, and the union of two sets A and B is the set $A \cup B$ which contains all the elements of A together with those elements of B not already included.

For instance, if $A = \{1, 2, 3, 4, 5\}$ and $B = \{2, 4, 6, 8, 10\}$

$$A \cup B = \{1, 2, 3, 4, 5, 6, 8, 10\}$$

In the Venn diagrams of Fig. 9.3 (a) and (b), the shaded region represents $A \cup B$. In Fig. 9.3 (b) A and B are disjoint sets.

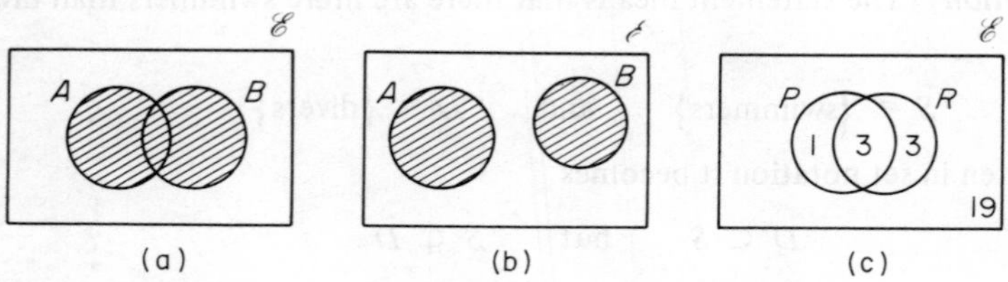

Fig 9.3

Example 9.3

If the universal set contains all the letters of the alphabet, $R = \{$letters in random$\}$ and $P = \{$letters in panama$\}$, draw a Venn diagram to represent these sets, entering the number of elements in the appropriate regions. State the number of elements in the sets (a) $P \cup R$, (b) $P \cup R'$, (c) $P \cup P'$.

Solution The Venn diagram is as shown in Fig. 9.3 (c).

(a) $P \cup R = \{$p, a, n, m, r, d, o$\}$ and $\underline{n(P \cup R) = 7}$.

(b) The set R', the complement of R, has 20 elements, but one of them (the letter p) is in the set P. Therefore $n(P \cup R') = 4 + 19 = \underline{23}$.

(c) $P \cup P' = \mathscr{E}$, therefore $n(P \cup P') = 26$.

In Fig. 9.4 some general results are illustrated by Venn diagrams.

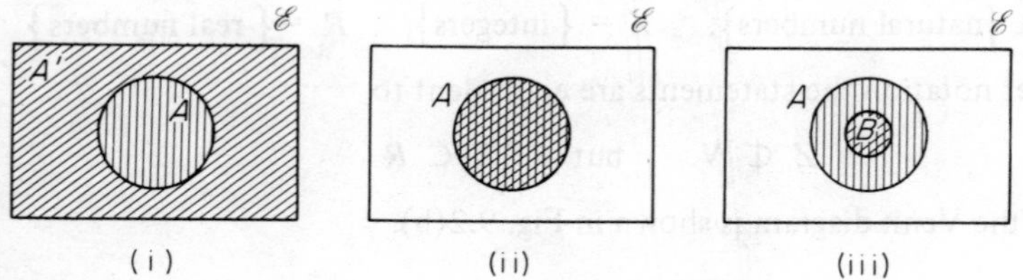

Fig 9.4

(i) $A \cup A' = \mathscr{E}$ (ii) $A \cup A = A$ (iii) $A \cup B = A \Rightarrow A \supset B$

(b) Intersection of sets

The symbol used for intersection is $\cap$, and the intersection of two sets A and B is defined as the set $A \cap B$ containing all the elements which are in both A and B. Sets having no common element are called disjoint.

In Example 9.3, $P \cap R = \{$a, n, m$\}$ since the letters a, n, m are in both words, and in Fig. 9.3(c) they are in the region of overlap of the two circles.

Intersecting sets are represented by overlapping areas in Venn diagrams, and those in Fig. 9.5 are illustrating some general statements about intersection of sets.

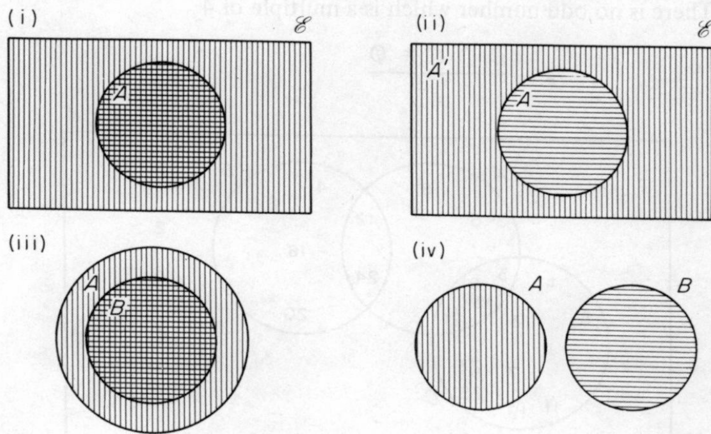

Fig 9.5

(i) $\mathscr{E} \cap A = A$ (ii) $A \cap A' = \emptyset$

(iii) $A \cap B = B \Rightarrow B \subset A$ (iv) $A \cap B = \emptyset \Rightarrow A$ and B are disjoint

Example 9.4

The universal set contains the natural numbers from 1 to 25 inclusive. If $A = \{\text{multiples of } 3\}$, $B = \{\text{multiples of } 4\}$, $C = \{\text{odd numbers}\}$, write down the elements of the sets (a) $A \cap B$, (b) $A \cap C'$, (c) $\mathscr{E} \cap A$, (d) $B \cap C$. Draw a Venn diagram to illustrate these sets and lists the elements of (i) A only, (ii) B only.

Solution (a) Numbers which are multiples of both 3 and 4 are multiplies of 12, and therefore

$$\underline{A \cap B = \{12, 24\}}$$

(b) Since the set C contains only odd numbers, the complement C' contains only even numbers, and the set $A \cap C'$ contains the even numbers which are multiples of 3

$$\underline{A \cap C' = \{6, 12, 18, 24\}}$$

(c) The universal set contains all the numbers from 1 to 25 so the elements of $\mathscr{E} \cap A$ are the elements of A

$$\underline{\mathscr{E} \cap A = \{3, 6, 9, 12, 15, 18, 21, 24\}}$$

(d) There is no odd number which is a multiple of 4

$$\underline{B \cap C = \emptyset}$$

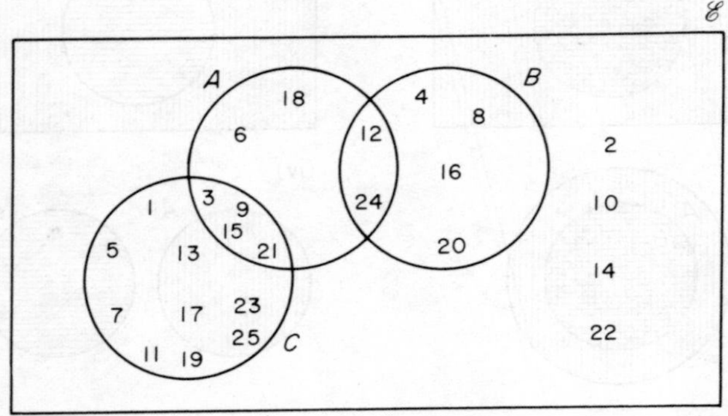

Fig 9.6

(i) The elements of A only are the elements of A that are in neither B nor C, that is $A \cap B' \cap C'$. They are $\underline{6 \text{ and } 18}$.
(ii) The elements of B only are the elements of $B \cap A'$. They are $\underline{4, 8, 16 \text{ and } 20}$.

(c) Combining the numbers of elements in sets

$$n(A) + n(B) = n(A \cup B) + n(A \cap B)$$

This is true for the number of elements in any two sets A and B, and it is used for calculating probabilities in Chapter 20. If any three of the numbers are known in a particular problem, the fourth can be calculated directly, without using a Venn diagram.

For any three sets A, B and C the relation is given by

$$n(A) + n(B) + n(C) = n(A \cup B \cup C) - n(A \cap B \cap C) + n(A \cap B) + n(B \cap C) + n(A \cap C)$$

This last equation could be verified for the sets A, B and C of Example 9.4

$$n(A) = 8, \quad n(B) = 6, \quad n(C) = 13, \quad n(A \cup B \cup C) = 21,$$

$$n(A \cap B) = 2, \quad n(A \cap C) = 4$$

and the other two sets are empty

$$8 + 6 + 13 = 27 \quad \text{and also} \quad 21 + 2 + 4 = 27$$

Exercise 9.1

(1) How many of the proper subsets of the set $\{3, 6, 9, 12, 15\}$ contain (a) 3 elements, (b) 4 elements?

(2) If $n(A) = 8$, $n(B) = 12$, $n(A \cap B) = 5$, how many elements has the set $A \cup B$?

(3) Find the value of $n(B)$ given that $n(A) = 19$, $n(A \cap B) = 7$ and $n(A \cup B) = 38$.

(4) When $\mathscr{E} = \{\text{natural numbers} < 30\}$, $A = \{\text{even numbers}\}$, $B = \{\text{multiples of } 4\}$, $C = \{\text{prime numbers}\}$, find the value of (a) $n(A')$, (b) $n(A \cap C)$, (c) $n(A \cup B)$. List the elements of the sets $A \cap B$ and $A \cap C$.

(5) $\mathscr{E} = \{\text{positive integers less than } 13\}$, $A = \{1, 2, 6, 11\}$, $B = \{2, 4, 6, 8, 10, 12\}$, $C = \{3, 6, 9, 12\}$. List the elements of the sets (a) $A \cup C$, (b) C', (c) $A \cap B$, (d) $A' \cap B$, (e) $A' \cap B'$, (f) $(A \cup B) \cap C$, (g) $A \cup (B \cap C)$, (h) $(A \cup B)'$.

(6) $\mathscr{E} = \{\text{letters of the alphabet}\}$, $P = \{\text{m, i, s, t}\}$, $Q = \{\text{s, c, e, n, t}\}$, $R = \{\text{a, l, m, o, s, t}\}$. State the number of elements in the sets (a) $P \cap R$, (b) P', (c) $Q \cup R$, (d) $P' \cup R'$, (e) $P \cap Q \cap R$, (f) $P \cup Q \cup R$, (g) $(P \cup R)'$.

(7) Which of the following statements are *true* for the sets P, Q, R of Question 6, (a) $\{\text{m, i, s, t}\} \supset P$, (b) $\{\text{m, o, s, t}\} \subset P \cap R$, (c) $P \not\subset R$, (d) $Q \cup R \supset P$?

(8) $\mathscr{E} = \{\text{a, b, c, d, e, f, m, n, r, s}\}$, $X = \{\text{a, b, c, m, n}\}$, $Y = \{\text{a, d, r, s}\}$ and $Z = \{\text{m, n, s}\}$. Draw a Venn diagram illustrating these sets and list the elements of the sets (a) $X \cap Y$, (b) $Z \cup Y$, (c) X', (d) $Y' \cap Z$, (e) $Y' \cap X'$, (f) $(X \cup Y) \cap (Y \cup Z)$, (g) $(X \cap Z) \cup Y$.

9.5 THE SOLUTION OF PROBLEMS INVOLVING SETS

In solving problems Venn diagrams may be used and also the equations given in Section 9.4(c). Remember that a Venn diagram is only an illustration of particular sets.

Example 9.5

The universal set contains all the cards in a deck of 52 playing cards, $A = \{\text{aces}\}$, $D = \{\text{diamonds}\}$, $P = \{\text{picture cards}\}$. Draw a Venn diagram showing the number of elements in each region, and state the number in (a) at least one of the sets, (b) none of the sets, (c) exactly one of the sets.

Solution Since an ace is not a picture card, A and P are disjoint sets. $A \cap D$ has only one element, the ace of diamonds, and $P \cap D$ has three elements, king, queen and knave of diamonds. There are 3 aces in A only, 9 diamonds in D only and 9 picture cards in P only. All these numbers are entered in the diagram shown in Fig. 9.7(a).

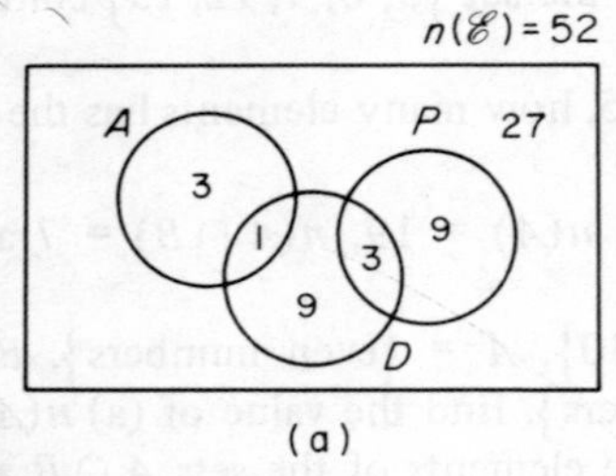

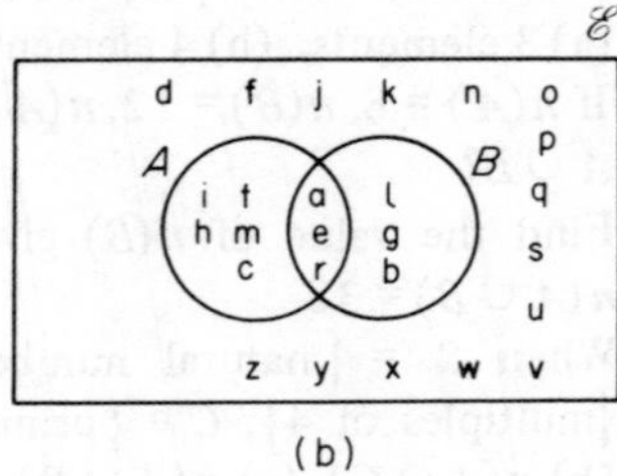

Fig 9.7

(a) The set of elements in at least one of A, D, P is the union $A \cup D \cup P$

$$n(A \cup D \cup P) = 3 + 1 + 3 + 9 + 9 = \underline{25}$$

(b) The elements in none of the sets are in the complement of the union

$$n(A \cup D \cup P)' = 52 - 25 = \underline{27}$$

(c) There are 3 + 9 + 9 elements in exactly one of the sets, that is $\underline{21}$.

Example 9.6

ℰ = {letters of the alphabet}, A = {letters in the word arithmetic}, B = {letters in the word algebra}. Illustrate these sets by a Venn diagram and state the number of elements in the sets (a) $A \cup B$, (b) $A \cup B'$, (c) $A \cap B$, (d) $A' \cap B'$. Verify that for these sets $n(A) + n(B) = n(A \cup B) + n(A \cap B)$.

Solution Remembering that each letter is only written once, they are entered in the regions of the Venn diagram [Fig. 9.7(b)].

(a) $n(A \cup B) = \underline{11}$.

(b) The set B' contains the 20 letters outside the region B, and there are 3 elements of A not included, the letters a, e, and r. $n(A \cup B') = \underline{23}$.

(c) $n(A \cap B) = \underline{3}$.

(d) The elements of $A' \cap B'$ are those outside the region $A \cup B$

$$n(A' \cap B') = 26 - 11 = \underline{15}$$

$$n(A) + n(B) = 8 + 6 = 14$$

$$n(A \cup B) + n(A \cap B) = 11 + 3 = 14$$

Example 9.7

Everyone in a group of 48 people bought at least one of three newspapers, A, B, and C. 30 people bought A, 25 bought B and 15 bought C, 7 people had both A and B, 6 had both B and C, while 16 bought B only. How many in the group bought (a) A and C but not B, (b) all three papers?

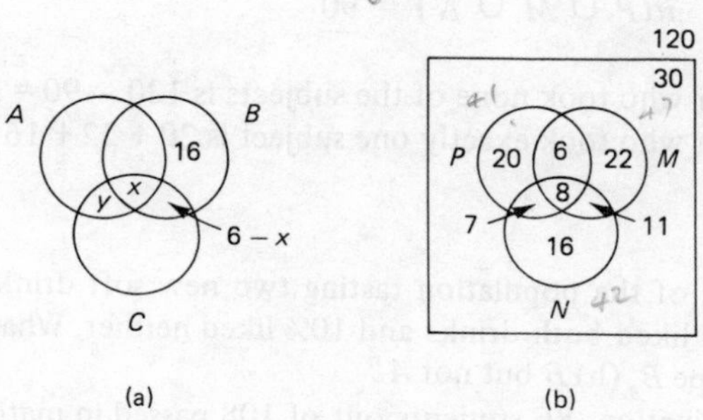

Fig 9.8

Solution Let the number $n(A \cap B \cap C)$ be x and $n(A \cap B' \cap C)$ by y as marked on the Venn diagram in Fig. 9.8(a), so that $n(A \cap C) = x + y$

$$n(A) + n(B) + n(C) = n(A \cup B \cup C) + n(A \cap B) + n(A \cap C) + n(B \cap C) - n(A \cap B \cap C)$$

(a) Substituting the given values

$$30 + 25 + 15 = 48 + 7 + x + y + 6 - x \Rightarrow y = 9$$

9 people bought A and C but not B.

(b) Considering the set B

$$16 + 7 + 6 - x = 25 \Rightarrow x = 4$$

4 people bought three papers.

Example 9.8

In a group of 120 students, 41 studied painting, 47 took music and 42 needlework. 14 students studied both painting and music, 15 took painting and needlework, and 19 took music and needlework, while 8 students studied all three subjects. Draw a Venn diagram and find how many of the students studied (a) at least one subject, (b) none of the subjects, (c) exactly one subject.

Solution The numbers are entered on the Venn diagram shown in Fig. 9.8(b). Start at the centre with the 8 who took three subjects. Then fill in the remaining 6 of the 14 in $P \cap M$, the remaining 7 of the 15 in $P \cap N$, and 11 of the 19 in $M \cap N$.

The number in P only is $41 - (6 + 8 + 7) = 20$. The number in M only is $47 - (6 + 8 + 11) = 22$. The number in N only is $42 - (7 + 8 + 11) = 16$.

(a) The number of students who studied at least one subject is given by

$$n(P \cup M \cup N) = \underline{90}$$

(b) The number who took none of the subjects is $120 - 90 = \underline{30}$.

(c) The number who took exactly one subject is $20 + 22 + 16 = \underline{58}$.

Exercise 9.2

(1) In a sample of the population tasting two new soft drinks 61% liked type A, 8% liked both drinks and 10% liked neither. What percentage liked (a) type B, (b) B but not A?

(2) In an examination, 86 students out of 108 passed in mathematics, 35 passed in French and 3 failed in both subjects. How many passed in both?

(3) From a sample of 70 young people who liked listening to music, 42 listened to cassette players, 42 to records and 40 to live music. 8 students listened to cassettes only, 10 to records only, 20 to records and cassettes and 8 to music from all three sources. Find the number of students who listened to live music only.

(4) Of the 56 customers eating lunch in a restaurant 12 had fish and vegetables, 27 had meat and vegetables, 8 had fish only, 4 had meat only, while 2 people had meat, fish and vegetables. If nobody had both fish and meat without vegetables how many customers had vegetables only?

(5) At a meeting of the 35 members of a society, 15 people drank lemonade, 17 drank shandy and 19 drank orange juice. Two of the members had all three drinks, five had both lemonade and shandy, and seven had shandy and orange juice. How many members drank (a) shandy only, (b) lemonade and orange juice but not shandy?

9.6 BINARY OPERATIONS

Sets are the simplest algebraic structures, and a definition of *binary operations* is included here because they are the next stage in the study of general algebraic structures. Since it is only an introduction to the subject, the examples and the examination questions are quite simple, involving substitution in algebraic expressions.

Any operation in mathematics, such as addition or multiplication, combines two elements to give a single element; for instance $2 + 11 = 13$

and $1/2 \times 3/4 = 3/8$. When the result of combining two elements of a set is a member of the same set, the operation is called a binary operation for that set.

Definition: a binary operation combines two elements of a set to give a single element of the same set.

For the set of natural numbers, addition and multiplication are both binary operations because the sum and the product of any two natural numbers are always natural numbers. Subtraction and division are *not* binary operations on this set, since the difference or quotient of two natural numbers is not always a natural number; $2 - 11 = -9$ and $5 \div 3 = 1\frac{2}{3}$.

Addition and subtraction are both binary operations on the set of integers and for the set of rational numbers all four arithmetic operations are binary operations.

In algebra a general operation is usually represented by an asterisk $*$ but other symbols can be used.

Example 9.9
If $x * y$ means $2x + 3y$ what is the value of $-4 * 5$?

Solution The values $x = -4$ and $y = 5$ are substituted in the expression $2x + 3y$.

$$-4 * 5 = 2(-4) + 3(5) = -8 + 15 = \underline{7}$$

Example 9.10
A binary operation $*$ is defined on the set of integers by

$$x * y = x^2 + xy + y^2$$

Find the value of (a) $-2 * 7$, (b) $(-3 * 2) * (4 * -2)$

Solution

(a) $x * y = x^2 + xy + y^2$

Substitute -2 for x and 7 for y

$$\begin{aligned} -2 * 7 &= (-2)^2 + -14 + 7^2 \\ &= 4 - 14 + 49 \\ &= \underline{39} \end{aligned}$$

(b)
$$\begin{aligned} -3 * 2 &= 9 - 6 + 4 \\ &= 7 \\ 4 * -2 &= 16 - 8 + 4 \\ &= 12 \\ 7 * 12 &= 49 + 84 + 144 \\ &= \underline{277} \end{aligned}$$

Example 9.11
The binary operation $*$ is defined on the set of integers by

$$p * q = 2pq + 3$$

Find the value of x such that (a) $x * 3 = 15$, (b) $4 * (3 * 2x) = 123$.

Solution

(a) $p * q = 2pq + 3$

$x * 3 = 15 \Rightarrow 6x + 3 = 15$

Therefore $\underline{x = 2}$.

(b) $4 * (3 * 2x) = 4 * (12x + 3)$

$= 8(12x + 3) + 3$

$= 96x + 27$

$96x + 27 = 123 \Rightarrow \underline{x = 1}$.

Example 9.12
If $a \otimes b$ represents the remainder when the product ab is divided by 13, show that $\otimes$ is a binary operation on the set $\{1, 3, 9\}$.

Solution The product of all the possible pairs from the set is calculated and divided by 13 to obtain the remainder

$1 \otimes 1 = 1$, $1 \otimes 3 = 3 \otimes 1 = 3$, $1 \otimes 9 = 9 \otimes 1 = 9$,

$3 \otimes 3 = 9$, $3 \otimes 9 = 9 \otimes 3 = 1$, $9 \otimes 9 = 3$,

$(81 = 13 \times 6 + 3)$

The set $\{1, 3, 9\}$ is closed under the operation, and therefore $\otimes$ is a binary operation for the set.

Exercise 9.3

(1) If $x * y$ represents the smallest natural number divisible by both x and y, evaluate (a) $3 * 4$, (b) $8 * 32$, (c) $50 * 10$.

(2) $a * b$ represents the product ab divided by 5, so that $2 * 10 = 4$. (a) Evaluate $4 * 5$. (b) Find integer values of x such that (i) $3 * x = 54$, (ii) $(x + 1) * 3 = 9$, (iii) $2x * x = 10$.

(3) If $x * y$ is defined as the remainder when $3x + y$ is divided by 9, evaluate (a) $5 * 3$, (b) $8 * -2$, (c) $4 * (3 * 4)$.

(4) If the operation $*$ is defined on the set of rational numbers by $x * y = x + y - 2xy$ solve the equation $x * 4 = (x * 5) - (x * 3)$.

(5) $p * q$ denotes $(p + q)/q$. (a) Find the value of x when $x * 6 = 4$, (b) Solve the equation $2 * x = x * 32$.

(6) If $a * b$ denotes $a^2 - ab + 2$, find the value of $2 * (3 * 2)$. What is the value of x if $x * 3 = 0$?

9.7 OPERATION TABLES

When an operation is defined on a finite set which has a small number of elements, as in Example 9.12, the results are conveniently displayed in an *operation table*.

Table 9.1

$\otimes$	1	3	9
1	1	3	9
3	3	9	1
9	9	1	3

Table 9.2

$\oplus$	0	1	2	3	4
0	0	1	2	3	4
1	1	2	3	4	0
2	2	3	4	0	1
3	3	4	0	1	2
4	4	0	1	2	3

Table 9.1 shows the result $a \otimes b$ for all pairs in the set $\{1, 3, 9\}$ as given in Example 9.12. By convention the first element, a, indicates the row and the second element the column.

Since the table contains only the numbers 1, 3 and 9, $\otimes$ is a binary operation for this set; the set is said to be *closed* for the operation $\otimes$.

(a) Identity elements

The identity element for a binary operation $*$ is the one leaving every element of the set unchanged, and it is usually denoted by e. In algebraic form, $a * e = e * a = a$ for every element a in the set.

0 is the identity element for addition of real numbers, and 1 is the identity element for multiplication. For the union of sets the empty set $\emptyset$ is the identity element, since $A \cup \emptyset = A$ for every set A.

It can be readily seen from an operation table whether there is an identity element. In Table 9.1, the row labelled 1 is the same as the column heading, and 1 is the identity element. 0 is the identity element in Table 9.2.

(b) Inverse elements

The inverse of an element a is written a^{-1}. If e is the identity element for a binary operation $*$, $a * a^{-1} = a^{-1} * a = e$ for every element in the set.

When an operation table has been constructed the inverse of an element can be found by looking along the appropriate row to see which column contains the identity element. In Table 9.1, 9 is the inverse of 3 and 3 is the inverse of 9; they are a pair of inverse elements. The identity element is always self-inverse.

Example 9.13

Construct an operation table for the operation $\oplus$ defined on the set $\{0, 1, 2, 3, 4\}$ by: $a \oplus b$ is the remainder when the sum $a + b$ is divided by 5. State whether $\oplus$ is a binary operation and name the identity element and the inverse of each element if there is one. Use the table to evaluate (a) $(1 \oplus 3) \oplus (2 \oplus 4)$, (b) $(3 \oplus 4) \oplus 4$.

Solution The results are shown in Table 9.2, and since it contains only the numbers 0, 1, 2, 3, 4, ⊕ is a binary operation.

The identity element is 0 because the row labelled 0 is unchanged. 1 and 4 are a pair of inverse elements and so are 2 and 3.

(a) From the table, 1 ⊕ 3 = 4 and 2 ⊕ 4 = 1. Therefore

$$(1 \oplus 3) \oplus (2 \oplus 4) = 4 \oplus 1 = \underline{0}$$

(b) $3 \oplus 4 = 2,\ (3 \oplus 4) \oplus 4 = 2 \oplus 4 = \underline{1}$

Exercise 9.4

Table A

*	2	3	4
2	3	2	4
3	2	3	4
4	3	4	2

Table B

⊕	3	5	7	9
3	5	7	3	9
5	9	7	5	3
7	3	7	2	5
9	9	7	3	5

Table C

*	*a*	*b*	*c*	*d*
a	*d*	*c*	*b*	*a*
b	*a*	*c*	*d*	*b*
c	*d*	*b*	*a*	*c*
d	*a*	*b*	*c*	*d*

Table D

○	4	5	6
4	4	6	5
5	6	5	4
6	5	4	6

(1) For each of the operation Tables A, B, C and D shown above
   (i) state whether the operation is a binary operation
   (ii) name the identity element if there is one
   (iii) name the inverse of each element if it exists.

(2) For Table C, (i) show that $(b * c) * a = c * (c * d)$. (ii) Which element is equal to $(a * b) * (c * d)$?

(3) From Table D determine the value of x when (a) $(5 \circ 6) \circ x = 6$, (b) $(6 \circ 4) \circ (5 \circ 4) = x$.

(4) With reference to Table B solve the equations (a) $x \oplus 5 = 7$, (b) $(x \oplus 3) \oplus 9 = 9$, (c) $5 \oplus 9 = x \oplus 7$.

CHAPTER 10

MATRICES

A set is a collection of elements in no particular order, but a *matrix* is a mathematical unit in which the elements are numbers arranged in rows and columns. The same numbers arranged in a different order would give a different matrix, and matrices are used to present numerical information in a compact form.

10.1 THE SIZE OF A MATRIX

Matrices are usually represented by capital letters and when specifying the size of a matrix the number of rows is given first and then the number of columns

$$P = \begin{pmatrix} 1 & 3 & 5 \\ 2 & 4 & 1 \end{pmatrix} \text{ is two by three (written } 2 \times 3)$$

since it has 2 rows and 3 columns. The first row, for instance, is $(1 \;\; 3 \;\; 5)$ and the third column is $\begin{pmatrix} 5 \\ 1 \end{pmatrix}$

$$M = \begin{pmatrix} 1 & 2 \\ 3 & -4 \\ 5 & 6 \\ -7 & 8 \end{pmatrix} \text{ is } 4 \times 2$$

and

$$N = \begin{pmatrix} 5 & 0 & 9 \\ 1 & 1 & 1 \\ 2 & 4 & 1 \end{pmatrix} \text{ is } 3 \times 3$$

10.2 OPERATIONS DEFINED ON MATRICES

Like sets, matrices are not single numbers so the ordinary arithmetic operations cannot be used, and matrix addition and matrix multiplication are defined in a special way.

(a) Addition and subtraction

Two matrices are added or subtracted by combining the corresponding elements, and this is only possible when the two matrices are the same size.

If

$$A = \begin{pmatrix} a_{11} & a_{12} & a_{13} \\ a_{21} & a_{22} & a_{23} \end{pmatrix} \qquad \text{and } B = \begin{pmatrix} b_{11} & b_{12} & b_{13} \\ b_{21} & b_{22} & b_{23} \end{pmatrix}$$

Then the matrix

$$A + B = \begin{pmatrix} a_{11} + b_{11} & a_{12} + b_{12} & a_{13} + b_{13} \\ a_{21} + b_{21} & a_{22} + b_{22} & a_{23} + b_{23} \end{pmatrix}$$

This shows the double subscript method of describing the position of an element in the general algebraic form of a matrix. For instance, a_{13} is the element in the first row and third column of A and b_{22} is in row 2 and column 2 of the matrix B.

Example 10.1

Find the matrices $P + Q$ and $P - Q$ when

$$P = \begin{pmatrix} 1 & 3 & 5 \\ 2 & 4 & 1 \end{pmatrix} \qquad \text{and } Q = \begin{pmatrix} 2 & -1 & 4 \\ -3 & 2 & 5 \end{pmatrix}$$

Solution

$$P + Q = \begin{pmatrix} 1 + 2 & 3 + -1 & 5 + 4 \\ 2 + -3 & 4 + 2 & 1 + 5 \end{pmatrix} = \begin{pmatrix} 3 & 2 & 9 \\ -1 & 6 & 6 \end{pmatrix}$$

$$P - Q = \begin{pmatrix} 1 - 2 & 3 - -1 & 5 - 4 \\ 2 - -3 & 4 - 2 & 1 - 5 \end{pmatrix} = \begin{pmatrix} -1 & 4 & 1 \\ 5 & 2 & -4 \end{pmatrix}$$

(b) Multiplication by a single number

Multiplication by a number can be regarded as repeated addition, so that $2P = P + P$ and $3Q = Q + Q + Q$. For the matrices P and Q of Example 10.1

$$2P = \begin{pmatrix} 1+1 & 3+3 & 5+5 \\ 2+2 & 4+4 & 1+1 \end{pmatrix} = \begin{pmatrix} 2 & 6 & 10 \\ 4 & 8 & 2 \end{pmatrix}$$

each element of P is multiplied by 2

$$3Q = \begin{pmatrix} 2 \times 3 & -1 \times 3 & 4 \times 3 \\ -3 \times 3 & 2 \times 3 & 5 \times 3 \end{pmatrix} = \begin{pmatrix} 6 & -3 & 12 \\ -9 & 6 & 15 \end{pmatrix}$$

each element of Q being multiplied by 3.

In general, when a matrix is multiplied by a number, each element of the matrix is multiplied by the number and this applies to negative numbers and fractions as well as natural numbers.

Example 10.2

When $A = \begin{pmatrix} 2 & 3 \\ -1 & 4 \end{pmatrix}$ and $B = \begin{pmatrix} 5 & -3 \\ 4 & 7 \end{pmatrix}$, find the matrix (a) $2A - B$, (b) $2B - 3A$.

Solution

(a) $$2A = \begin{pmatrix} 4 & 6 \\ -2 & 8 \end{pmatrix}, \quad 2A - B = \begin{pmatrix} 4-5 & 6--3 \\ -2-4 & 8-\ \ 7 \end{pmatrix} = \underline{\begin{pmatrix} -1 & 9 \\ -6 & 1 \end{pmatrix}}$$

(b) $$2B = \begin{pmatrix} 10 & -6 \\ 8 & 14 \end{pmatrix}, \quad 3A = \begin{pmatrix} 6 & 9 \\ -3 & 12 \end{pmatrix}$$

$$2B - 3A = \begin{pmatrix} 10-6 & -6-9 \\ 8--3 & 14-12 \end{pmatrix} = \underline{\begin{pmatrix} 4 & -15 \\ 11 & 2 \end{pmatrix}}$$

(c) Equal matrices

Two matrices are equal when the corresponding element in each matrix has the same numerical value.

Example 10.3

If $M = \begin{pmatrix} 2 & 4 \\ x & 3 \end{pmatrix}$, $N = \begin{pmatrix} 1 & y \\ 5 & 2 \end{pmatrix}$, and $2M + N = \begin{pmatrix} 5 & 10 \\ 11 & 8 \end{pmatrix}$ calculate the value of x and y.

Solution

$$2M + N = \begin{pmatrix} 4+1 & 8+y \\ 2x+5 & 6+2 \end{pmatrix} = \begin{pmatrix} 5 & 8+y \\ 2x+5 & 8 \end{pmatrix}$$

If

$$\begin{pmatrix} 5 & 8+y \\ 2x+5 & 8 \end{pmatrix} = \begin{pmatrix} 5 & 10 \\ 11 & 8 \end{pmatrix}$$

then by comparing corresponding elements

$$2x + 5 = 11 \Rightarrow \underline{x = 3}$$

$$8 + y = 10 \Rightarrow \underline{y = 2}$$

Exercise 10.1

(1) If $A = \begin{pmatrix} 2 & 0 \\ 3 & 5 \end{pmatrix}, B = \begin{pmatrix} -1 & -3 \\ 2 & -1 \end{pmatrix}, C = \begin{pmatrix} -2 & 3 \\ 4 & -1 \end{pmatrix}$ find the matrix
(a) $A + B$, (b) $A - B$, (c) $A + B + C$.

(2) Given that $K = \begin{pmatrix} 4 & 1 \\ -2 & 5 \end{pmatrix}$ and $L = \begin{pmatrix} -2 & 6 \\ 3 & -1 \end{pmatrix}$, find
(a) $3K$, (b) $2L$, (c) $K + 3L$, (d) $2L - K$, (e) $\frac{1}{2}K$.

(3) Find the matrix A such that $A + \begin{pmatrix} 2 & -4 \\ -1 & -3 \end{pmatrix} = \begin{pmatrix} 2 & 4 \\ 3 & -5 \end{pmatrix}$.

(4) Given that

$$A = \begin{pmatrix} 2 & -3 \\ -1 & 4 \\ 3 & 2 \end{pmatrix}$$

find a matrix X satisfying the equation (a) $2X = A$, (b) $X = 2X - A$.

(5) Find the value of x and y given that (a) $A = \begin{pmatrix} 2 & x \\ -1 & 4 \end{pmatrix}$,

$$B = \begin{pmatrix} -5 & 3 \\ 2 & y \end{pmatrix} \text{ and } 3A + B = \begin{pmatrix} 1 & 6 \\ -1 & 8 \end{pmatrix},$$

(b) $\begin{pmatrix} 1 & -2y \\ 2 & y \end{pmatrix} + \begin{pmatrix} 3 & x \\ 2 & -2x \end{pmatrix} = \begin{pmatrix} 4 & 7 \\ 4 & -5 \end{pmatrix}$.

(d) Matrix multiplication

Matrix multiplication is not defined in the same way as matrix addition, but in a special way which is valid only when the number of columns in the first matrix is the same as the number of rows in the second. This is because each element of the product matrix is obtained by multiplying one row of the first matrix by one column of the second, term by term. The method is shown in the following examples.

If A is $m \times n$ and B is $n \times p$ the product AB is $m \times p$ and the two matrices A and B are said to be *compatible* or to conform for multiplication. For example, if A is 2×3 and B is 3×4, then AB is 2×4 but the matrix product BA cannot be evaluated since B has 4 columns and A has only 2 rows.

Suppose

$$A = \begin{pmatrix} a_{11} & a_{12} & a_{13} \\ a_{21} & a_{22} & a_{23} \end{pmatrix} \quad \text{and } B = \begin{pmatrix} b_{11} & b_{12} \\ b_{21} & b_{22} \\ b_{31} & b_{32} \end{pmatrix}$$

the matrix product AB exists and is 2×2, call it

$$\begin{pmatrix} c_{11} & c_{12} \\ c_{21} & c_{22} \end{pmatrix}$$

The element in the first row and first column of AB is the product of the first row of A and the first column of B

$$c_{11} = (a_{11} \quad a_{12} \quad a_{13}) \times \begin{pmatrix} b_{11} \\ b_{21} \\ b_{31} \end{pmatrix} = a_{11}b_{11} + a_{12}b_{21} + a_{13}b_{31}$$

$$c_{12} = (a_{11} \quad a_{12} \quad a_{13}) \times \begin{pmatrix} b_{12} \\ b_{22} \\ b_{32} \end{pmatrix} = a_{11}b_{12} + a_{12}b_{22} + a_{13}b_{32}$$

and similarly for the other elements.

In matrix multiplication the order must always be specified. In the product AB, A is said to be *postmultiplied* by B and B is *premultiplied* by A.

Example 10.4

Find the product matrix (i) AB, (ii) BA when $A = \begin{pmatrix} 2 & 3 \\ -1 & 4 \end{pmatrix}, B = \begin{pmatrix} 5 & -3 \\ 4 & 7 \end{pmatrix}$.

Solution Since A and B are both 2×2, the product matrices are both 2×2.

(i) Suppose AB is $\begin{pmatrix} a_{11} & a_{12} \\ a_{21} & a_{22} \end{pmatrix}$

$$a_{11} = (2 \quad 3) \times \begin{pmatrix} 5 \\ 4 \end{pmatrix} = 2 \times 5 + 3 \times 4 = 22$$

$$a_{12} = (2 \quad 3) \times \begin{pmatrix} -3 \\ 7 \end{pmatrix} = 2 \times -3 + 3 \times 7 = 15$$

$$a_{21} = (-1 \quad 4) \times \begin{pmatrix} 5 \\ 4 \end{pmatrix} = -1 \times 5 + 4 \times 4 = 11$$

$$a_{22} = (-1 \quad 4) \times \begin{pmatrix} -3 \\ 7 \end{pmatrix} = -1 \times -3 + 4 \times 7 = 31$$

$$AB = \begin{pmatrix} 22 & 15 \\ 11 & 31 \end{pmatrix}$$

(ii) Suppose BA is $\begin{pmatrix} b_{11} & b_{12} \\ b_{21} & b_{22} \end{pmatrix}$

$$b_{11} = (5 \quad -3) \times \begin{pmatrix} 2 \\ -1 \end{pmatrix} = 5 \times 2 + -3 \times -1 = 13$$

$$b_{12} = (5 \quad -3) \times \begin{pmatrix} 3 \\ 4 \end{pmatrix} = 5 \times 3 + -3 \times 4 = 3$$

$$b_{21} = (4 \quad 7) \times \begin{pmatrix} 2 \\ -1 \end{pmatrix} = 4 \times 2 + 7 \times -1 = 1$$

$$b_{22} = (4 \quad 7) \times \begin{pmatrix} 3 \\ 4 \end{pmatrix} = 4 \times 3 + 7 \times 4 = 40$$

$$BA = \begin{pmatrix} 13 & 3 \\ 1 & 40 \end{pmatrix}$$

After a little practice the products can be written down without the preliminary working, as shown in the next Example.

Example 10.5

$$P = \begin{pmatrix} 1 & 3 & 5 \\ 2 & 4 & 6 \end{pmatrix}, \qquad Q = (3 \quad -2 \quad 5), \qquad R = \begin{pmatrix} 7 & -2 \\ 4 & 9 \\ -1 & 3 \end{pmatrix}$$

State which pairs of matrices are compatible and find the products where possible.

Solution
P is 2×3, Q is 1×3; they are not compatible.
Q is 1×3, P is 2×3; they are not compatible.
P is 2×3, R is 3×2; PR exists and is 2×2.
R is 3×2, P is 2×3; RP exists and is 3×3.
Q is 1×3, R is 3×2; QR exists and is 1×2.
R is 3×2, Q is 1×3; they are not compatible.
Only three of the six products can be evaluated, PR, QR, and RP

$$\begin{pmatrix} 1 & 3 & 5 \\ 2 & 4 & 6 \end{pmatrix} \times \begin{pmatrix} 7 & -2 \\ 4 & 9 \\ -1 & 3 \end{pmatrix} = \begin{pmatrix} 7+12+-5 & -2+27+15 \\ 14+16+-6 & -4+36+18 \end{pmatrix}$$

Therefore

$$PR = \underline{\begin{pmatrix} 14 & 40 \\ 24 & 50 \end{pmatrix}}$$

$$(3 \quad -2 \quad 5) \times \begin{pmatrix} 7 & -2 \\ 4 & 9 \\ -1 & 3 \end{pmatrix} = (21+-8+-5 \quad -6+-18+15)$$

Therefore

$$\underline{QR = (8 \quad -9)}$$

$$\begin{pmatrix} 7 & -2 \\ 4 & 9 \\ -1 & 3 \end{pmatrix} \times \begin{pmatrix} 1 & 3 & 5 \\ 2 & 4 & 6 \end{pmatrix} = \begin{pmatrix} 7+-4 & 21+-8 & 35+-12 \\ 4+18 & 12+36 & 20+54 \\ -1+6 & -3+12 & -5+18 \end{pmatrix}$$

$$\underline{RP = \begin{pmatrix} 3 & 13 & 23 \\ 22 & 48 & 74 \\ 5 & 9 & 13 \end{pmatrix}}$$

Exercise 10.2

(1) If A is a $3 \times m$ matrix and B is a 2×5 matrix, for what value of m does the matrix (a) AB, (b) A^2, (c) BA exist?

(2) $A = \begin{pmatrix} 2 & -1 \\ 3 & -4 \end{pmatrix}, B = \begin{pmatrix} -1 & 3 \\ 2 & -5 \end{pmatrix}, C = \begin{pmatrix} 1 \\ -2 \end{pmatrix}, D = (-2 \quad 13).$

If the matrices are compatible find the products (a) AC, (b) AB, (c) BC, (d) CD, (e) DC, (f) C^2, (g) A^2, (h) D^2.

(3) Evaluate the product (a) $(-1 \quad 2 \quad -3)\begin{pmatrix} 4 \\ -3 \\ 2 \end{pmatrix}$, (b) $\begin{pmatrix} 2 & 1 & 3 \\ 1 & -1 & 2 \\ 3 & 2 & -1 \end{pmatrix}\begin{pmatrix} 2 & 3 & -1 \\ -1 & 1 & 1 \\ 1 & -2 & 2 \end{pmatrix}$

(4) Find the matrix product (a) $\begin{pmatrix} 1 & -2 \\ 3 & 4 \end{pmatrix}\begin{pmatrix} x \\ y \end{pmatrix}$, (b) $\begin{pmatrix} -2 & x \\ 3 & y \end{pmatrix}\begin{pmatrix} 3 \\ -4 \end{pmatrix}$.

(5) Solve the matrix equations (a) $\begin{pmatrix} 6 \\ -9 \end{pmatrix} = 3\begin{pmatrix} x \\ y \end{pmatrix}$, (b) $\begin{pmatrix} x \\ y \end{pmatrix} = \begin{pmatrix} 2 & -1 \\ 3 & 1 \end{pmatrix}\begin{pmatrix} 1 \\ 4 \end{pmatrix}$,

(c) $\begin{pmatrix} 4 & 5 \\ -3 & 2 \end{pmatrix}\begin{pmatrix} x \\ y \end{pmatrix} = \begin{pmatrix} 11 \\ 9 \end{pmatrix}$.

10.3 SOME SPECIAL MATRICES

(a) Zero matrix

A zero matrix has every element zero, and the addition of a zero matrix to any matrix of the same size leaves the matrix unchanged. We say the zero matrix is the identity element for addition of matrices of that size. For a 2 × 3 matrix

$$\begin{pmatrix} 1 & 2 & 3 \\ 4 & 5 & 6 \end{pmatrix} + \begin{pmatrix} 0 & 0 & 0 \\ 0 & 0 & 0 \end{pmatrix} = \begin{pmatrix} 1 & 2 & 3 \\ 4 & 5 & 6 \end{pmatrix}$$

(b) Square matrix

A square matrix has the same number of rows as it has columns.

(c) Unit matrix

A unit matrix is a square matrix with the elements on the leading diagonal 1 and all other elements 0. The unit matrix of any size is the identity element for multiplication of matrices of that size, and it is usually given the symbol I_n

$$I_2 = \begin{pmatrix} 1 & 0 \\ 0 & 1 \end{pmatrix} \qquad \text{and } I_3 = \begin{pmatrix} 1 & 0 & 0 \\ 0 & 1 & 0 \\ 0 & 0 & 1 \end{pmatrix}$$

For any 2×2 matrix

$$\begin{pmatrix} a & b \\ c & d \end{pmatrix} \begin{pmatrix} 1 & 0 \\ 0 & 1 \end{pmatrix} = \begin{pmatrix} a & b \\ c & d \end{pmatrix}$$

(d) Inverse matrix

The inverse of a square matrix A is denoted by A^{-1}, and the product of a square matrix and its inverse is the corresponding unit matrix. Thus for 2-square matrices

$$AA^{-1} = A^{-1}A = \begin{pmatrix} 1 & 0 \\ 0 & 1 \end{pmatrix}$$

10.4 TWO BY TWO MATRICES

(a) Determinants

The determinant of a square matrix is a single number associated with the matrix, and for a 2-square matrix $\begin{pmatrix} a & b \\ c & d \end{pmatrix}$ the determinant is the numerical value of $ad - cb$.

Thus if $A = \begin{pmatrix} 4 & 2 \\ 1 & 3 \end{pmatrix}$, the determinant is $(4 \times 3) - (1 \times 2)$ and $\det A$ or $|A| = 10$.

If the determinant of a matrix is zero the matrix is called *singular*.

Example 10.6

Calculate the determinant of the matrix (a) $\begin{pmatrix} 5 & 2 \\ -1 & 3 \end{pmatrix}$, (b) $\begin{pmatrix} 4 & -3 \\ 1 & -2 \end{pmatrix}$, (c) $\begin{pmatrix} 2 & 3 \\ -2 & 3 \end{pmatrix}$, (d) $\begin{pmatrix} 1.6 & 0.5 \\ 8 & 2.5 \end{pmatrix}$, (e) $\begin{pmatrix} 1 & 0 \\ 0 & 1 \end{pmatrix}$.

Solution

(a) $\det \begin{pmatrix} 5 & 2 \\ -1 & 3 \end{pmatrix} = (5 \times 3) - (-1 \times 2) = 15 + 2 = \underline{17}$

(b) $\det \begin{pmatrix} 4 & -3 \\ 1 & -2 \end{pmatrix} = (4 \times -2) - (1 \times -3) = -8 + 3 = \underline{-5}$

(c) $\det \begin{pmatrix} 2 & 3 \\ -2 & 3 \end{pmatrix} = (2 \times 3) - (-2 \times 3) = 6 + 6 = \underline{12}$

(d) $\det\begin{pmatrix}1.6 & 0.5\\ 8 & 2.5\end{pmatrix} = (1.6 \times 2.5) - (8 \times 0.5) = 4 - 4 = \underline{0}$

(e) $\det\begin{pmatrix}1 & 0\\ 0 & 1\end{pmatrix} = 1 - 0 = \underline{1}$

(b) Inverse matrices

(i) From the definition A matrix and its inverse must be the same size. If

$$A = \begin{pmatrix}3 & 1\\ 4 & 2\end{pmatrix} \text{ and } A^{-1} = \begin{pmatrix}a & b\\ c & d\end{pmatrix}$$

then by the definition

$$\begin{pmatrix}3 & 1\\ 4 & 2\end{pmatrix}\begin{pmatrix}a & b\\ c & d\end{pmatrix} = \begin{pmatrix}1 & 0\\ 0 & 1\end{pmatrix}$$

Multiplication leads to two pairs of simultaneous equations

$$3a + c = 1 \qquad 3b + d = 0$$

$$4a + 2c = 0 \qquad 4b + 2d = 1$$

These could be solved as shown in Section 8.4 to obtain values for *a, b, c* and *d* and hence the inverse matrix A^{-1}.

(ii) Short method There is a shorter method which applies only to 2-square matrices:

(1) Interchange the elements on the leading diagonal
(2) Change the sign of the other two elements
(3) Divide by the determinant.

If

$$A = \begin{pmatrix}a & b\\ c & d\end{pmatrix}, \qquad A^{-1} = \frac{1}{ad - cb}\begin{pmatrix}d & -b\\ -c & a\end{pmatrix}$$

A singular matrix has no inverse since the determinant is zero.

Example 10.7

Find the inverse of the matrix if it exists of (a) $\begin{pmatrix}3 & 1\\ 4 & 2\end{pmatrix}$, (b) $\begin{pmatrix}2 & 7\\ -1 & 4\end{pmatrix}$,

(c) $\begin{pmatrix}-2 & 7\\ -1 & 4\end{pmatrix}$, (d) $\begin{pmatrix}-5 & -2\\ 10 & 4\end{pmatrix}$.

Solution

(a) $\det\begin{pmatrix}3 & 1\\ 4 & 2\end{pmatrix} = 2,$ and the inverse is $\frac{1}{2}\begin{pmatrix}2 & -1\\ -4 & 3\end{pmatrix}$

Check

$$\frac{1}{2}\begin{pmatrix}2 & -1\\ -4 & 3\end{pmatrix}\begin{pmatrix}3 & 1\\ 4 & 2\end{pmatrix} = \frac{1}{2}\begin{pmatrix}6-4 & 2-2\\ -12+12 & -4+6\end{pmatrix}$$

$$= \frac{1}{2}\begin{pmatrix}2 & 0\\ 0 & 2\end{pmatrix} = \begin{pmatrix}1 & 0\\ 0 & 1\end{pmatrix}$$

(b) $\det\begin{pmatrix}2 & 7\\ -1 & 4\end{pmatrix} = 15$

the inverse matrix is

$$\frac{1}{15}\begin{pmatrix}4 & -7\\ 1 & 2\end{pmatrix}$$

or

$$\begin{pmatrix}4/15 & -7/15\\ 1/15 & 2/15\end{pmatrix}$$

(c) $\det\begin{pmatrix}-2 & 7\\ -1 & 4\end{pmatrix} = -8 + 7 = -1$

The inverse matrix is

$$-1\begin{pmatrix}4 & -7\\ 1 & -2\end{pmatrix} = \begin{pmatrix}-4 & 7\\ -1 & 2\end{pmatrix}$$

(d) $\det\begin{pmatrix}-5 & -2\\ 10 & 4\end{pmatrix} = -20 + 20 = \underline{0}$

The matrix has no inverse.

(c) Matrix equations

(i) Inverse matrices can be used to solve matrix equations. For instance, if $AX = B$, premultiplying both sides of the equation by A^{-1} gives $A^{-1}AX = A^{-1}B$. But $A^{-1}A = \mathrm{I}$ and $\mathrm{I}X = X$, where I is the unit matrix. Therefore $X = A^{-1}B$.

Example 10.8

Solve the matrix equation $AX = B$ when $A = \begin{pmatrix} 3 & 2 \\ 1 & 4 \end{pmatrix}$ and $B = \begin{pmatrix} 5 \\ 4 \end{pmatrix}$.

Solution

$$A^{-1} = \frac{1}{12-2}\begin{pmatrix} 4 & -2 \\ -1 & 3 \end{pmatrix} = \frac{1}{10}\begin{pmatrix} 4 & -2 \\ -1 & 3 \end{pmatrix}$$

$$X = A^{-1}B = \frac{1}{10}\begin{pmatrix} 4 & -2 \\ -1 & 3 \end{pmatrix}\begin{pmatrix} 5 \\ 4 \end{pmatrix} = \frac{1}{10}\begin{pmatrix} 12 \\ 7 \end{pmatrix} = \underline{\begin{pmatrix} 1.2 \\ 0.7 \end{pmatrix}}$$

Example 10.9

Find a matrix X such that $AX = B$ when $A = \begin{pmatrix} 4 & 2 \\ 5 & 3 \end{pmatrix}$, $B = \begin{pmatrix} 2 & 6 \\ 1 & 10 \end{pmatrix}$.

Solution

$$A^{-1} = \frac{1}{2}\begin{pmatrix} 3 & -2 \\ -5 & 4 \end{pmatrix} \qquad X = A^{-1}B = \frac{1}{2}\begin{pmatrix} 3 & -2 \\ -5 & 4 \end{pmatrix}\begin{pmatrix} 2 & 6 \\ 1 & 10 \end{pmatrix}$$

$$X = \frac{1}{2}\begin{pmatrix} 4 & -2 \\ -6 & 10 \end{pmatrix} = \underline{\begin{pmatrix} 2 & -1 \\ -3 & 5 \end{pmatrix}}$$

(ii) Simultaneous linear equations can also be solved using matrix equations. Two simultaneous linear equations in the form

$$ax + by = p$$

$$cx + dy = q$$

can be written as a matrix equation

$$\begin{pmatrix} a & b \\ c & d \end{pmatrix}\begin{pmatrix} x \\ y \end{pmatrix} = \begin{pmatrix} p \\ q \end{pmatrix}$$

This equation can be solved for the matrix $\begin{pmatrix} x \\ y \end{pmatrix}$ as shown in the previous section, by premultiplying both sides of the matrix equation by the inverse of the matrix of coefficients.

Example 10.10

Find the inverse of the matrix $\begin{pmatrix} 2 & -3 \\ 4 & 1 \end{pmatrix}$ and hence the solution of the simultaneous equations

$$2x - 3y = 9$$
$$4x + \ \ y = 11$$

Solution The inverse of the matrix $\begin{pmatrix} 2 & -3 \\ 4 & 1 \end{pmatrix}$ is

$$\underline{\frac{1}{14}\begin{pmatrix} 1 & 3 \\ -4 & 2 \end{pmatrix}}$$

When written in matrix form the equations are

$$\begin{pmatrix} 2 & -3 \\ 4 & 1 \end{pmatrix}\begin{pmatrix} x \\ y \end{pmatrix} = \begin{pmatrix} 9 \\ 11 \end{pmatrix}$$

and premultiplying both sides of this equation by the inverse matrix gives

$$\begin{pmatrix} x \\ y \end{pmatrix} = \frac{1}{14}\begin{pmatrix} 1 & 3 \\ -4 & 2 \end{pmatrix}\begin{pmatrix} 9 \\ 11 \end{pmatrix} = \frac{1}{14}\begin{pmatrix} 42 \\ -14 \end{pmatrix} = \begin{pmatrix} 3 \\ -1 \end{pmatrix}$$

The solution of the simultaneous equations is $\underline{x = 3, y = -1}$.

This is a simple example of a general method of solving simultaneous equations in several variables, and computer programs are available for obtaining inverse matrices. When the number of equations is greater than about 20, other methods are preferred, and for two equations the elimination method used in Section 8.4 is probably faster and more straightforward.

If the matrix of coefficients is singular there is no solution satisfying both equations and they are not consistent. For example, the equations

$$2x + y \ \ = 7$$
$$4x + 2y = 12$$

are not consistent, and the determinant of $\begin{pmatrix} 2 & 1 \\ 4 & 2 \end{pmatrix}$ is zero.

Exercise 10.3

(1) Calculate the determinant of

(a) $\begin{pmatrix} 2 & 3 \\ 1 & 4 \end{pmatrix}$, (b) $\begin{pmatrix} 3 & -4 \\ 1 & 2 \end{pmatrix}$, (c) $\begin{pmatrix} 0 & 1 \\ -1 & 1 \end{pmatrix}$, (d) $\begin{pmatrix} 2a & -b \\ 4a & 3b \end{pmatrix}$.

(2) For what value of x is the matrix singular?

(a) $\begin{pmatrix} 2 & 3 \\ x & 4 \end{pmatrix}$, (b) $\begin{pmatrix} -6 & -1 \\ 3 & x \end{pmatrix}$, (c) $\begin{pmatrix} 2a & 2b \\ 3a & x \end{pmatrix}$

(3) Find the inverse matrix of

(a) $\begin{pmatrix} 4 & 3 \\ 1 & 7 \end{pmatrix}$, (b) $\begin{pmatrix} 2 & -1 \\ 3 & -4 \end{pmatrix}$, (c) $\begin{pmatrix} 1 & -6 \\ 2 & 5 \end{pmatrix}$, (d) $\begin{pmatrix} -1 & 12 \\ 0 & -14 \end{pmatrix}$.

(4) If $A = \begin{pmatrix} 2 & 4 \\ -3 & 1 \end{pmatrix}$ and $B = \begin{pmatrix} -5 & -3 \\ 6 & 4 \end{pmatrix}$ find the inverse of B and hence solve the matrix equation $A = BX$.

(5) If $M = \begin{pmatrix} 3 & 2 \\ 1 & 1 \end{pmatrix}$ find the matrix N such that the product MN is the unit matrix.

(6) If $A = \begin{pmatrix} 1 & 2 \\ 3 & 1 \end{pmatrix}$ and $\mathrm{I} = \begin{pmatrix} 1 & 0 \\ 0 & 1 \end{pmatrix}$ find a number k such that $A^2 = 2A + k\mathrm{I}$.

(7) Write down the two simultaneous linear equations that can be solved from the matrix equation $\begin{pmatrix} 2 & 4 \\ 1 & -3 \end{pmatrix} \begin{pmatrix} x \\ y \end{pmatrix} = \begin{pmatrix} 10 \\ -5 \end{pmatrix}$.

(8) Solve the simultaneous equations

(a) $3y - 4x = 18, 4y + x = 5$

(b) $6a + 2b = 11, 2a - 5b = -19$.

10.5 THE USE OF A MATRIX TO PRESENT INFORMATION

A matrix provides a method of presenting a large amount of numerical information (data) in a convenient and compact form

$$M = \begin{pmatrix} 350 & 170 & 80 \\ 100 & 90 & 50 \\ 120 & 110 & 70 \\ 150 & 120 & 90 \end{pmatrix}, N = \begin{pmatrix} 23 \\ 16 \\ 45 \end{pmatrix}, P = (2.00 \quad 2.20 \quad 1.90 \quad 2.10)$$

For example, suppose the matrix M represents the stock of 4 different articles maintained in each branch of 3 chain stores. If P represents the price of each article in pounds sterling, then the matrix PM would be a 1×3 matrix showing the total value of the stock in each of the 3 branches.

Further, if the 3×1 matrix N gives the number of branches of each store in a particular region, the product MN is a 4×1 matrix showing the total stock of each of the 4 articles in that region. The product PMN is a single number giving the total value of the stock in the region.

Exercise 10.4

(1) The matrix A shows the number of cups of tea, coffee and soup respectively supplied by a vending machine on 3 days. The price of the drinks per cup is: tea 10 p, coffee 15 p, soup 12 p. (a) Form a 3×1 matrix of prices and use it to find the amount of cash taken on each of the 3 days. (b) What information would be obtained by premultiplying A by $(1 \quad 1 \quad 1)$?

$$A = \begin{pmatrix} \text{Tea} & \text{Coffee} & \text{Soup} \\ 87 & 51 & 21 \\ 43 & 78 & 34 \\ 55 & 89 & 12 \end{pmatrix}$$

(2)

Departure	Comfort	Excursion	Economy
10.00	15	35	60
12.30	12	40	70
16.00	8	50	34
Price (£)	70	35	30

The table shows the number of passengers on three flights to Paris in 3 classes and the price of a ticket for each class. Construct three matrices to represent this information, and use them to obtain (a) the total number of passengers on the 12.30 flight, (b) the revenue from each flight.

(3) Three brands of soft drinks A, B, and C are sold at 11 p, 13 p and 16 p respectively in four shops P, Q, R, S. On one day, 16, 20 and 6 bottles of brand A were sold in Q, R, and S respectively, while the sales of brand B were 20, 4, and 12 in shops P, Q, and S. 8 bottles of brand C were sold in each of shops P and R and 7 bottles in S.

Construct a 4×3 matrix M from the sales data, and a matrix L of prices, such that ML gives the total revenue from the drinks in each shop. (a) If N is the matrix $(1 \quad 1 \quad 1 \quad 1)$ which matrix product gives the total number sold of each brand of drink? (b) What information is given by the matrix NML?

PROGRESS TEST 2

A

(1) Simplify (a) $2x^2y^3 \times -3xy^4$, (b) $2x^2y + 3xy^2 - x^2y$, (c) $4(x + 2y) + 3(2x + y)$, (d) $2(3y + x) - 3(y - x)$, (e) $(m + n)^2 - 3(m - n)^2$. **(2)** Simplify (a) $(3x/y) + (x/3y)$, (b) $(2t^3/3x) \div (t/x)$, (c) $[3(x^2 + y)/4] - [(x^2 - y)/3]$, (d) $(2x/5y) - [(3 - 2x)/4y]$. **(3)** Given that $m = 2, x = -1, y = -2$, evaluate (a) $3mx + y^2$, (b) $(m - x)^2$, (c) $x^2/(2mx + y)$, (d) $(1/m) - (1/x) + (1/y)$. **(4)** Solve (a) $3z - 2z + 1 = 6$, (b) $2(x - 3) - 4(2 - x) = -1$, (c) $(x + 1)/3 + (3x + 1)/3 = 2$, (d) $(1/x) - 3x/(x + 1) = 1$, (e) $x^2 + x - 12 = 0$, (f) $3x^2 + 5x = 7$. **(5)** Solve (a) $2x - 6y = 10, 3x - 2y = 8$; (b) $t + 2s = 1, 0.1t + 6s = 8.8$; (c) $x + 2y = 3, xy = -2$. **(6)** Factorise (a) $xy - 2by - 4x + 8b$, (b) $18 - 2b^2$, (c) $x^2 - 2x - 8$, (d) $2y^2 - 12y + 18$. **(7)** Find the values of x which satisfy (a) $3(x - 2) \leqslant 5$, (b) $5 + \frac{1}{4}x \geqslant \frac{3}{8}x - 1$, (c) $3x^2 - 13x \leqslant 10$. **(8)** Make x the subject of the formula $y = u + (v/x)$. **(9)** $\mathscr{E} = \{1,2,3,4,5,6,7,8\}$, $A = \{1,3,5,7\}$, $B = \{2,4,6\}$, $C = \{1,3,5,7,8\}$. List the elements of the sets (a) C', (b) $A \cup C$, (c) $(A \cap C) \cup (B \cap A')$. **(10)** If $n(P) = 54$, $n(Q) = 27$ and $n(P \cap Q) = 12$, find $n(P \cup Q)$. **(11)** If $m * n$ denotes $2m - n^2$, find the value of (a) $5 * -3$, (b) $-3 * \sqrt{5}$, (c) $3 * (2 * 1)$. **(12)** $M = \begin{pmatrix} 2 & -3 \\ 1 & 2 \end{pmatrix}$, $N = \begin{pmatrix} -4 & 1 \\ 2 & -3 \end{pmatrix}$, $P = \begin{pmatrix} -4 \\ 1 \end{pmatrix}$; find the matrix (a) M^{-1}, (b) MN, (c) NP, (d) $MN - NM$, (e) X when $NX = P$.

B

(1) Simplify (a) $(2x - 3y) - 3(x + y)$, (b) $5/(x - 1) - 3/(x + 1)$, (c) $(x - 1)^2 - (x + 1)$. **(2)** Solve (a) $x/(x - 3) = -2$, (b) $3x^2 - 16x = 12$, (c) $4 = \sqrt{(x - 3)}$, (d) $(1/x) = (1/0.4) - (1/0.3)$. **(3)** Transpose the formula for y (a) $z = 3b\sqrt{(y + 2x)}$, (b) $2c = (y + b)/(y - c)$. **(4)** Factorise (a) $2x^2 - 8y^2$, (b) $(3x - 2)^2 - (3x - 2)$, (c) $4ay - 2yb - 6ax + 3xb$, (d) $3x^2 - 4x$, (e) $9a^2 + 24ab + 16b^2$. **(5)** Solve (a) $3y^2 + 6y = 5$, (b) $4x - 3y = 7, 2x + 2y = 5$, (c) $x^2 + y^2 = 5, x + y = -1$. **(6)** What number must be added to $9x^2 - 12x$ to make it a perfect square? **(7)** If $x = yz^2$ and z is increased by 12% while y is unchanged, find the percentage increase in x. **(8)** Find the matrix M if $3M + 2\begin{pmatrix} 1 & -3 \\ 2 & -7 \end{pmatrix} = \begin{pmatrix} 8 & 3 \\ 1 & -8 \end{pmatrix}$.

(9) Find the value of x and y if $\begin{pmatrix} 3 & x \\ y & 1 \end{pmatrix} \begin{pmatrix} -2 \\ 1 \end{pmatrix} = \begin{pmatrix} 12 \\ 1 \end{pmatrix}$. **(10)** If $p = -2$ and $q = 3$, evaluate (a) $\sqrt{(2p^2 + pq)}$, (b) $(2p + q)(q - p)$, (c) $q^2 - 3p^2$. **(11)** If x is an integer, $A = \{x : 13 - 3x \geqslant 2\}$ and $B = \{x : 2/3 \geqslant 1/x\}$ find the set $A \cap B$. **(12)** Solve (a) $5/x \geqslant 2/3$, (b) $x^2 - 5x + 6 \geqslant 0$.

INTRODUCTION TO GEOMETRY

The need to measure distances and directions precisely for erecting buildings and monuments stimulated the study of geometry in Greece and Egypt in the sixth century B.C. In the third century B.C. Euclid enlarged the knowledge of the subject and presented it as a continuous logical development in his books, which remained the main source of geometrical studies for the succeeding 2000 years.

11.1 POINTS, LINES AND PLANES

A point in geometry marks a position in space, and has no dimensions. A line joins two points and its length is a measure of the distance between points. The shortest line between any two points is called a straight line or line, all others are called curved lines or curves. A line has only one dimension, length.

A plane has two dimensions and is defined by two intersecting lines.

11.2 ANGLES

The difference in direction of two straight lines is described by the size of the angle between them, and angles are measured in degrees. The symbol used is $^\circ$ as in 30° or 45°.

(a) Angles at a point

When a number of lines in the same plane meet at a point the angles between them add up to 360°, a complete revolution. An angle is described either by capital letters in terms of the lines containing it, or by a single letter placed between the lines, and the symbol used is $\angle$ before the named angle. In Fig. 11.1(a) $\angle AOD = 60^\circ$ and $\angle COD = 30^\circ$.

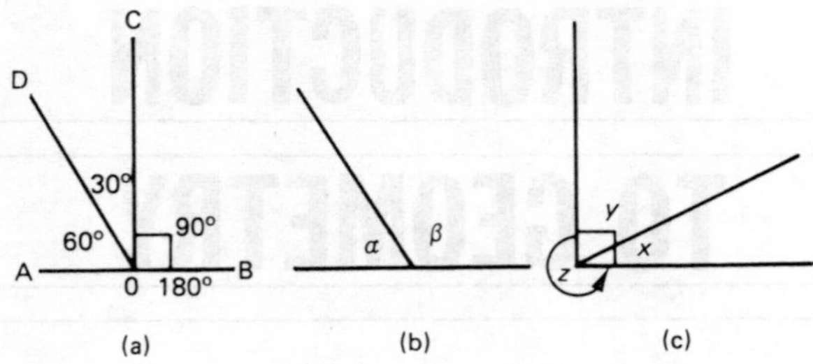

Fig 11.1

(b) Angles on a straight line

A half revolution, 180°, is called a straight angle and angles which add up to 180° are *supplementary* angles. In Fig. 11.1(a), ∠AOB is a straight angle, and ∠AOD is the supplement of ∠DOB. In Fig. 11.1(b) ∠α is the supplement of ∠β and ∠β is the supplement of ∠α.

(c) Right angles

A right angle is a quarter revolution, 90°, and is given the symbol ∟. Angles which add up to 90° are called *complementary* angles. In Fig. 11.1(a), ∠BOC is a right angle, and ∠AOD is the complement of ∠COD.

(d) Other angles

Acute angles are smaller than 90°. *Obtuse angles* are between 90° and 180°. *Reflex angles* are greater than 180°. In Fig. 11.1(b) ∠α is acute, ∠β is obtuse and in Fig. 11.1(c) ∠z is reflex.

11.3 EQUAL ANGLES FORMED BY INTERSECTING LINES

(a) Vertically opposite angles

When two straight lines cross they form two pairs of equal angles which are called vertically opposite angles [Fig. 11.2(a)]. If both pairs are right angles the lines are said to be *perpendicular* to each other [Fig. 11.2(b)], and the symbol used is ⊥, AB ⊥ MN.

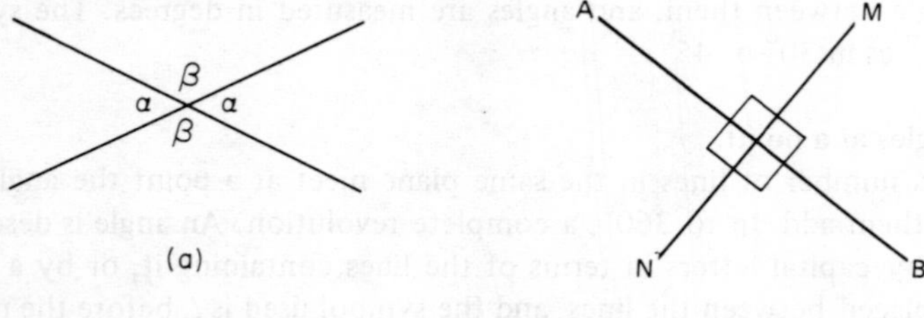

Fig 11.2

(b) Alternate and corresponding angles

Straight lines in the same plane which have the same direction are called *parallel*, the distance between them is constant. Lines that are parallel are marked with arrows in diagrams, and the shortest distance between two parallel lines is a line perpendicular to both. In Fig. 11.3(a) AB || CD and AC = BD.

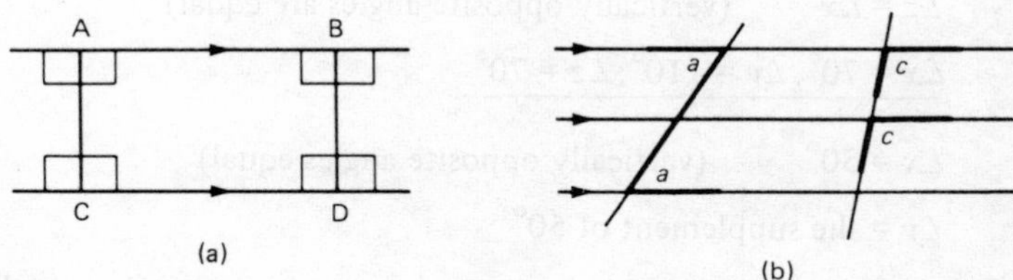

Fig 11.3

Since the parallel lines have the same direction they make equal angles with a *transversal*, that is any other line that crosses them. When parallel lines are crossed by a transversal 'Z' shapes and 'F' shapes are formed as well as 'X' shapes. In Fig. 11.3(b) the angles marked *a* are in the corners of a 'Z' and on 'alternate' sides of the transversal. The angles marked *c* are in 'corresponding' positions relative to the parallel lines, and are in the corners of an 'F' shape.

Alternate angles are equal.
Corresponding angles are equal.

Example 11.1

Calculate the angles marked with letters in Fig. 11.4.

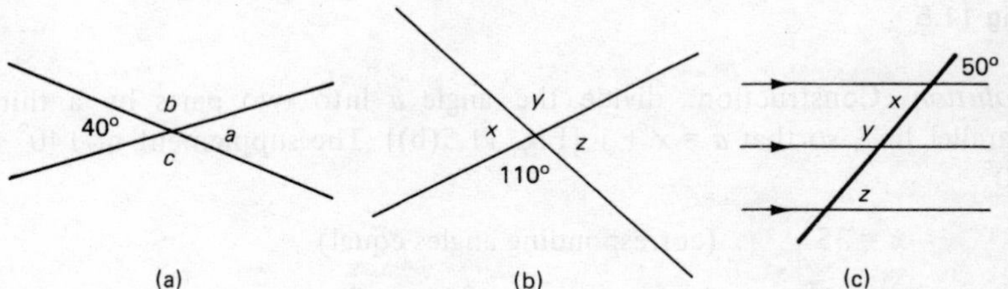

Fig 11.4

Solution

(a) $\angle a = 40^\circ$ (vertically opposite angles are equal)

$\angle b = 180^\circ - 40^\circ$ (adjacent angles on a straight line are supplementary)

$= 140^\circ$

$\angle c = \angle b$ (vertically opposite angles are equal)

$\underline{\angle a = 40^\circ, \angle b = 140^\circ, \angle c = 140^\circ}$

(b) $\angle y = 110^\circ$ (vertically opposite angles are equal)

$\angle x = 70^\circ$ (supplement of 110°)

$\angle z = \angle x$ (vertically opposite angles are equal)

$\underline{\angle x = 70^\circ, \angle y = 110^\circ, \angle z = 70^\circ}$

(c) $\angle x = 50^\circ$ (vertically opposite angles equal)

$\angle y =$ the supplement of 50°

$= 130^\circ$ (corresponding angles equal)

$\angle z = \angle x = 50^\circ$ (alternate angles equal)

$\underline{\angle x = 50^\circ, \angle y = 130^\circ, \angle z = 50^\circ}$

Example 11.2

Find the size of the angle marked *a* in Fig. 11.5(a).

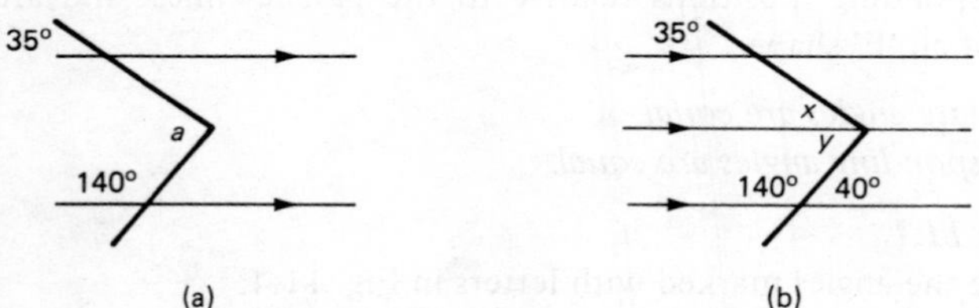

Fig 11.5

Solution Construction: divide the angle *a* into two parts by a third parallel line, so that $a = x + y$ [Fig. 11.5(b)]. The supplement of 140° is 40°

$x = 35^\circ$ (corresponding angles equal)

$y = 40^\circ$ (alternate angles are equal)

$a = x + y = \underline{75^\circ}$

Exercise 11.1

Calculate the size of the angles marked by letters in Fig. 11.6 (1) to (16).

(17) Of the three angles at a point, one is double the second angle which is treble the third angle. What is the size of (a) the smallest, (b) the largest angle?

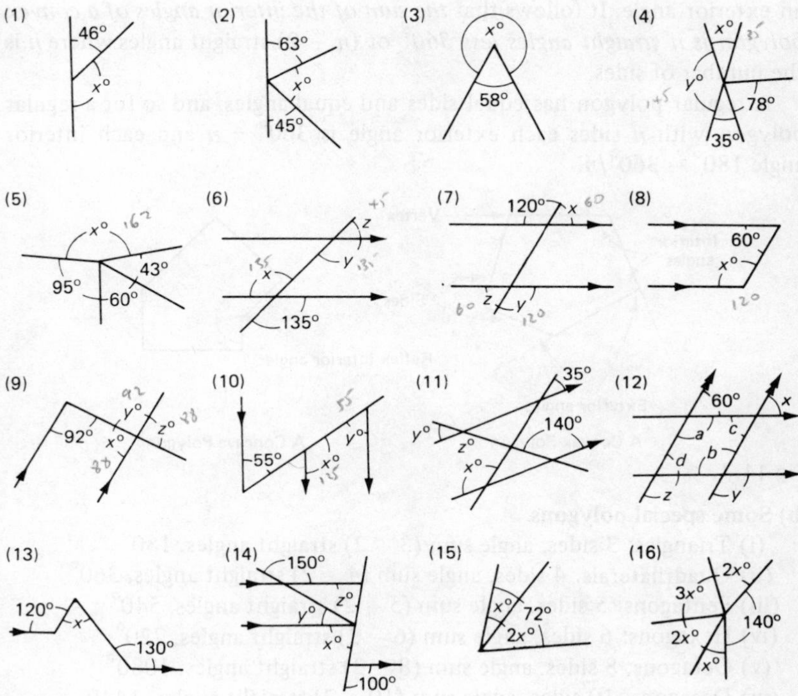

Fig 11.6

11.4 POLYGONS

Geometrical figures are the shapes formed by intersecting lines and planes, plane figures are formed by lines in the same plane. A *polygon* is a plane figure bounded by straight lines called the *sides* of the polygon and any two adjacent sides meet at a *vertex*. We shall consider only convex polygons in which the angle between adjacent sides is less than a straight angle.

(a) Angle sum of polygons

(i) Exterior angles

An exterior angle is formed by extending one of the sides of a polygon as in Fig. 11.7. When the sides of a polygon are traversed in order, returning to the starting point, a complete revolution has been made. It follows that *the sum of the exterior angles of any convex polygon is 360°*.

(ii) Interior angles

A polygon with n sides has n interior angles, each one is the supplement of an exterior angle. It follows that *the sum of the interior angles of a convex polygon is n straight angles less 360°* or $(n - 2)$ straight angles where n is the number of sides.

A regular polygon has equal sides and equal angles, and so for a regular polygon with n sides each exterior angle is $360° \div n$ and each interior angle $180° - 360°/n$.

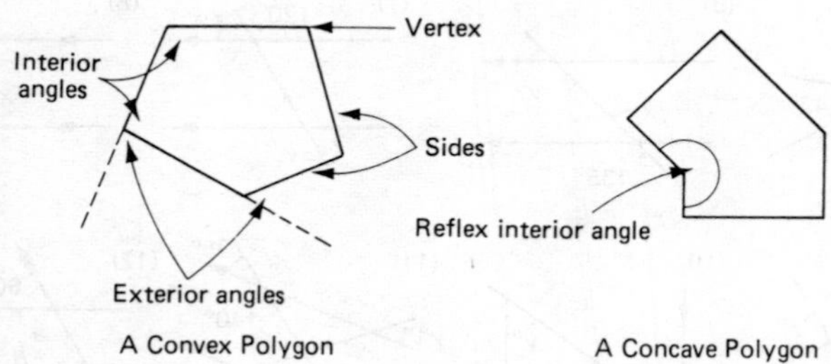

Fig 11.7

(b) Some special polygons

(i) Triangles: 3 sides, angle sum $(3 - 2)$ straight angles, 180°
(ii) Quadrilaterals: 4 sides, angle sum $(4 - 2)$ straight angles, 360°
(iii) Pentagons: 5 sides, angle sum $(5 - 2)$ straight angles, 540°
(iv) Hexagons: 6 sides, angle sum $(6 - 2)$ straight angles, 720°
(v) Octagons: 8 sides, angle sum $(8 - 2)$ straight angles. 1080°
(vi) Decagons: 10 sides, angle sum $(10 - 2)$ straight angles, 1440°.

Example 11.3

(a) Calculate the size of each interior angle of a regular polygon with 20 sides.

Solution Each exterior angle is $360°/20 = 18°$. Each interior angle is therefore $180° - 18° = \underline{162°}$.

(b) How many sides has a regular polygon with each interior angle 156°?

Solution Suppose there are n sides, then $(n - 2) \times 180° = n \times 156°$

$$180n - 360 = 156n \Rightarrow 24n = 360 \Rightarrow n = 15$$

The number of sides is $\underline{15}$.

Example 11.4

(a) How many sides has a polygon whose interior angles total 4320°?

Solution

$$(n - 2) \times 180° = 4320° \Rightarrow n - 2 = 4320/180 = 24$$

The number of sides is 26.

(b) A hexagon has interior angles x°, $2x^\circ$, $x + 50^\circ$, 130°, 140°, 150°. Calculate the value of x, and hence the size of the two smallest angles of the polygon.

Solution The angle sum of a hexagon is 720°

$$x + 2x + x + 50 + 130 + 140 + 150 = 720 \Rightarrow 4x = 250$$

$$x = 62\tfrac{1}{2}$$

The two smallest angles are $62\frac{1}{2}^\circ$ and $112\frac{1}{2}^\circ$.

11.5 TRIANGLES

Any two sides of a triangle are together greater than the third side; the longest side is always opposite the greatest angle, and equal sides are opposite equal angles.

(a) Types of triangle

(i) Equilateral: these are the regular triangles with equal sides and each angle 60°

(ii) Isosceles: they have two sides and two angles equal

(iii) Scalene: all three sides are different

(iv) Acute angled triangles have three angles less than 90°

(v) Obtuse angled triangles have one angle greater than 90°

(vi) Right angled triangles have one angle of 90°.

Example 11.5

In an isosceles triangle ABC, ∠ABC = 30°. What is the size of the other angles (a) when AB = BC, (b) when AB = AC?

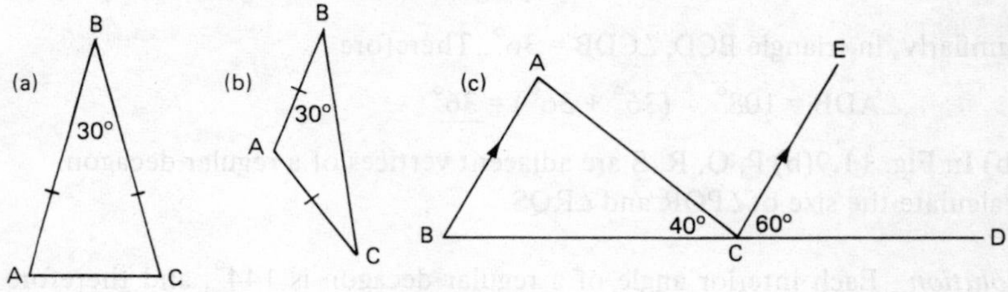

Fig 11.8

Solution The diagrams are shown in Fig. 11.8(a) and (b).

(a) ∠BAC + ∠BCA = 180° − 30° = 150°,

Since AB = BC, ∠BAC = ∠BCA = 75°.

(b) When AB = AC, ∠ACB = ∠ABC = 30°, ∠BAC = 180° − 60° = 120°.

Example 11.6

In Fig. 11.8(c), the triangle ABC has BC extended to D and CE drawn parallel to BA, $\angle ACB = 40^\circ$ and $\angle ECD = 60^\circ$. Calculate $\angle BAC$ and $\angle ABC$.

Solution $\angle ACE = 180^\circ - 40^\circ - 60^\circ = 80^\circ$ (angles on a straight line).
Since BA and CE are parallel, $\angle BAC = \angle ACE$ (alternate angles).
Therefore $\underline{\angle BAC = 80^\circ}$. Also, $\angle ABC = \angle ECD$ (corresponding angles).
Therefore $\underline{\angle ABC = 60^\circ}$.

Example 11.6 illustrates two important properties of triangles:

(i) The angle sum is 180°
(ii) An exterior angle is equal to the sum of the interior opposite angles. The exterior angle $ACD = \angle CAB + \angle CBA$.

Example 11.7

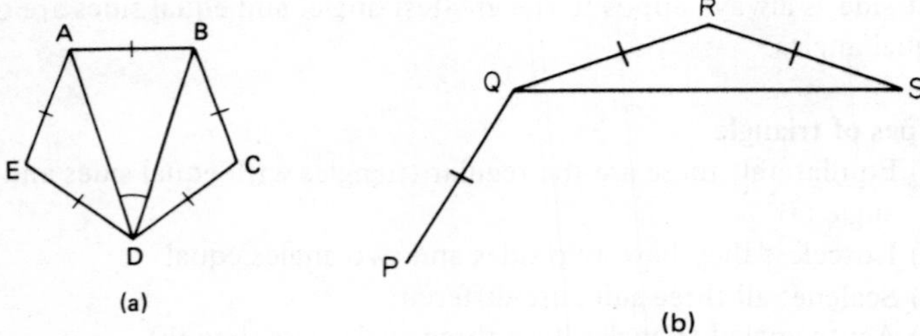

Fig 11.9

(a) Calculate $\angle ADB$ in a regular pentagon ABCDE [Fig. 11.9(a)].

Solution Each interior angle is 108°. In the isosceles triangle AED

$\angle EAD = \angle EDA = 36^\circ$ (angle sum of a triangle is 180°)

Similarly, in triangle BCD, $\angle CDB = 36^\circ$. Therefore

$$\angle ADB = 108^\circ - (36^\circ + 36^\circ) = \underline{36^\circ}$$

(b) In Fig. 11.9(b) P, Q, R, S are adjacent vertices of a regular decagon. Calculate the size of $\angle PQR$ and $\angle RQS$.

Solution Each interior angle of a regular decagon is 144°, and therefore $\underline{\angle PQR = 144^\circ}$.

Triangle QRS is isosceles, since RQ = RS. Hence

$$\angle RQS + \angle RSQ = 180^\circ - 144^\circ = 36^\circ$$

Therefore $\underline{\angle RQS = 18^\circ}$.

(b) Pythagoras' theorem

In a right angled triangle the side opposite the right angle is called the *hypotenuse* and Pythagoras' theorem states

> 'In any right angled triangle the square on the hypotenuse is equal to the sum of the squares on the other two sides.'

This is an important property of right angled triangles which is widely used in both geometry and trigonometry.

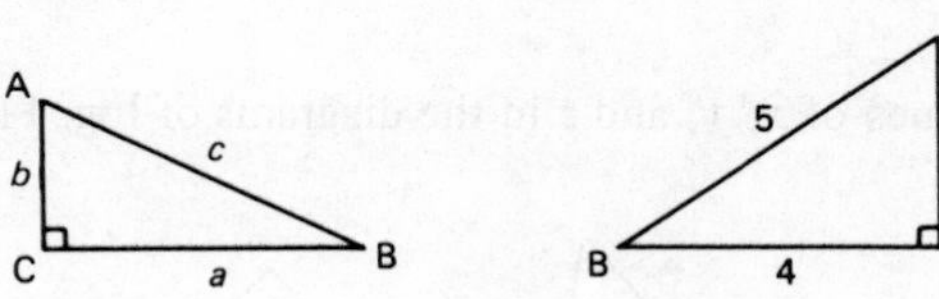

Fig 11.10

In mathematical notation, if the length of the hypotenuse is c units

$$c^2 = a^2 + b^2 \qquad \text{(Fig. 11.10)}$$

If $a = 4$ cm, $b = 3$ cm, then $c = \sqrt{(4^2 + 3^2)}$ cm = 5 cm.

The converse is also true, and if in a triangle ABC

$$AB^2 = AC^2 + BC^2$$

then the angle at C is a right angle.

Example 11.8

Calculate the value of x in each of the triangles in Fig. 11.11.

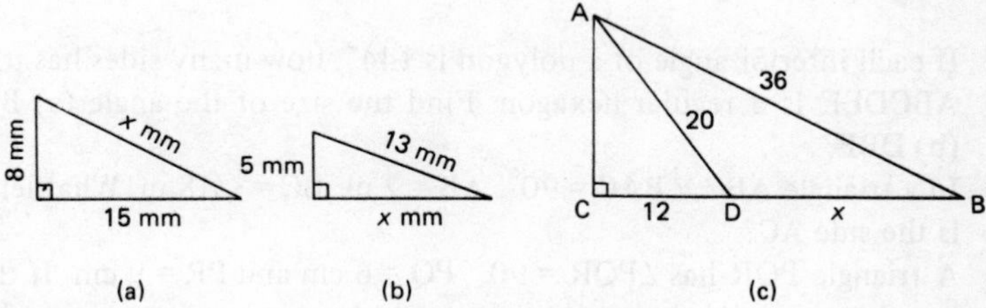

Fig 11.11

Solution

(a) $x^2 = 8^2 + 15^2 = 289 \qquad x = \sqrt{289} = \underline{17}$

(b) $13^2 = 5^2 + x^2 \Rightarrow x^2 = 169 - 25 = 144 \qquad x = \underline{12}$

(c) In triangle ACD, $AC^2 = AD^2 - CD^2$. Therefore

$$AC = \sqrt{(400 - 144)} = \sqrt{256} = 16 \text{ units}$$

In triangle ACB, $AB^2 = AC^2 + CB^2$, and substituting the numerical values gives

$$36^2 = 16^2 + (12 + x)^2$$

$$(12 + x)^2 = 1040$$

$$12 + x = 32.25$$

$x = 20.2$ units correct to 3 sig. fig.

Exercise 11.2

Calculate the values of x, y, and z in the diagrams of Fig. 11.12 (1) to (7).

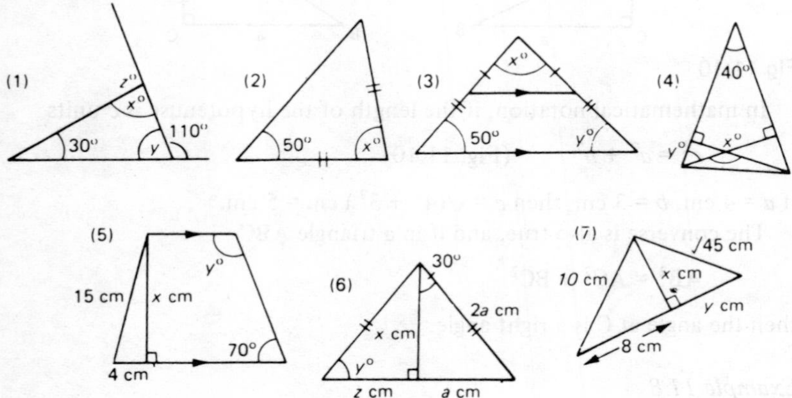

Fig 11.12

(8) If each interior angle of a polygon is 144°, how many sides has it?

(9) ABCDEF is a regular hexagon. Find the size of the angle (a) BCD, (b) DBE.

(10) In a triangle ABC, ∠BAC = 90°, AB = 2 m, BC = $\sqrt{18}$ m. What length is the side AC?

(11) A triangle PQR has ∠PQR = 90°, PQ = 6 cm and PR = 9 cm. If these lengths are given to the nearest centimetre, find the greatest and the least possible value of the length QR, correct to 3 sig. fig.

11.6 SIMILAR AND CONGRUENT TRIANGLES

(a) Similar triangles

In general, two polygons are said to be similar when they have exactly the same shape, and for triangles it is sufficient that they have the same angles.

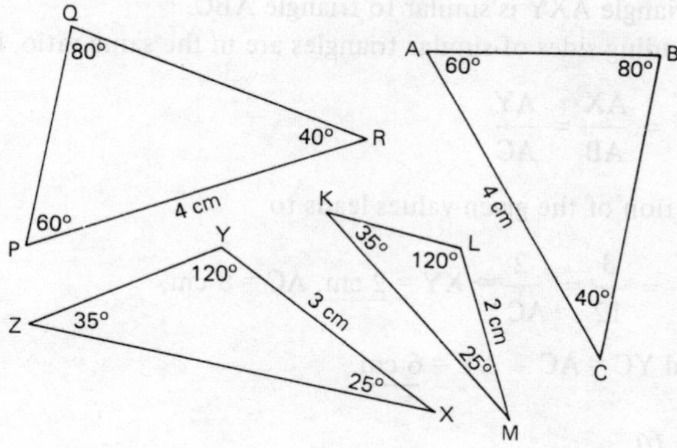

Fig 11.13

In Fig. 11.13 the triangles ABC and PQR are similar to each other because they have the same angles 40°, 60° and 80°. Triangles XYZ and MLK are also similar to each other.

A second condition is that the sides containing equal angles are in the same ratio, and either of these conditions is sufficient to prove similarity.

Example 11.9

In Fig. 11.14 XY is parallel to BC, AX = 3 cm, AY = 2 cm, XB = 9 cm and BC = 8 cm. Show that ΔAXY is similar to ΔABC and calculate the lengths XY and YC.

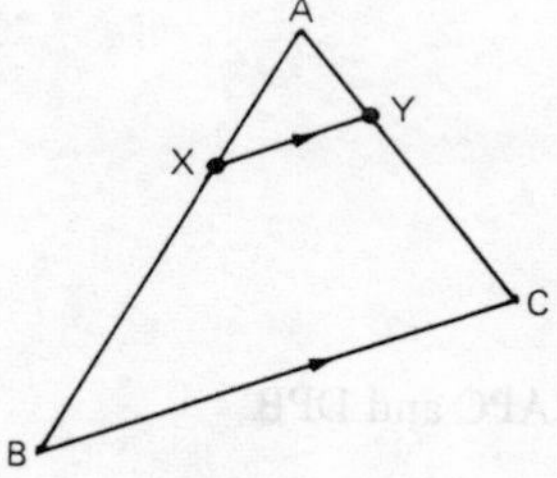

Fig 11.14

Solution

Since XY ∥ BC

$\angle AXY = \angle ABC$ (corresponding angles)

$\angle AYX = \angle ACB$ (corresponding angles)

and

$\angle XAY = \angle BAC$ (common angle)

Therefore triangle AXY is similar to triangle ABC.

Corresponding sides of similar triangles are in the same ratio, hence

$$\frac{XY}{BC} = \frac{AX}{AB} = \frac{AY}{AC}$$

and substitution of the given values leads to

$$\frac{XY}{8} = \frac{3}{12} = \frac{2}{AC} \Rightarrow XY = \underline{2\text{ cm}},\ AC = 8\text{ cm},$$

$$\text{and } YC = AC - AY = \underline{6\text{ cm}}$$

Example 11.10
In Fig. 11.15, two lines AB and CD intersect at P. If AP = 4 cm, PB = 3 cm, CP = 6 cm and PD = 2 cm, what is the ratio of DB to AC?

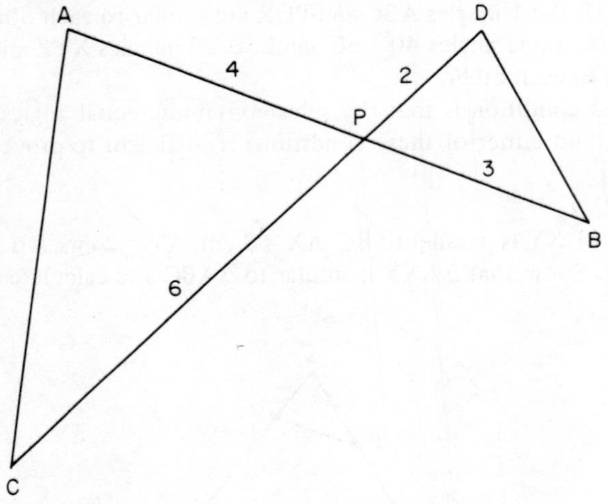

Fig 11.15

Solution In the triangles APC and DPB

$\angle APC = \angle DPB$ (vertically opposite angles)

$$\frac{AP}{CP} = \frac{4\text{ cm}}{6\text{ cm}} = \frac{2}{3}, \text{ and also } \frac{DP}{BP} = \frac{2}{3}$$

Therefore ΔDPB is similar to ΔAPC because sides containing equal angles are in the same ratio. Hence

$$\underline{\frac{DB}{AC} = \frac{DP}{AP} = \frac{1}{2}}$$

(b) Congruent triangles

Two figures are said to be congruent when they are exactly the same size as well as the same shape, and the symbol for congruent is ≡, meaning identical or equal in all respects. In Fig. 11.13 ΔPQR is congruent to ΔABC because they are similar, and corresponding sides PR and AC are equal.

There are four different conditions for triangles to be congruent, and each is sufficient. They are illustrated in Fig. 11.16.

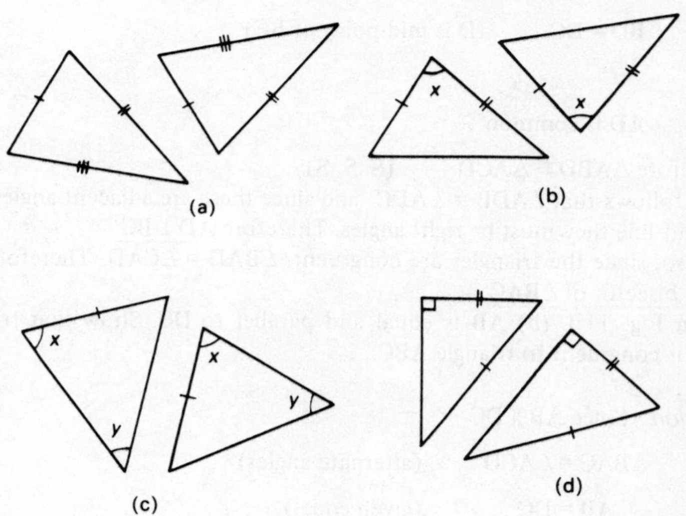

Fig 11.16

(a) Three sides equal (S-S-S)
(b) Two sides and the included angle equal (S-A-S)
(c) Two angles and a corresponding side equal (A-A-S)
(d) A right angle, the hypotenuse and one other side equal (R-H-S).

Example 11.11

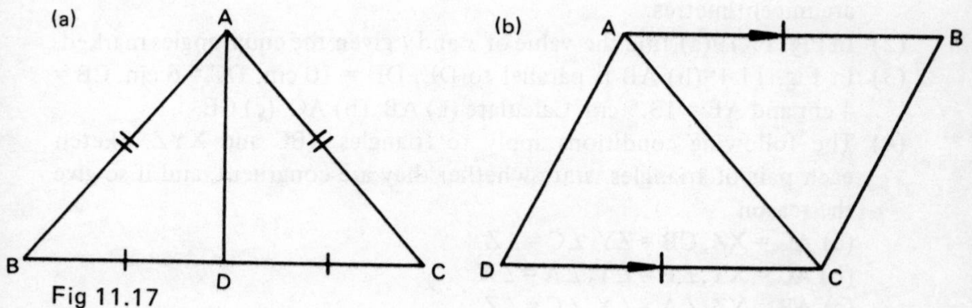

Fig 11.17

(a) In Fig. 11.17(a) given that ABC is an isosceles triangle and D is the mid-point of BC, show that the median AD is perpendicular to BC and bisects BAC.

Solution A median of a triangle is a line joining a vertex to the mid-point of the opposite side.

In the triangles ABD and ACD

AB = AC (sides of isosceles triangle)

BD = DC (D is mid-point of BC)

and

AD is common

Therefore $\triangle ABD \equiv \triangle ACD$ (S S S).

It follows that $\angle ADB = \angle ADC$, and since these are adjacent angles on a straight line they must be right angles. Therefore AD ⊥ BC.

Also, since the triangles are congruent, $\angle BAD = \angle CAD$. Therefore AD is the bisector of $\angle BAC$.

(b) In Fig. 11.17(b) AB is equal and parallel to DC. Show that triangle ADC is congruent to triangle ABC.

Solution Since AB || DC

$\angle BAC = \angle ACD$ (alternate angles)

AB = DC (given equal)

and

AC is common to both triangles

Therefore $\triangle ABC \equiv \triangle ADC$ (S A S).

Exercise 11.3

(1) Calculate the lengths marked *a* and *b* in Fig. 11.18. The measurements are in centimetres.
(2) In Fig. 11.19(a) find the value of *s* and *t* given the equal angles marked.
(3) In Fig. 11.19(b) AB is parallel to DE, DE = 10 cm, DC = 6 cm, CB = 4 cm and AE = 13.5 cm. Calculate (a) AB, (b) AC, (c) CE.
(4) The following conditions apply to triangles ABC and XYZ. Sketch each pair of triangles, state whether they are congruent, and if so give the reason.
(a) AC = XZ, CB = ZY, $\angle C = \angle Z$,
(b) AC = XY, $\angle B = \angle Y$, $\angle A = \angle X$,
(c) AB = YZ, $\angle A = \angle X$, $\angle C = \angle Z$,

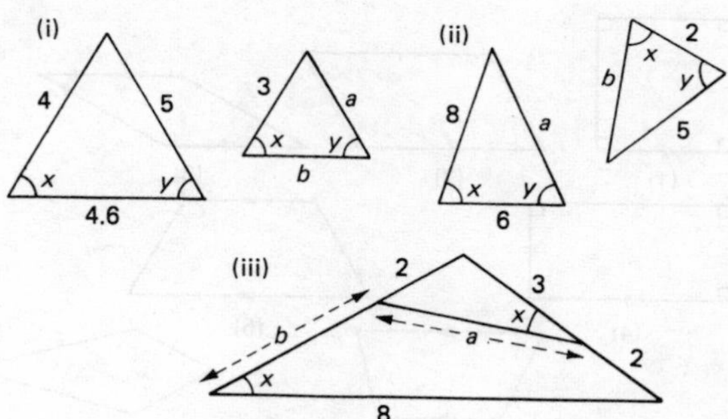

Fig 11.18

(d) AC = XY, ∠B = ∠Z, ∠A = ∠X,
(e) CB = ZY, AB = XY, ∠C = ∠Z,
(f) ∠B = ∠Y = 90°, AC = XZ, ∠A = ∠X.

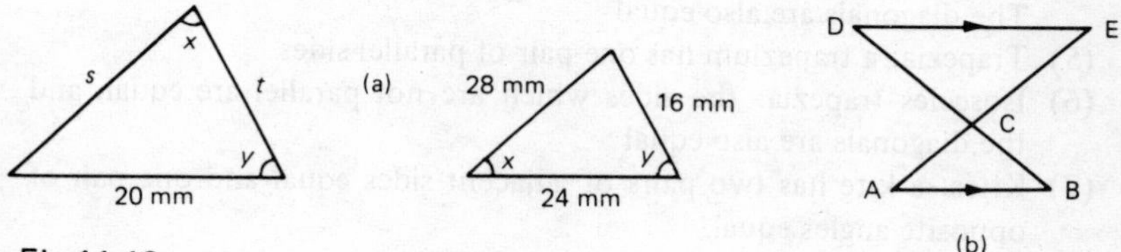

Fig 11.19

(5) Triangle PQR has ∠PQR = 90°, ∠QRP = 60° and PQ = 80 mm. PR is extended to S so that ∠RSQ = 30°. What is the length of QS? If QR = $80/\sqrt{3}$ mm what is the length of (a) RP, (b) RS?
(6) Triangle RST has a right angle at R and the perpendicular from R to ST meets ST at Z. Prove triangles RZT and SRT are similar, and calculate the lengths RZ and RS if TZ = 4 cm and RT = 5 cm.

11.7 QUADRILATERALS

Quadrilaterals are polygons with four sides, the sum of the interior angles is 360°, and the two lines joining opposite vertices are called diagonals.

Special quadrilaterals

The important types of quadrilateral are shown in Fig. 11.20.

(1) Squares: a square is a regular quadrilateral with all sides equal and interior angles of 90°. The diagonals are also equal
(2) Parallelograms: opposite sides are equal and parallel, opposite angles are equal, adjacent angles are supplementary. The diagonals bisect each other
(3) Rhombi: a rhombus is a parallelogram with four equal sides. The

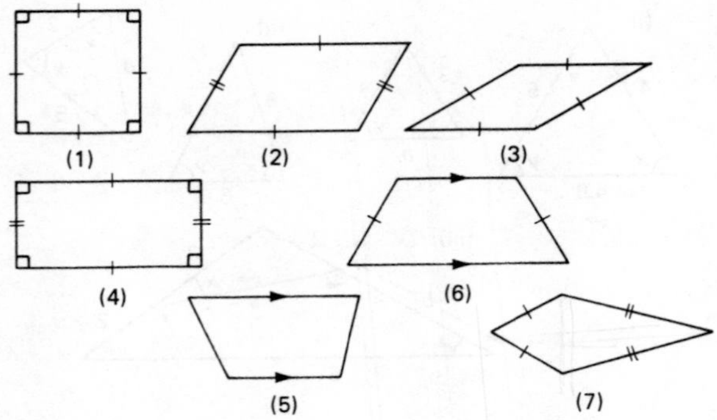

Fig 11.20

diagonals bisect each other at right angles

(4) Rectangles: a rectangle is a parallelogram with four right angles. The diagonals are also equal
(5) Trapezia: a trapezium has one pair of parallel sides
(6) Isosceles trapezia: the sides which are not parallel are equal, and the diagonals are also equal
(7) Kites: a kite has two pairs of adjacent sides equal and one pair of opposite angles equal.

Example 11.12
Find the size of the angles marked x, y, z in Fig. 11.21(a) and (b).

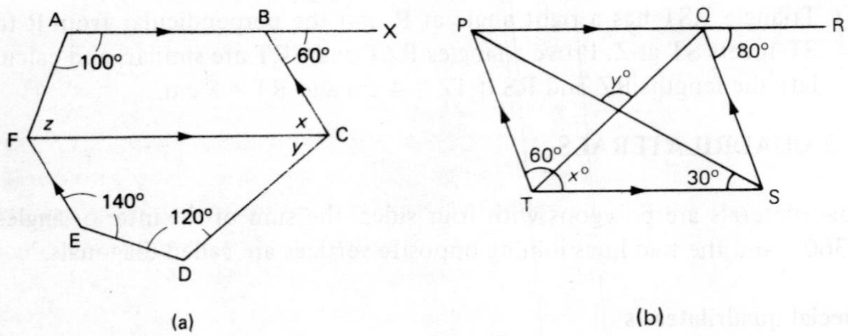

Fig 11.21

Solution (a) In the quadrilateral ABCF

$x = 60^\circ$ (alternate angles BX ∥ FC)

$\angle ABC = 120^\circ$ (supplement of $\angle CBX$)

Therefore

$z = 360° - (120° + 100° + 60°)$ (angle sum of quadrilateral 360°)

$= 80°$

In the quadrilateral CDEF

$\angle CFE = x = 60°$ (alternate angles EF || CB)

Therefore

$y = 360° - (60° + 140° + 120°)$ (angle sum of quadrilateral 360°)

$= 40°$

Answer: $x = 60°, y = 40°, z = 80°$.

(b) $\angle PQS = 100°$ (supplement of $\angle RQS$)

$\angle TQS = \angle PTQ = 60°$ (alternate angles PT || QS)

$\angle PQT = 100° - 60° = 40°$

$\angle STQ = \angle PQT = 40° \Rightarrow x = 40°$

$\angle QPS = \angle PST = 30°$ (alternate angles PQ || TS)

$y = 180° - (30° + 40°)$ (angle sum of triangle 180°)

$= 110°$

Answer: $x = 40°, y = 110°$.

11.8 SYMMETRICAL PROPERTIES OF POLYGONS

(a) Line symmetry

(i) Regular polygons

A line or axis of symmetry is a mirror line, it divides a plane figure into two identical parts, each the mirror image of the other. Since one vertex is not distinguished from another, a regular polygon with n sides has line symmetry of order n. When n is odd, as in the equilateral triangle, each vertex is joined to the mid-point of the opposite side [Fig. 11.22(a)].

When the number of sides is even, the lines of symmetry join opposite vertices and also mid-points of opposite sides, as in the square shown in Fig. 11.22(b).

A circle is the limiting regular polygon with an infinite number of lines of symmetry called diameters [Fig. 11.22(c)].

(ii) Irregular polygons

Some irregular polygons have line symmetry, but the order is always less

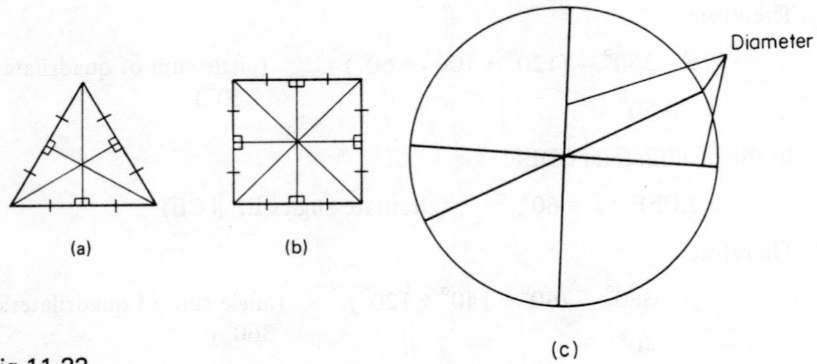

Fig 11.22

than for the regular figure.

Isosceles triangles, isosceles trapezia and kites have one line of symmetry each, rectangles and rhombi have two each, as illustrated by Fig. 11.23.

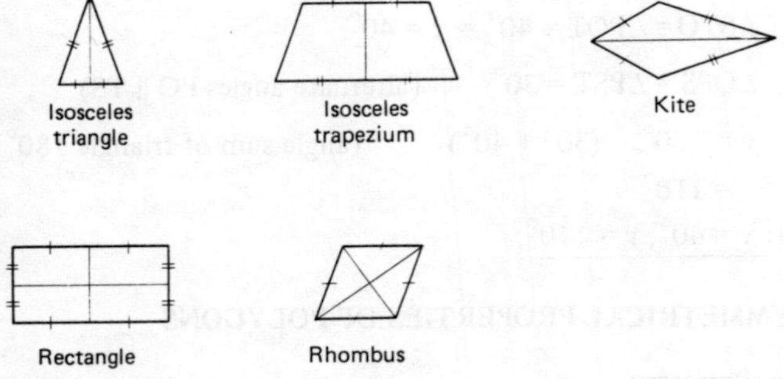

Fig 11.23

(b) Rotational symmetry

The centre of symmetry of a regular polygon is the point of intersection of the axes of symmetry, and the polygon can be rotated about the centre of symmetry in the plane so that the vertices change places. For example, an equilateral triangle has three rotations which leave its position in the plane unaltered, $120°$, $240°$ and $360°$. In general, a regular n-sided polygon has rotational symmetry of order n.

(c) Point symmetry

If the position of a polygon in the plane is unchanged by rotation through $180°$, it is said to have point symmetry about the centre of rotation. Regular polygons with an even number of sides have point symmetry about the centre, and parallelograms have point symmetry about the intersection of the diagonals.

Example 11.13

$\mathscr{E}$ = {quadrilaterals with at least one line of symmetry}

P = {quadrilaterals with point symmetry},

R = {quadrilaterals with rotational symmetry of order 4}

L = {quadrilaterals with at least 2 lines of symmetry}

Name the elements of the sets (a) $P \cap L$, (b) $R \cap L$, (c) L'.

Solution (a) All parallelograms have point symmetry, but only squares, rectangles and rhombi have 2 or more lines of symmetry

$P \cap L$ = {squares, rectangles, rhombi}

(b) Only squares have rotational symmetry of order four

$R \cap L = R$ = {squares}

(c) The set L' contains the quadrilaterals with exactly one line of symmetry

L' = {isosceles trapezia, kites}

Exercise 11.4

(1) A quadrilateral KLMN has diagonals intersecting at P, LP = PK and PM = PN. Show that triangle LPM is congruent to triangle KPN and find the size of ∠LKP if ∠PMN = 67°.

(2) Calculate the length of a side of a rhombus with diagonals of length (a) 6 cm and 5 cm, (b) 50 mm and 70 mm.

(3) What is th length of the sides of a square with diagonals (a) 90 mm, (b) 10 m?

(4) Find the value of x and name the special quadrilateral ABCD whose angles taken in order are (a) $x°$, $2x°$, $x°$, $2x°$ $\frac{1}{2}x°$, $2x°$, $1\frac{1}{2}x°$, $2x°$.

(5) ABCD is a parallelogram with ∠ADB = 40° and ∠BDC = 30°. CE is a line parallel to DB cutting AB produced (extended) at E. Calculate (a) ∠DAB, (b) ∠CBE, (c) ∠BEC, and prove that CE = DB and that BC bisects ED.

(6) Which of the quadrilaterals has/have *exactly one* line of symmetry (a) a parallelogram, (b) an isosceles trapezium, (c) a rhombus, (d) a kite?

CHAPTER 12

GEOMETRICAL CONSTRUCTIONS

A locus in plane geometry is a line or curve joining all points in the plane which satisfy the given conditions. For example, a circle is the locus of all the points in a plane which are equidistant from a fixed point in the plane; the fixed point is the centre, the fixed distance is the radius and the locus is the circumference of the circle. An arc is part of the circumference.

For the constructions described in this chapter, a ruler, a pair of compasses and sharp pencils are needed: parallel rulers or set squares are used for drawing parallel lines, and protractors for measuring angles. In examinations, all construction lines must be shown, and it is worth noting that marks are awarded for neatness and accuracy of construction as well as the correct method. The answers to the exercises in this chapter are given to the nearest 0.5 unit.

12.1 TO CONSTRUCT THE BISECTOR OF A GIVEN ANGLE

The given angle XOY is shown in Fig. 12.1(a).

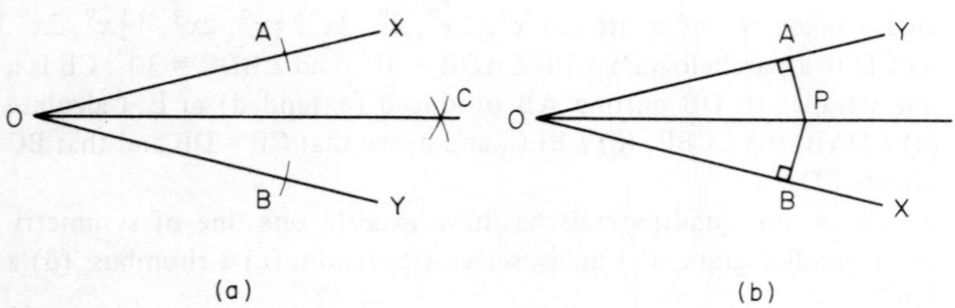

Fig 12.1

Method of construction

(i) Place the point of the compasses at O and with any convenient radius draw arcs to cut OX at A and OY at B

(ii) With centre at A, draw an arc between X and Y
(iii) Repeat with centre B and the same radius, so that the arcs cut at C
(iv) Join OC.

Comments
(i) The line OC is the bisector of the acute angle XOY. If the line CO is extended to D, then OD is the bisector of the reflex angle XOY
(ii) It can be proved that triangles OAC and OBC are congruent (S-S-S) and hence that $\angle AOC = \angle BOC$
(iii) In Fig. 12.1(b), the distance of the point P on the bisector from the lines OX and OY is the perpendicular distance PA and PB. Since triangles OPA and OPB are congruent (R-H-S) these are equal, and it follows that *the locus of points in a plane which are equidistant from two straight lines is the bisector of the angle between the lines*.

12.2 TO CONSTRUCT THE PERPENDICULAR BISECTOR OF A GIVEN STRAIGHT LINE

The given line XY is shown in Fig. 12.2(a).

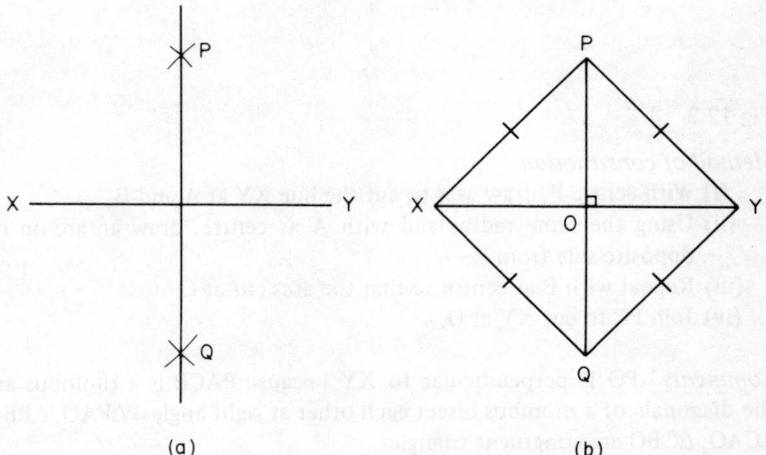

Fig 12.2

Method of construction
(i) With centre X and radius more than half the distance XY, draw arcs above and below the line XY
(ii) With centre Y and the *same* radius, draw arcs to cut the other arcs at P and Q
(iii) Join P and Q.

Comments

(i) The line PQ is the perpendicular bisector of XY

(ii) If PQ and XY intersect at O [Fig. 12.2(b)], then O is the point of intersection of the diagonals of the rhombus XPYQ

(iii) *the locus in a plane of points equidistant from two given points is the perpendicular bisector of the line joining the two points.*

12.3 TO CONSTRUCT A LINE PERPENDICULAR TO A GIVEN LINE FROM A GIVEN POINT NOT ON THE LINE

The given line XY and the point P are shown in Fig. 12.3.

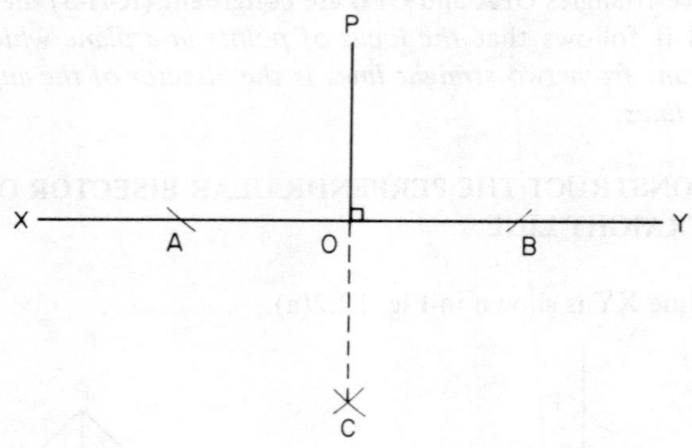

Fig 12.3

Method of construction

(i) With centre P, draw arcs to cut the line XY at A and B

(ii) Using the same radius and with A as centre, draw an arc on the opposite side from P

(iii) Repeat with B as centre so that the arcs cut at C

(iv) Join PC to cut XY at O.

Comments PO is perpendicular to XY because PACB is a rhombus and the diagonals of a rhombus bisect each other at right angles. ΔPAO, ΔPBO, ΔCAO, ΔCBO are congruent triangles.

Exercise 12.1

(1) Draw a line AB of length 70 mm and from A draw another line AX such that angle BAX is 54°. From B construct a perpendicular to AX meeting AX at the point Z. Measure the length of BZ.

(2) Draw a line of length 85 mm and construct the perpendicular bisector of this line AB to cut it at O. With centre A and radius 65 mm, draw an arc to cut the perpendicular bisector at P. Measure PO and angle PAO.

(3) Using a protractor make an angle KLM of 70°. Construct the bisector of this angle using ruler and compasses, and mark on it the point H, 92 mm from L. Construct the perpendicular HG from H to the line KL and measure the lengths HG and GL.

(4) Draw two straight lines POQ and ROS, using a protractor to make the angle POR 64°. Use ruler and compasses to construct the bisectors of the angles POR and ROQ and measure the angle between the bisectors.

12.4 TO CONSTRUCT AN ANGLE EQUAL TO A GIVEN ANGLE

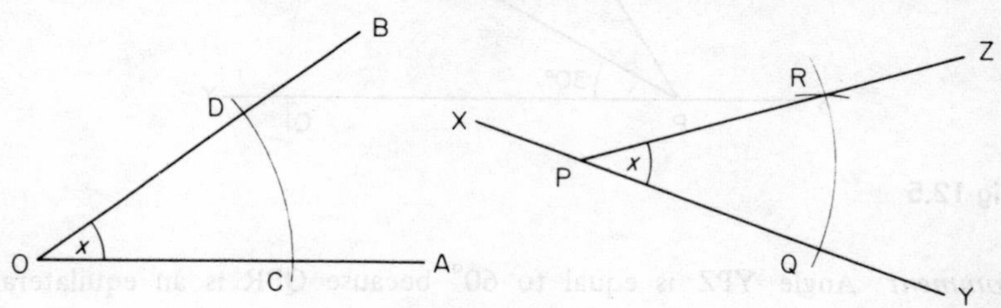

Fig 12.4

Figure 12.4 shows the given angle AOB and a given point P on a straight line XY.

Method of construction

(i) With centre O and any convenient radius draw an arc to cut OA at C and OB at D

(ii) With centre P and the *same* radius draw an arc to cut the line XY at Q

(iii) With centre Q and radius equal to CD, draw an arc to cut the arc through Q at R

(iv) Join PR and extend it to Z.

Comment ∠YPZ = ∠AOB because the triangles COD and QPR are congruent (S-S-S).

12.5 TO CONSTRUCT ANGLES OF 60° AND 30°

Fig. 12.5 shows a given line XY and a point P.

(a) Method of construction for 60°

(i) With P as centre and any convenient radius draw an arc to cut the line XY at Q

(ii) With centre at Q and the *same* radius, draw an arc to cut the first arc at R

(iii) Join PR and extend it to Z.

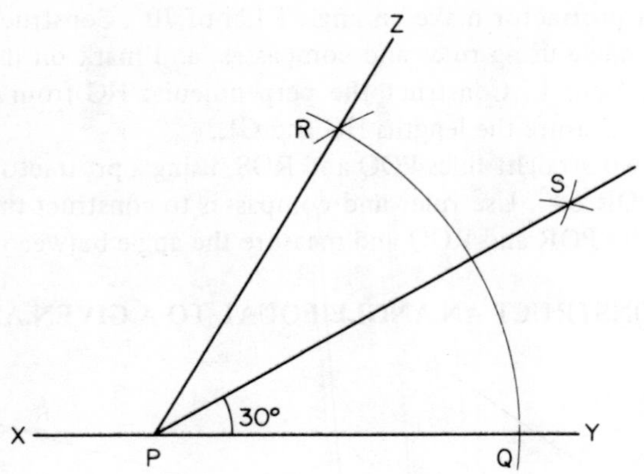

Fig 12.5

Comment Angle YPZ is equal to 60° because QPR is an equilateral triangle.

(b) Method of construction for 30°
Use the method described in Section 12.1 to draw the bisector PS of the angle QPR, and the angle YPS is then equal to 30°.

12.6 TO CONSTRUCT AN ANGLE OF 45°

Methods of construction

(a) Construct an angle of 90° as in Section 12.2 or 12.3 and then bisect it using the method of Section 12.1 to obtain an angle of 45°

(b) This is an alternative method when the angle is required at one end of the given line. Fig. 12.6 shows the given line XY and it is required to construct angle YXZ of 45°.

(i) With centre X and any convenient radius, draw a long arc to cut the line XY at A

(ii) With centre A and the same radius make a small arc cutting the first arc at B

(iii) With centre B and the same radius, draw a long arc to cut the first arc at C

(iv) With centre C and the same radius draw an arc cutting the second long arc at D

(v) Join DX and construct the bisector XZ of the right angle YXD.

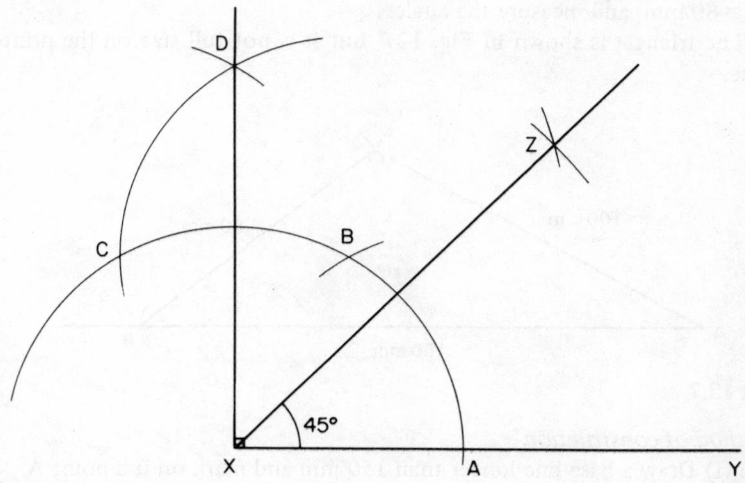

Fig 12.6

Comment

(i) Since the radius was the same each time, ∠AXB and ∠BXC are both 60° and XBDC is a rhombus. Hence ∠BXD = 30° and ∠AXD = 90°

(ii) Angles of 15°, $22\frac{1}{2}$° and also combinations such as 75° and 105° can be constructed using ruler and compasses.

Exercise 12.2

In this exercise the only instruments needed are compasses and a ruler.

(1) Draw a line AB of length 55 mm and construct an angle BAC of 120° so that the length AC is 62 mm. Join BC and measure its length.

(2) Construct an angle XYZ of 105° with YX = 93 mm and YZ = 70 mm. Measure the length of XZ.

(3) Draw a line PQ of length 82 mm and construct an angle QPR of $67\frac{1}{2}$° making the length of PR 76 mm. From R construct a line perpendicular to PQ to meet PQ at S, and measure RS.

12.7 TO CONSTRUCT TRIANGLES OF GIVEN DIMENSIONS

There are four different sets of measurements which can specify a particular triangle, and they are the four conditions for congruency stated in Section 11.6(b).

(a) The lengths of three sides are given

Instruction Construct a triangle ABC having AB = 150 mm, AC = 100 mm,

BC = 80 mm, and measure the angles.

The triangle is shown in Fig. 12.7 but it is not full size on the printed page.

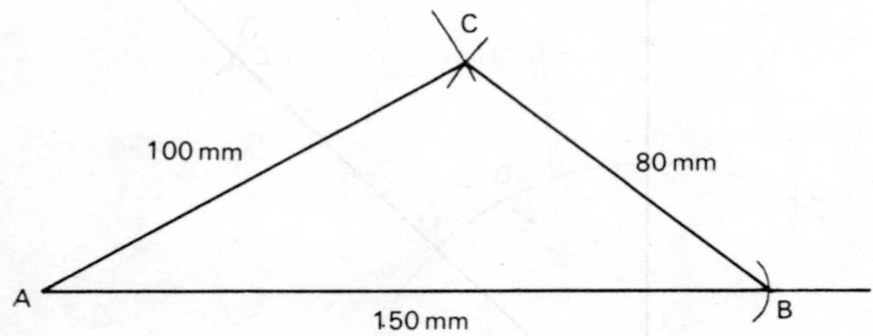

Fig 12.7

Method of construction

(i) Draw a base line longer than 150 mm and mark on it a point A

(ii) With centre A and radius 150 mm draw an arc to cut the base line at B

(iii) With centre A and radius 100 mm draw an arc in the approximate position of C

(iv) With centre B and radius 80 mm draw an arc to cut the second arc at C

(v) Join AC and BC to form the triangle ABC.

Measurement Measured with a protractor to the nearest half degree the angles are

$$\angle A = 29.5^\circ, \angle B = 38^\circ, \angle C = 112.5^\circ$$

(b) Two sides and the included angle are given

Instruction Construct a triangle ABC having AB = 95 mm, AC = 72 mm, and angle BAC = 35°. Measure the length of BC.

The triangle is not printed full size in Fig. 12.8.

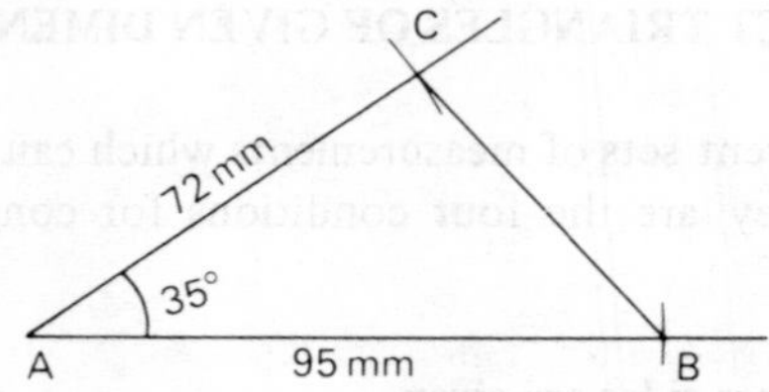

Fig 12.8

Method of construction

(i) Draw a base line longer than 95 mm and mark on it a point A
(ii) With centre A and radius 95 mm draw an arc to cut the base line at B
(iii) Using a protractor make an angle at A of 35° with AB, drawing the line longer than 72 mm
(iv) With centre A and radius 72 mm draw an arc to cut this line at C
(v) Join B and C and measure the length BC.

Measurement By measurement with a ruler to the nearest half millimetre, BC = 55 mm.

(c) One side and two angles are given

Instruction Construct a triangle ABC having BC = 105 mm, ∠ABC = 52°, and ∠ACB = 47°. Measure the length of AB and AC.

The required triangle is shown in Fig. 12.9, not full size.

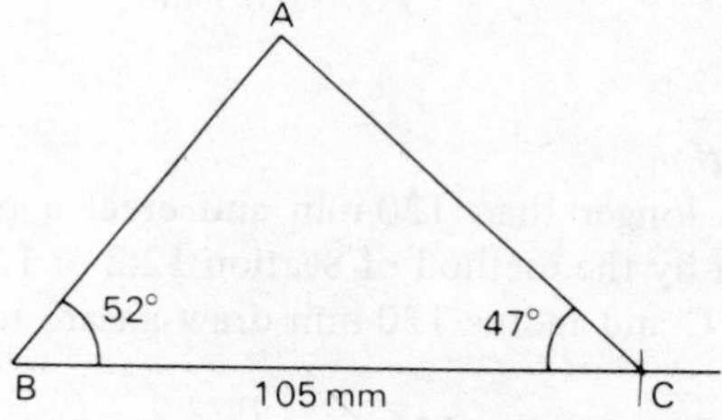

Fig 12.9

Method of construction

(i) Draw a base line longer than 105 mm and mark on it a point B
(ii) With centre B and radius 105 mm draw an arc to cut the base line at C
(iii) Using a protractor, make an angle at B of 52° with BC and an angle at C of 47° with CB
(iv) Extend the lines if necessary to intersect at A, and measure the lengths AB and AC.

Measurement Measured with a ruler to the nearest half millimetre, AB = 78 mm, AC = 84 mm.

(d) The hypotenuse and one other side are given

Instruction Construct a triangle ABC having a right angle at C, hypotenuse AB = 130 mm and the side BC = 120 mm. Measure angle BAC and the length of the side AC.

The triangle is shown, not full size, in Fig. 12.10.

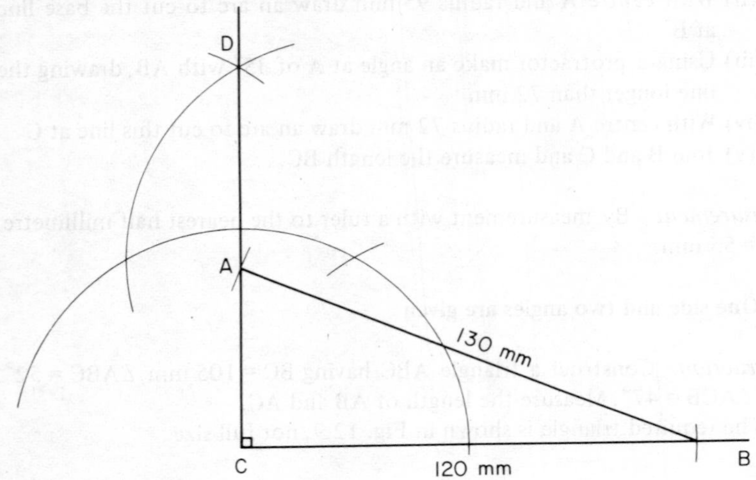

Fig 12.10

Method of construction

(i) Draw a line longer than 120 mm and erect a perpendicular at a point C on it by the method of Section 12.2 or 12.6

(ii) With centre C and radius 120 mm draw an arc to cut the base line at B

(iii) With centre B and radius 130 mm, draw an arc to cut the perpendicular at A. Join A and B to form the triangle ABC.

Measurement Measured with a ruler AC = 50 mm and by measurement with a protractor, ∠BAC = $67\frac{1}{2}°$.

Exercise 12.3

Construct triangles having the dimensions given.

(1) XY = 74 mm, YZ = 100 mm, XZ = 124 mm. Measure ∠YXZ.

(2) AB = 40 mm, ∠BAC = 48°, ∠ABC = 65°. Measure AC and BC.

(3) QR = 60 mm, PQ = 48 mm, ∠ PQR = 54°. Measure PR.

(4) Construct a right angled triangle having the longest side 84 mm and one angle 38°. Measure the other two sides.

(5) Construct a right angled triangle ABC with AB = 40 mm, AC = 60 mm, and ∠ABC = 90°. Bisect the hypotenuse and draw a circle with the mid point O as centre and OC as radius. Mark any point P on the circle on the opposite side from B, join AP and BP and measure the angles APB and ACB.

(6) A triangle PQR has $\angle QPR = 90°$, PR = 80 mm and PQ = 64 mm. A point T inside the triangle is equidistant from PR and PQ. If the distance QT is 50 mm, construct the triangle using ruler and compasses, mark the position of the point T and measure the length of RT.

(7) Construct a triangle ABC having AC = 72 mm, AB = 50 mm and BC = 93 mm. Bisect the angles ABC and ACB letting the bisectors intersect at O. From O, draw a line perpendicular to BC to meet BC at Z. Measure OZ.

(8) Draw any triangle XYZ and construct the perpendicular bisectors of the sides to intersect at a point O. With centre O and radius OX draw a circle. This circle is called the circumscribed circle of the triangle and it should pass through the vertices X, Y and Z.

12.8 TO CONSTRUCT QUADRILATERALS OF GIVEN DIMENSIONS

(a) Parallel lines

In technical drawing, sets of hinged parallel rulers are available for drawing parallel lines, but there is a very simple method using a ruler and a set square which is accepted by some Examining Boards.

Fig. 12.11 shows a line XY and a point P and it is required to draw through P a line parallel to XY.

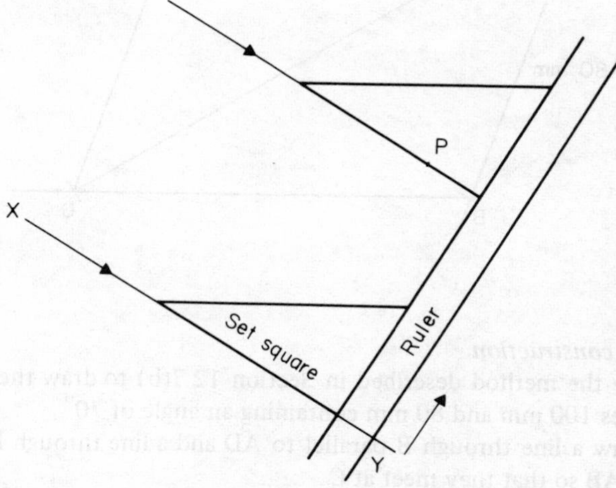

Fig 12.11

Method of construction

(i) Place one edge of a set square along the line XY

(ii) Place the edge of the ruler along a different edge of the set square

(iii) Hold the ruler firmly and slide the set square along until the edge that was along XY passes through P

(iv) Draw a line through P parallel to XY.

(b) To construct squares and rectangles with given dimensions

Method of construction
The general method is to draw a line perpendicular to a given line as in Section 12.6 and mark off the lengths of two adjacent sides. The remaining sides can be drawn parallel as shown in Fig. 12.11.

(c) To construct parallelograms having given dimensions

Instruction Construct a parallelogram ABCD having AB = 80 mm, AD = 100 mm and ∠BAD = 70°. Measure the length of the diagonal AC.

The parallelogram, not full size, is shown in Fig. 12.12.

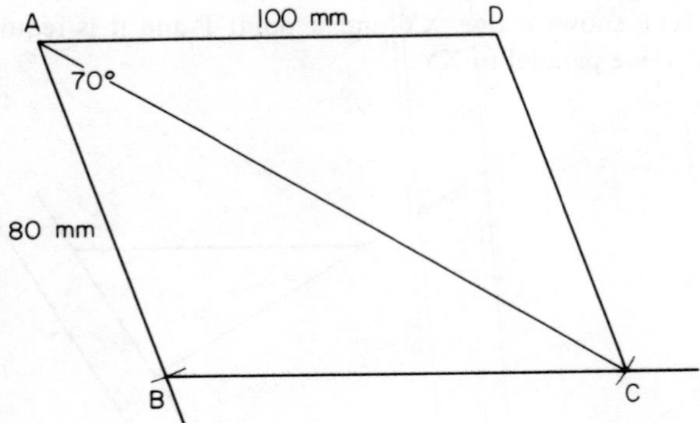

Fig 12.12

Method of construction

(i) Use the method described in Section 12.7(b) to draw the adjacent sides 100 mm and 80 mm containing an angle of 70°

(ii) Draw a line through B parallel to AD and a line through D parallel to AB so that they meet at C

(iii) Measure the length of AC.

Measurement AC = 148 mm.

Exercise 12.4

(1) Construct a parallelogram ABCD with sides AB = 68 mm, BC = 54 mm and diagonal AC = 50 mm, and measure BD. From D construct a line perpendicular to the extension of BC and measure the length of this normal.

(2) Construct a rhombus having diagonals of length 60 mm and 75 mm and measure the length of a side.

(3) Construct a kite ABCD having AB of length 50 mm, BC 80 mm and $\angle BAD = 55^\circ$. Measure BD and $\angle BCA$.

(4) Construct a square ABCD with sides 55 mm, bisect DC at X and measure AX, BX and the angle DAX. Construct an equilateral triangle BCE and measure the length of the median from E.

CHAPTER 13

PERIMETER, AREA AND VOLUME

13.1 PERIMETERS OF PLANE FIGURES

The perimeter of a polygon is the total length of its sides, so that a regular polygon with n sides of length l cm has perimeter nl cm. There is no general formula for the perimeter of irregular polygons but when the dimensions are known the perimeter is easily calculated.

(a) Triangles
A triangle with sides of length a, b and c units has a perimeter of $a + b + c$ units.

(b) Parallelograms
A parallelogram having adjacent sides of length a and b units has perimeter $2a + 2b$ units.

(c) Circles
The perimeter of a circle is its circumference and for every circle the circumference is in the same constant ratio to the diameter. This constant value is an irrational number, given the symbol π (pi), and the usual approximations are $3\frac{1}{7}$ or 3.142.

For every circle having diameter d units, $C/d = \pi$ or $C = \pi d$ and since the radius of a circle is half the diameter, $C = 2\pi r$.

Example 13.1
(a) Calculate the circumference of a circle (i) when the radius is 14 cm, (ii) when the diameter is 183 mm.

Solution

(i) $C = 2\pi r = 2 \times \dfrac{22}{7} \times 14 \text{ cm} = \underline{88 \text{ cm}}$

(ii) $C = \pi d = 3.142 \times 183$ mm = $\underline{575 \text{ mm}}$

(b) What is the radius of a circle which has a circumference of 352 m?

Solution

$$C = 2\pi r \Rightarrow r = \frac{C}{2\pi} = \frac{352}{2} \times \frac{7}{22} = \underline{56 \text{ m}}$$

13.2 AREAS OF PLANE FIGURES

The area of a polygon is the quantity of the plane surface bounded by the sides of the polygon; it has two dimensions and is measured in square units, km^2, m^2, cm^2, or mm^2. A practical unit for large areas such as building or agricultural land is the hectare (ha), which is 10 000 m^2.
Units of area

$$1 \text{ km}^2 = 10^6 \text{ m}^2, \qquad 1 \text{ ha} = 10^4 \text{ m}^2$$

$$1 \text{ m}^2 = 10^4 \text{ cm}^2, 1 \text{ cm}^2 = 100 \text{ mm}^2$$

(a) Rectangles
The area of a rectangle having sides of length a units and b units is given by the formula $A = ab$ square units. This is easily seen when the sides are marked in units of length as in Fig. 13.1.

A rectangle 3 cm by 6 cm has area 18 cm^2 and a square with sides of length 3 cm has area 9 cm^2.

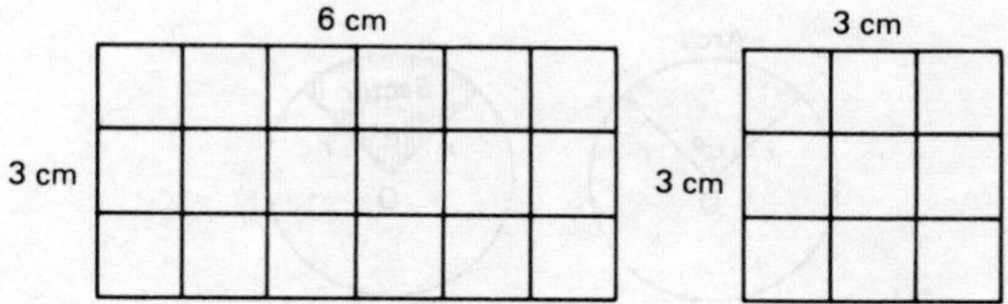

Fig 13.1

(b) Parallelograms
The area of parallelogram is the product of the length of one side and the perpendicular distance to the opposite side [Fig. 13.2(a)]

$$A = l_1 h_1 = l_2 h_2$$

(c) Triangles
Any parallelogram is cut by a diagonal into two congruent triangles, and so the area of a triangle is half the area of a parallelogram with the same base and height [Fig. 13.2(b)]

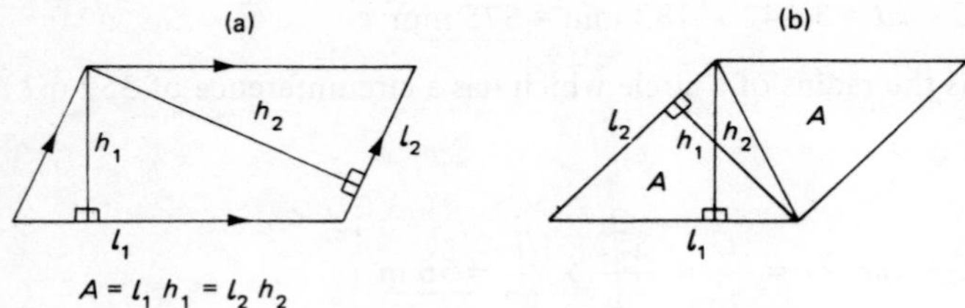

Fig 13.2

$$A = \tfrac{1}{2}l_1h_1 = \tfrac{1}{2}l_2h_2$$

(d) Trapezia

A trapezium can be considered as two triangles of the same height. If the parallel sides are of length l_1 and l_2 and the perpendicular distance between them is h, then the area of the trapezium is

$$A = \tfrac{1}{2}l_1h + \tfrac{1}{2}l_2h \text{ or } \tfrac{1}{2}h(l_1 + l_2)$$

(e) Circles

The area of a circle with radius r units is given by the formula $A = \pi r^2$ square units.

(i) Area of a sector

An arc is part of the circumference of a circle and a sector is a part of a circle bounded by an arc and two radii.

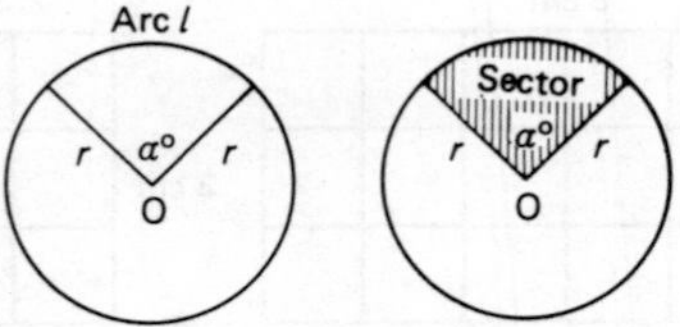

Fig 13.3

Fig. 13.3 shows a sector of a circle with radius r, including an angle $a°$. The angle $a°$ is *subtended* at the centre of the circle by the arc of length l, and the whole circumference subtends a complete revolution 360°. The area A of the sector is the same proportion of the area of the circle as the arc l is of the whole circumference. Hence

$$\frac{A}{\pi r^2} = \frac{l}{2\pi r} = \frac{a}{360}$$

(ii) Area of a segment
A straight line joining two points on the circumference of a circle is called a chord, and every chord of a circle cuts the area into two parts called *segments* [Fig. 13.4(a)].

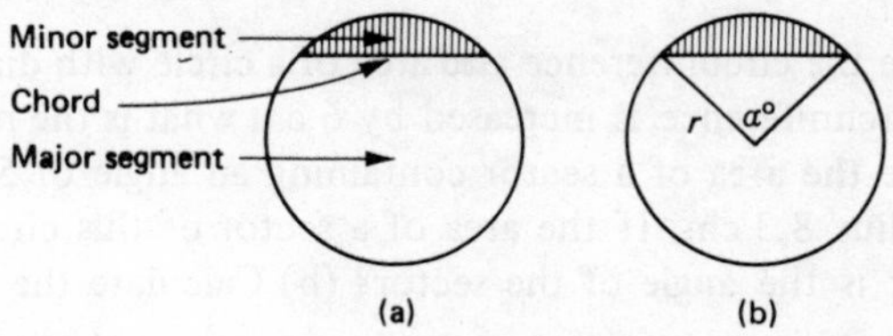

Fig 13.4

The area of a segment is the difference in area of a sector and an isosceles triangle [Fig. 13.4(b)].

Example 13.2
Calculate the radius of the circle of which the area of the quarter circle is 628 mm². ($\pi = 3.14$.)

Solution

$$A = \pi r^2 = 4 \times 628 \text{ mm}^2 \Rightarrow r^2 = \frac{4 \times 628}{3.14} = 800 \text{ mm}^2$$

The radius $r = \sqrt{800}$ = 28.3 mm.

Exercise 13.1
Where necessary, take $\pi = 3.142$.

(1) The area of a triangle ABC is 69.6 cm² and AB = 16 cm. Calculate the distance of C from AB. If the distance from A to CB is 24 cm, what is the length of CB?
(2) Find the area of an equilateral triangle with sides of length 24 cm.
(3) Calculate the length of a side of an equilateral triangle of which the area is $25\sqrt{3}$ cm².
(4) The area of a rectangle is 4392 mm² and one side is 122 mm. What is the length of the adjacent side? If both sides are increased by 20% calculate the percentage increase in area.
(5) The rectangular floor of a room is 7.1 m by 4.8 m. Calculate the least number of square carpet tiles with sides 250 mm which would cover the floor.
(6) Calculate the area of each of the figures (a) a rectangle 28 mm by 42 mm, (b) a parallelogram with opposite sides 75 cm and their distance apart 38 cm, (c) a circle with radius 6 cm, (d) a circle with

diameter 38 mm, (e) a trapezium with parallel sides 23 cm and 49 cm, 16 cm apart.

(7) If the diagonals of a rhombus are 15 cm and 8 cm, what is its area?

(8) A rectangular building plot is 160 m by 40 m. What is its area in hectares?

(9) Calculate the circumference and area of a circle with diameter 13 cm. If the circumference is increased by 6 cm what is the new diameter?

(10) Calculate the area of a sector containing an angle of 50° in a circle with radius 8.3 cm. If the area of a sector of this circle is 72 cm^2, (a) what is the angle of the sector? (b) Calculate the length of the arc.

(11) A tyre of a car has a diameter of 550 mm. If there is no slipping, how many revolutions does the wheel make when the car travels a distance of 8 km?

(f) Irregular figures

Other polygons and composite figures are divided into triangles, quadrilaterals and parts of circles and the areas added together.

Example 13.3

(a) Calculate the area of the pentagon in Fig. 13.5(a); the lengths are in cm.

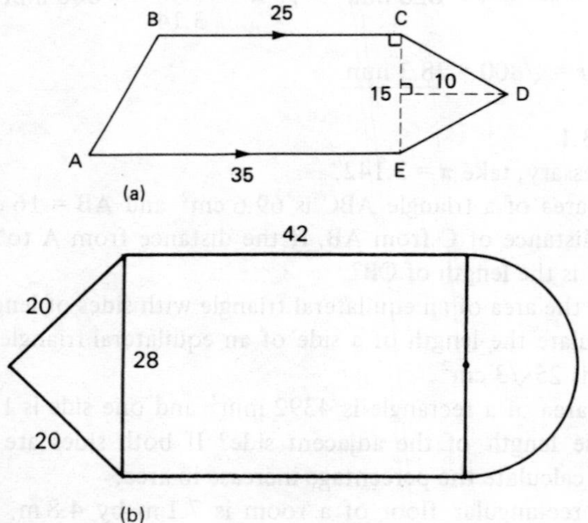

Fig 13.5

Solution

$$\text{Area of trapezium ABCE} = \tfrac{1}{2} \times 15 \times (25 + 35) = 450 \text{ cm}^2$$

$$\text{Area of triangle CDE} = \tfrac{1}{2} \times 15 \times 10 = 75 \text{ cm}^2$$

Therefore the area of the pentagon ABCDE is $\underline{525 \text{ cm}^2}$.

(b) Calculate the area of the plane figure shown in Fig. 13.5(b) which is made up of a right angled triangle, a rectangle and a semicircle. The measurements are given to the nearest millimetre and π is $3\frac{1}{7}$.

Solution

$$\text{Area of right angled triangle: } \tfrac{1}{2} \times 20 \times 20 = \quad 200 \text{ mm}^2$$

$$\text{Area of rectangle: } 28 \times 42 \qquad = 1176 \text{ mm}^2$$

$$\text{Area of semicircle: } \frac{1}{2} \times \frac{22}{7} \times 14^2 \qquad = \quad \underline{308 \text{ mm}^2}$$

$$1684 \text{ mm}^2$$

Therefore total area is $\underline{1684 \text{ mm}^2}$.

Example 13.4

(a) Calculate (i) the shaded area between the two circles in Fig. 13.6(a) when $d_1 = 17$ cm and $d_2 = 7$ cm, (ii) the value of d_2 when $d_1 = 4$ mm and the difference in area is 750 mm².

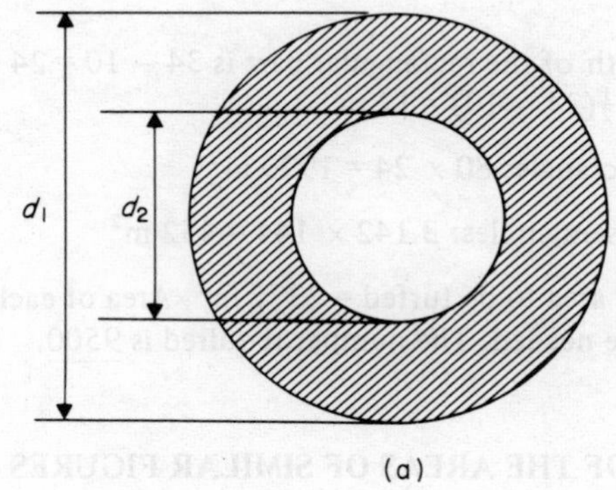

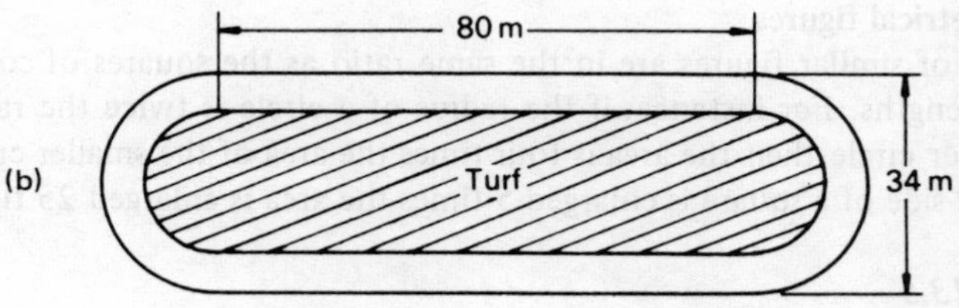

Fig 13.6

Solution
(i) The difference in area of the two circles is

$$\pi r_1^2 - \pi r_2^2$$

When

$$r_1 = \frac{d_1}{2} = \frac{17}{2} \text{ cm and } r_2 = \frac{7}{2} \text{ cm}$$

the required area is 3.142 $(8.5^2 - 3.5^2)$ = $\underline{189 \text{ cm}^2}$

(ii) $$3.142\left[20^2 - \left(\frac{d_2}{2}\right)^2\right] = 750 \Rightarrow \left(\frac{d_2}{2}\right)^2 = 400 - \frac{750}{3.142}$$

$$\frac{d_2}{2} = \sqrt{161.3} = 12.7 \text{ mm}$$

Therefore d_2 = $\underline{25.4 \text{ mm}}$.
(b) Fig. 13.6(b) represents a running track 5 m wide enclosing a plot of land in the shape of a rectangle with semicircular ends. Calculate to the nearest hundred the number of turfs, each 1 m by $\frac{1}{4}$ m, which would cover the plot.

Solution The width of the rectangular part is 34 – 10 = 24 m. The radius of the semicircle is $\frac{1}{2}$(34 – 10) = 12 m

Area of rectangle: 80 × 24 = 1920 m^2.

Area of 2 semicircles: 3.142 × 144 = 452 m^2

Therefore the total area to be turfed = 2372 m^2. Area of each turf is $\frac{1}{4}$ m^2, and so 2372 × 4 are needed. The number required is $\underline{9500}$.

13.3 THE RATIO OF THE AREAS OF SIMILAR FIGURES

(a) Geometrical figures

The areas of similar figures are in the same ratio as the squares of corresponding lengths. For instance, if the radius of a circle is twice the radius of a smaller circle then the area is four times the area of the smaller circle, and if the side of a square is enlarged 5 times the area is enlarged 25 times.

Example 13.5
The areas of two circles are in the ratio 16 : 9. Calculate the radius of the smaller circle when the radius of the larger circle is 24 mm.

Solution Since the areas are in the ratio 16 : 9 corresponding lengths are in the ratio $\sqrt{16} : \sqrt{9} = 4 : 3$ and the radius of the smaller circle is $\frac{3}{4}$ the radius of the larger, or 18 mm.

Example 13.6

(a) Calculate the area of ΔABC in Fig. 13.7 when AB = 18 cm, AX = 3 cm and area AXY = 8 cm^2.

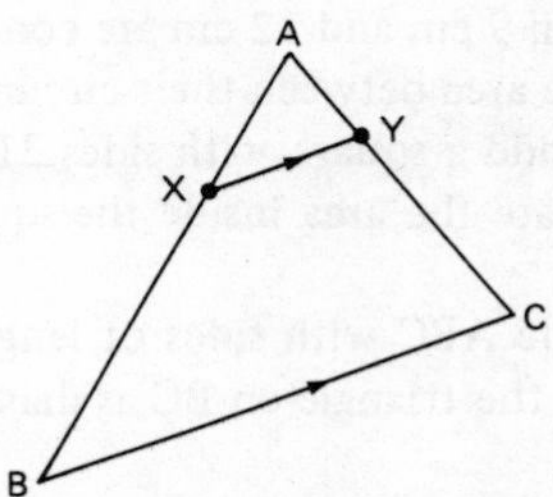

Fig 13.7

Solution Triangles AXY and ABC are similar (3 angles the same). Therefore area AXY : area ABC = AX^2 : AB^2 = 1 : 36

$$\text{Area ABC} = 36 \times 8\ \text{cm}^2 = \underline{288\ \text{cm}^2}$$

(b) Calculate the area of the trapezium XYCB when AY = 2 cm, YC = 4 cm and the area of ΔABC = 180 cm^2.

Solution

$$\text{AC} = 2 + 4 = 6\ \text{cm},\ \text{AY}^2 : \text{AC}^2 = 2^2 : 6^2 = 1 : 9$$

Therefore, area AXY = (1/9) area ABC = 20 cm^2 and area of trapezium XYCB = 180 − 20 = 160 cm^2.

(b) Scale drawings, plans and maps

Photographic enlargements and plans and maps drawn accurately to scale are also similar figures, and corresponding areas may be calculated when the scale factor is known.

Example 13.7

On a map a lake was represented by an area 30 mm^2, and the longest part was 8 mm. If the corresponding length was actually 240 m what was the area of the lake?

Solution 8 mm : 240 m = 8 mm : 240 000 mm and so the scale of the map was 1 : 30 000 or 1 : 3×10^4.

Corresponding areas are in the ratio $1:(3\times10^4)^2$ or $1:9\times10^8$ so that each square millimetre on the map represents 9×10^8 mm^2. An area of 30 mm^2 on the map represents 2.7×10^{10} mm^2 and therefore the area of the lake was 2.7 ha since 1 ha $=10^{10}$ mm^2.

Exercise 13.2

(1) Calculate the area of a path of width 80 cm surrounding a rectangular lawn 25 m by 9 m.

(2) Two circles with radii 9 cm and 12 cm are concentric (have the same centre). Calculate the area between their circumferences. ($\pi=3.142$.)

(3) A circle is drawn inside a square with sides 21 cm so that it touches all four sides. Calculate the area inside the square but *not* inside the circle. ($\pi=22/7$.)

(4) An equilateral triangle ABC with sides of length 75 mm has a semicircle drawn outside the triangle on BC as diameter. What is the area of the figure?

(5) Twenty circular discs each having a diameter of 14 mm are arranged in a single layer on a rectangular plate 147 mm by 35 mm. Calculate the area of the plate left uncovered. ($\pi=22/7$.)

(6) Calculate the area of each of the shapes in Fig. 13.8; all the measurements are in centimetres.

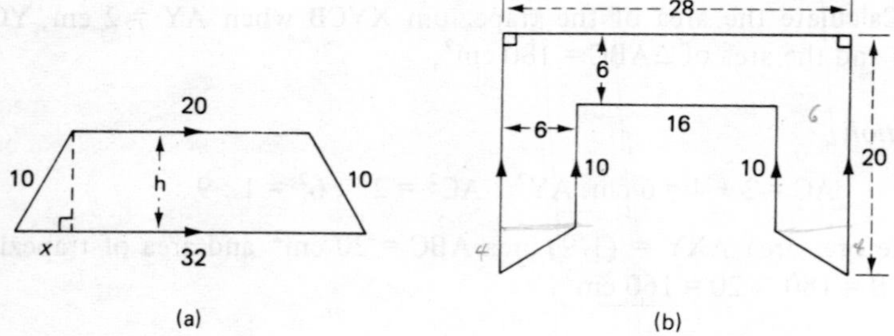

Fig 13.8

(7) A triangle ABC has AB = 60 mm and AC = 90 mm and D and E are the mid-points of AB and AC respectively. Calculate the ratio of the areas of triangles ADE and ABC. If P is a point on BC such that BP = PC and DE = 50 mm, find the lengths (a) BC, (b) DP, (c) EP.

(8) Find the ratio of the areas (a) AGF : ABCD, (b) BFC : ADF in the diagram of Fig. 13.9.

(9) Find the area in hectares represented by 30 mm^2 on a map 1 : 15 000.

(10) A window is drawn on a scale diagram as a rectangle 25 mm by 35 mm. If the shorter sides of the actual window are 1.7 m calculate (a) the longer sides, (b) the area of the glass.

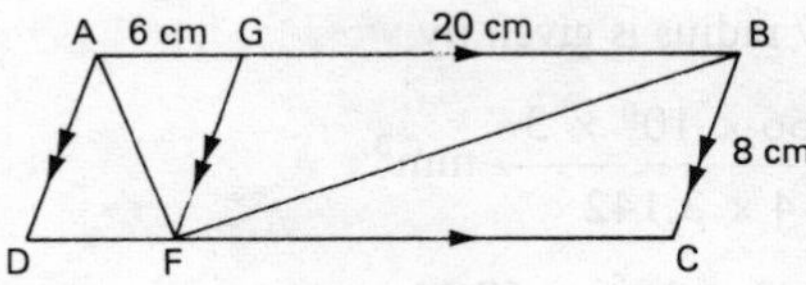

Fig 13.9

13.4 VOLUMES OF GEOMETRICAL SOLIDS

A geometrical solid, or polyhedron, is a three-dimensional figure consisting of intersecting polygons called faces, in different planes. Two faces of a polyhedron meet in a line called an edge; two edges meet in a corner or vertex.

(a) Units of volume
Since a solid has three dimensions, the units of volume are cubic: mm^3, cm^3, m^3 and km^3. $1\ km^3 = 10^9\ m^3$. $1\ m^3 = 10^6\ cm^3$ or $10^9\ mm^3$. A convenient unit for fluids is the litre (l) which is $1\ dm^3$ or $1000\ cm^3$.

(b) Regular polyhedra
All the faces of a regular polyhedron are identical regular polygons at the same distance from the centre of the solid.

A *sphere* is the limiting form, when the number of faces becomes so great that each one is a point, every point on the surface of a sphere being at the same distance (the radius) from the centre.

The volume of a sphere with radius r units is given by the formula

$$V = \frac{4}{3}\pi r^3 \text{ cubic units}$$

Example 13.8
Calculate the volume of a sphere with radius 65 mm. What is the new radius if the volume is increased by 10%?

Solution

$$V = \frac{4}{3}\pi r^3 = 65^3 \times 3.142 \times 4 \div 3 = \underline{1.15 \times 10^6\ mm^3}$$

New volume

$$110\%\ V = 1.10 \times 1.15 \times 10^6 = 1.266 \times 10^6\ mm^3$$

Therefore the new radius is given by

$$r^3 = \frac{1.266 \times 10^6 \times 3}{4 \times 3.142} \text{ mm}^3$$

$$r = \sqrt[3]{(3.022 \times 10^5)} = 67.11 \text{ mm}$$

The new radius is 67 mm to nearest millimetre.

(c) Right prisms

A prism has a polygon as base and uniform cross section, and in right prisms the plane of the base is perpendicular (or normal) to the length or height of the prism. The volume of a right prism with cross sectional area A and height h is given by the formula

$$V = Ah \text{ cubic units}$$

A cylinder is the limiting form of prism with a circular base and a curved surface, and its volume is given by

$$V = Ah = \pi r^2 h$$

A cube, or regular hexahedron, is a square prism with six square faces. A cuboid is a rectangular right prism.

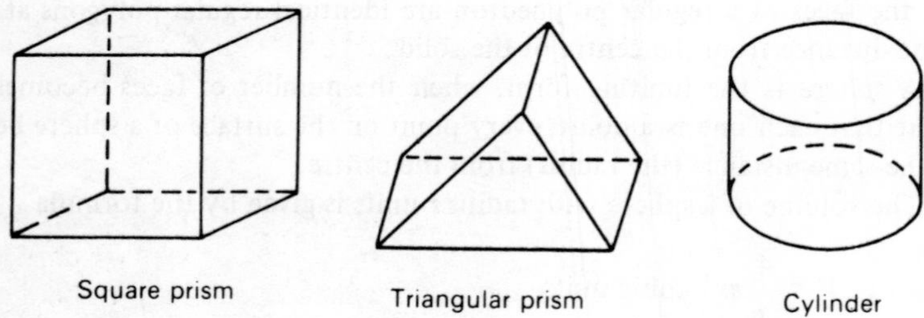

Fig 13.10

Example 13.9

A metal water pipe has internal diameter 21 mm and external diameter 28 mm. Calculate (a) the volume of water leaving the pipe in 1 minute when the rate of flow along the pipe is half a metre per second, (b) the volume of metal in 1 metre of the pipe.

Solution Fig. 13.11 shows a cross section of the pipe.

(a) 0.50 m/s = 30 000 mm/minute

$$\text{Area of cross section} = \pi r_1{}^2 = 3.142 \times 10.5^2 \text{ mm}^2$$

Therefore

$$\text{the volume of water} = 3.142 \times 10.5^2 \times 30\,000 \text{ mm}^3$$
$$= \underline{1.039 \times 10^7 \text{ mm}^3 \text{ or } 10.4 \text{ litres}}$$

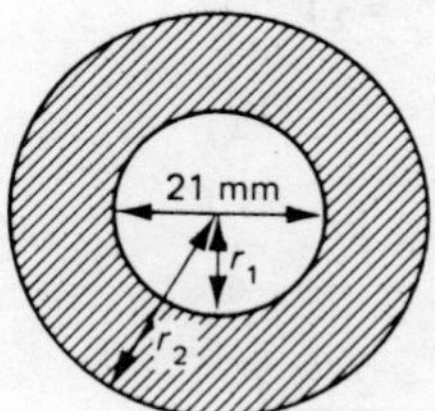

Fig 13.11

(b) Area of outside circle $= \pi {r_2}^2 = 3.142 \times 14^2 \text{ mm}^2$

Area of inside circle $= \pi {r_1}^2 = 3.142 \times 10.5^2 \text{ mm}^2$

Therefore

$$\text{area of cross section of the metal} = \pi {r_2}^2 - \pi {r_1}^2$$

and

$$\text{volume of metal in 1 m of pipe} = 1000\,\pi({r_2}^2 - {r_1}^2)$$
$$= 3142(14^2 - 10.5^2) \text{ mm}^3$$
$$= \underline{2.7 \times 10^5 \text{ mm}^3}$$

Example 13.10

Calculate (a) the volume of a right prism of length 7 cm having a triangular base with sides of length 3 cm, 4 cm and 5 cm, (b) the percentage change in the volume of a right cylinder when the height is trebled and the diameter is halved.

Solution

(a) A triangle having sides in the ratio 3 : 4 : 5 has a right angle.

Area of the base $A = \frac{1}{2} \times 3 \times 4 = 6 \text{ cm}^2$, height $h = 7$ cm

Therefore

$$\text{the volume } V = Ah = 6 \times 7 = \underline{42 \text{ cm}^3}$$

(b) Suppose the original height is h_1 and diameter d_1, then

$$\text{volume } V = \pi \left(\frac{d_1}{2}\right)^2 h_1 \text{ or } \frac{\pi {d_1}^2 h_1}{4}$$

The new height $h_2 = 3h_1$ and diameter $d_2 = \frac{1}{2}d_1$ and so

$$\text{the new volume } V_2 = \frac{\pi d_2{}^2 h_2}{4} = \frac{\pi d_1{}^2 \times 3h_1}{4 \times 4}$$

$$= \tfrac{3}{4}V$$

$$\text{Change in volume} = V - V_2$$

$$= V - \frac{3\,V}{4}$$

$$= \underline{25\%\ V}$$

(d) Right pyramids

A pyramid tapers to a point called the *apex* and so the area of cross section is not constant. In a right pyramid the normal from the apex to the base passes through the centre of symmetry of the base, and the length of this normal is the height of the pyramid.

The general formula for the volume of a right pyramid is

$$V = \tfrac{1}{3}Ah \text{ cubic units}$$

where A is the area of the base and h is the height.

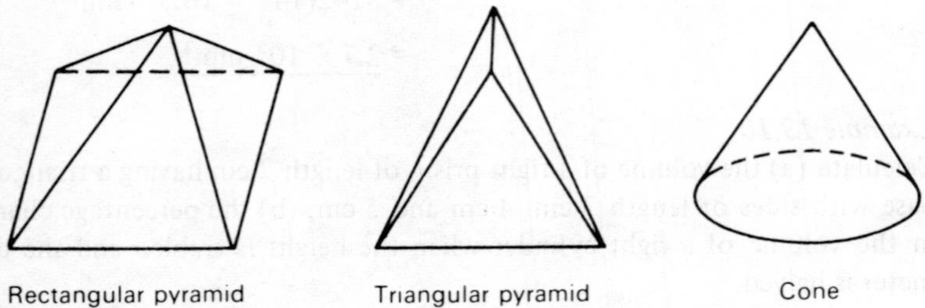

Fig 13.12

A cone is a pyramid with a circular base and a curved surface. The volume of a cone with a base of radius r units and a height h is

$$V = \tfrac{1}{3}\pi r^2 h \text{ cubic units}$$

Example 13.11

Fig. 13.13 represents a plane section through the axis of symmetry of a compound solid formed of a cone, two cylinders and a hemisphere. Find an expression in terms of x and π for the volume of the solid.

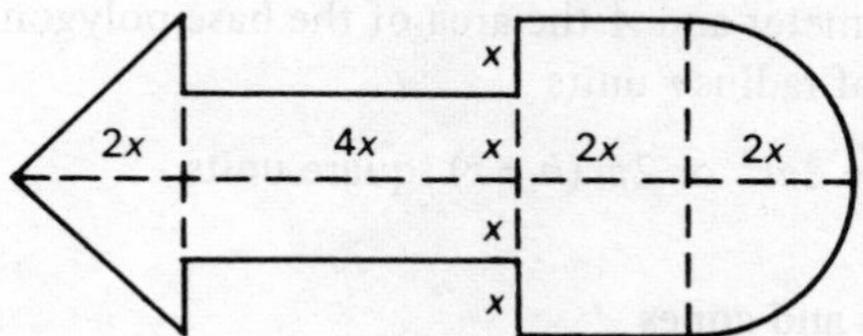

Fig 13.13

Solution

Volume of cone; height $2x$, radius $2x$ $\quad \frac{1}{3}\pi(2x)^3 = \frac{8}{3}\pi x^3$

Volume of cylinder; height $4x$, radius x $\quad \pi x^2(4x) = 4\pi x^3$

Volume of cylinder; height $2x$, radius $2x$ $\quad \pi(2x)^2\ 2x = 8\pi x^3$

Volume of hemisphere; radius $2x$ $\quad \frac{2}{3}\pi(2x)^3 = \frac{16}{3}\pi x^3$

The total volume of the solid is $\underline{20\pi x^3}$.

Example 13.12

Calculate the ratio of the volume of a pyramid with a square base of side 10 cm and height 6 cm and a rectangular prism 5 cm by 4 cm by 9 cm.

Solution

Volume of pyramid $\quad \frac{1}{3} \times 100 \times 6 = 200\ \text{cm}^3$

Volume of prism $\quad 5 \times 4 \times 9 = 180\ \text{cm}^3$

Ratio of volumes = 200 : 180 = $\underline{10 : 9}$.

13.5 SURFACE AREAS OF GEOMETRICAL SOLIDS

The surface area of a polyhedron is the sum of the areas of the faces, and there is no general formula.

(a) Spheres

The surface area of a sphere with a radius of r units is

$S = 4\pi r^2$ square units

(b) Right prisms and cylinders

The general formula for the surface area of a right prism of height h units is

$S = Ph + 2A$

where P is the perimeter and A the area of the base polygon.

For a cylinder of radius r units

$$S = 2\pi rh + 2\pi r^2 \text{ or } 2\pi r(h + r) \text{ square units}$$

(c) Right pyramids and cones

A pyramid has triangular faces and the area of these is calculated as in Section 13.2(c) when the dimensions are given.

The *slant edge* is the distance from the apex to one corner of the base and the *slant height* is the height of one of the triangular faces.

In a right cone the slant height, l, is the distance from the apex to a point on the circumference of the base, and it is related to the height of the cone by the formula $l^2 = r^2 + h^2$ (using Pythagoras' theorem). The curved surface of a cone has area πrl and the total area is $S = \pi rl + \pi r^2$ or $\pi r(r + l)$.

Example 13.13

A right cone has a curved surface area of 85.5 cm². Calculate (a) the curved surface area of a cone which has three times the radius and twice the slant height, (b) the radius of a sphere which has the same curved surface area as the second cone, (c) the height of a cylinder which has the same radius and surface area as the sphere.

Solution

(a) Suppose the slant height is l cm and the radius r cm. Then

$$\pi rl = 85.5 \text{ cm}^2 \text{ and } \pi \times 3r \times 2l = 6\pi rl = \underline{513 \text{ cm}^2}$$

(b) Surface of a sphere $4\pi r^2 = 513 \text{ cm}^2 \Rightarrow r^2 = (513/4\pi) \text{ cm}^2$

$$r^2 = 40.82 \text{ cm}^2 \Rightarrow \underline{r = 6.39 \text{ cm}}$$

(c) Surface area of cylinder $2\pi r(r + h) = 4\pi r^2$

Hence

$$r + h = 2r \Rightarrow h = r = \underline{6.39 \text{ cm}}$$

Exercise 13.3

If necessary take $\pi = 3.142$ in this exercise.

(1) Calculate the volume in m³ of ready mixed concrete needed to construct the path described in Exercise 13.2 Question (1) if the path is to be 7 cm thick, assuming there is no contraction.

(2) What is the greatest length of gold foil 10 mm wide and 4×10^{-3} mm thick that can be beaten from 2 mm³ of the metal?

(3) Calculate the volume and surface area of (a) a sphere of diameter 2.5 m, (b) a cone of slant height 9.5 cm and base diameter 4.2 cm, (c) a solid cylinder of diameter 28 mm and height 65 mm.

(4) How many litres of liquid will fill a cylindrical tank of internal diameter 0.8 m and length 1.7 m? Calculate the height of a rectangular tank 0.7 m by 1.1 m which has the same capacity.

(5) Tape of thickness 0.05 mm is wound tightly on the core of a circular spool. If the core has a diameter of 2 cm and the core and tape together a diameter of 14 cm, calculate the length, in metres, of the tape.

(6) A child's toy consists of a cylinder of length 18 cm and diameter 5 cm, with a hemisphere of the same diameter on one end and a cone of the same diameter and height 6 cm on the other end. Calculate (a) the volume, (b) the surface area of the toy.

(7) Find the volume of a right prism of length 60 cm having as cross section (a) a right angled triangle with sides 45 mm, 60 mm and 75 mm, (b) a regular hexagon with sides of length 40 mm.

13.6 RATIOS OF AREAS AND VOLUMES OF SIMILAR SOLIDS

(a) Geometrical figures

If two similar three-dimensional figures have a corresponding length in the ratio $a : b$, then corresponding areas are in the ratio $a^2 : b^2$ and corresponding volumes are in the ratio $a^3 : b^3$.

This is illustrated in Fig. 13.14 by two cubes of sides 2 cm and 4 cm. The areas of the faces are in the ratio 4 cm^2 : 16 cm^2 or $2^2 : 4^2$ and the volumes of the cubes are in the ratio 8 cm^3 : 64 cm^3 or $2^3 : 4^3$.

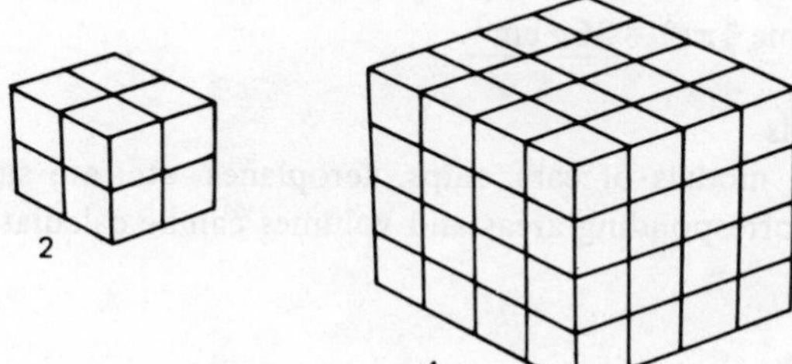

Fig 13.14

All cubes are similar to each other and so are all spheres, because they are regular solids, but solids which are not regular are similar only if corresponding lengths are in the same ratio. Thus a cylinder with a radius of 4 cm and height 8 cm is similar to a cylinder with a radius of 2 cm and height 4 cm, it is *not* similar to one with a radius of 2 cm and height 8 cm.

Example 13.14
A cylinder with a diameter of 6 cm has a height of 9 cm. Calculate (a) the radius and curved surface area of a similar cylinder with a height of 12 cm, (b) the volume of a similar cylinder which has a diameter of 9 cm.

Solution
(a) Corresponding lengths are in the same ratio, therefore $r : 3 = 12 : 9$ and the radius is $\underline{4 \text{ cm}}$.

$$\text{Curved surface area } \pi dh = 3.142 \times 8 \times 12 = \underline{302 \text{ cm}^2}$$

(b) $h : 9 = 9 : 6 \Rightarrow h = 13.5$ cm. Volume of a cylinder of radius 4.5 cm and height 13.5 cm is

$$3.142 \times 4.5^2 \times 13.5 = \underline{859 \text{ cm}^3}$$

Example 13.15
A cone with a base area of K cm^2 has a volume of V cm^3. Find in terms of K the base area of a similar cone with a volume of $27V$ cm^3.

Solution Ratio of volumes is $V : 27V = 1 : 3^3$. Ratio of corresponding areas $1 : 3^2 = K : 9K$. Therefore the required area is $\underline{9K \text{ cm}^2}$.

Example 13.16
A metal sphere of radius 12 cm is formed without wastage into 64 identical spheres. Calculate in terms of π the surface area and volume of each of the small spheres.

Solution Volumes are in the ratio 1 : 64 and radii in the ratio $1 : \sqrt[3]{64} = 1 : 4$. The radius of each small sphere is 12/4 = 3 cm. Surface area $4\pi r^2 = \underline{36\pi \text{ cm}^2}$; volume $\frac{4}{3}\pi r^3 = \underline{36\pi \text{ cm}^3}$.

(b) Scale models
Accurate scale models of cars, ships, aeroplanes, etc. are similar to the originals and corresponding areas and volumes can be calculated when the scale is known.

Example 13.17
A scale model 10 cm long is made of a car of length 5 m. (a) If the car has a window area of 2.75 m^2, what is the corresponding area on the model? (b) If the luggage space on the model car is 4 cm^3, what is the capacity in the real car.

Solution
(a) Ratio of lengths 10 cm : 5 m = 1 : 50. Ratio of areas $1 : 50^2 = 1 : 2500$.

2.75 m^2 = 27 500 cm^2 on car corresponds to 27 500/2 500 cm^2

= 11 cm^2 on the model

(b) Ratio of volumes 1 : 50^3 = 1 : 125 000. 4 cm^3 corresponds to 500 000 cm^3 = 0.5 m^3.

The capacity is 0.5 m^3.

(c) Frusta of cones and pyramids

When a small cone or pyramid is removed from the top of a similar solid the portion remaining is called a frustum. The volume of a frustum of a cone is the difference in the volumes of two similar cones.

Example 13.18

Calculate the height, volume and total surface area of the frustum remaining when a cone of slant height 25 cm is removed from the top of a cone of slant height 40 cm by cutting in a plane parallel to the base, which is of radius 24 cm. (See Fig. 13.15.)

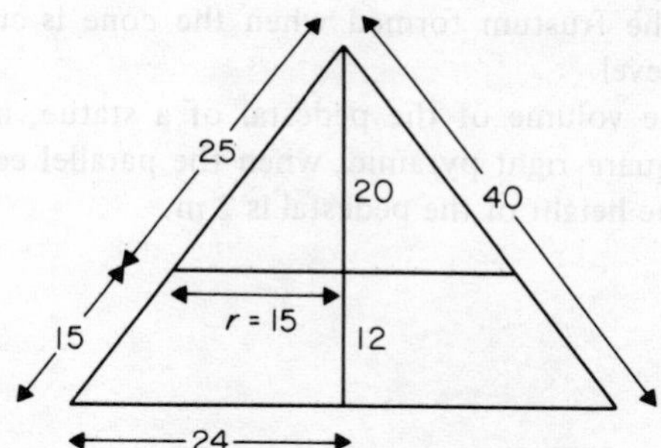

Fig 13.15

Solution If the base of the smaller cone has radius r, $r : 24 = 25 : 40$ since similar figures have corresponding lengths in the same ratio. Therefore the radius is 15 cm and the height $\sqrt{(25^2 - 15^2)} = 20$ cm. The height of the larger cone is $\sqrt{(40^2 - 24^2)} = 32$ cm, and hence the height of the frustum is 12 cm.

$$\text{The volume of the frustum} = \tfrac{1}{3}\pi \times 24^2 \times 32 - \tfrac{1}{3}\pi \times 15^2 \times 20$$

$$= 14\,600 \text{ cm}^3$$

$$\text{The curved surface area} = \pi \times 24 \times 40 - \pi \times 15 \times 25 = 1838 \text{ cm}^2$$

$$\text{The plane surface area} = \pi \times 24^2 + \pi \times 15^2 = 2517 \text{ cm}^2$$

Hence the total surface area of the frustum = 4355 cm^2.

Exercise 13.4

(1) Calculate the ratios of (a) the surface areas, (b) the volumes of two cubes with sides of length 12 cm and 20 cm.

(2) If two similar hexagonal blocks have a corresponding edge of length 2.4 cm and 3.6 cm respectively, what is the ratio of their areas? Calculate the volume of the smaller block if the volume of the other is 12.5 cm^3.

(3) Two similar cones have radii 60 mm and 15 mm. If the height of the smaller is 140 mm calculate (a) the height of the other cone, (b) the ratio of their volumes.

(4) A model car is built to a scale of 1 : 130 and the fuel tank has a capacity of 12 mm^3. Calculate the capacity in litres of the fuel tank on the original car.

(5) A scale model of a ship is 90 cm long when the actual length is 300 m. If a window on the ship has an area of 6 m^2 what is the area in mm^2 of the window on the model?

(6) A cone 25 cm high has a base area of 150 cm^2. Find (a) the height above the base at which the area of the cross section is 80 cm^2, (b) the volume of the frustum formed when the cone is cut parallel to the base at this level.

(7) Calculate the volume of the pedestal of a statue, in the form of a truncated square right pyramid, when the parallel edges are 2 m and 1.2 m and the height of the pedestal is 2 m.

MAPPINGS AND FUNCTIONS, VARIATION

This chapter contains three topics, which although closely related, could be studied separately. Mappings and mapping diagrams (Sections 14.1 and 14.2) and the composition of functions (Section 14.3) are required by only a few Examining Boards. Function notation and the ability to evaluate functions for given values of the variable (Section 14.4) are used later in the book, while a knowledge of variation (Section 14.5) is essential in engineering and science, and is required by most Boards.

14.1 MAPPINGS

(a) Domain and range

If set A is $\{1, 2, 3, 4\}$ and set B is $\{2, 3, 4, 5\}$ for each element of set A there is an element in B which is one greater, and we say there is a *mapping* from set A to set B. Set A is called the *domain* of the mapping and set B the *range.*

The sets in a mapping may be infinite, like the set of real numbers, or finite. When the number of elements is small they can be represented by arrow diagrams, as in Fig. 14.1.

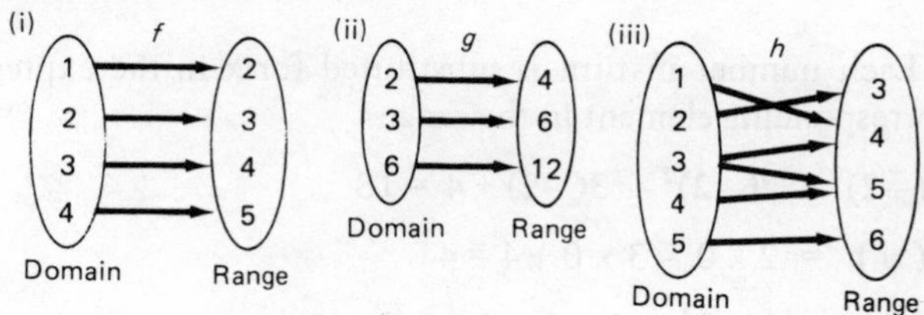

Fig 14.1

In algebra a mapping is represented by a single letter, usually f or g, and in Fig. 14.1(i) f represents the mapping 'adding 1'. We say 'f maps 1 to 2, 2 to 3, 3 to 4, and 4 to 5' and in mapping notation this is written $f\colon 1 \to 2, f\colon 2 \to 3$ and so on.

When defining a mapping the domain should be stated, and there are several ways of writing it, such as

$f\colon x \to x+1, \quad$ for the domain $\{1, 2, 3, 4\}$

$g\colon x \to 2x, \quad x \in \{2, 3, 6\}$

$f\colon x \to 3x+2, \{x : -4 < x < 4\}$

This is read as 'the set of values of x between -4 and $+4$'.

(b) Functions

A function is a special kind of mapping in which each element in the domain is mapped to one and only one element in the range, and most of the mappings in this book are functions. In Fig. 14.1 the mappings f and g are both functions, but the mapping h is *not* a function because the element 3 is mapped to two different elements in the range of h.

In function notation the mapping $f\colon x \to x^2 + 2$ would be written $f(x) = x^2 + 2$.

When $x = 1$ we write $f(1) = 1^2 + 2 = 3$, meaning 3 is the value of $f(x)$ when x takes the value 1.

In general $f(a) = a^2 + 2$ is the value of the function $f(x)$ when x takes the value a.

(c) Sequences

A sequence is a function defined on the natural numbers. If the nth term of a sequence S is given by $S(n)$ where n is a natural number, then the first term of the sequence is $S(1)$, the second term is $S(2)$, the fifteenth term is $S(15)$ and so on, (see Example 14.2).

Example 14.1

(a) Find the range of the function $f\colon x \to 2x^2 - 3x + 4$ for the domain $\{-2, 0, 2\}$.

Solution Each number in turn is substituted for x in the expression, to give the corresponding element in the range

$$f(-2) = 2(-2)^2 - 3(-2) + 4 = 18$$

$$f(0) = 2 \times 0 - 3 \times 0 + 4 = 4$$

$$f(2) = 2 \times 2^2 - 3 \times 2 + 4 = 6$$

The range is therefore the set $\underline{\{18, 4, 6\}}$.

(b) If the range of a function f is $\{f(x) : -2 \leqslant f(x) \leqslant 2\}$, and $f(x) = 3x + 7$, what is the domain?

Solution Since $3x+7$ increases as x increases, only the extreme values are calculated

$$3x+7=-2 \Rightarrow x=-3, \qquad 3x+7=2 \Rightarrow x=-1\tfrac{2}{3}$$

and the domain is $\underline{\{x : -3 \leqslant x \leqslant -1\tfrac{2}{3}\}}$

Example 14.2
(a) A sequence is defined by $S(n)=(-1)^n\ (n^2-3n)$ where n is a natural number. Find the first three terms and the tenth term of the sequence.

Solution

$$S(1)\ = (-1)^1\ (1^2-3)=(-1)(-2)=\underline{2}$$

$$S(2)\ = (-1)^2\ (2^2-6)=1\times -2=\underline{2}$$

$$S(3)\ = (-1)^3\ (3^2-9)=-1\times 0=\underline{0}$$

$$S(10)= (-1)^{10}\ (10^2-30)=1\times 70=\underline{70}$$

(b) Find the domain of the function $f(x)=4/x$ for the range $-4 \leqslant f(x) \leqslant -1/4$

Solution

$$\frac{4}{x}=-4 \Rightarrow x=-1, \frac{4}{x}=\frac{1}{-4} \Rightarrow x=-16$$

Since the value of $f(x)$ decreases steadily as x increases, the domain is $\underline{\{x : -16 \leqslant x \leqslant -1\}}$.

Example 14.3
A function f is defined by $f(x)=2x^2-2x-3$. Calculate the value of $f(0)$ and $f(-1)$ and hence obtain an estimate of the value of x between 0 and -1 for which $f(x)=0$.

Solution

$$f(0)=0-0-3=\underline{-3}, \qquad f(-1)=2\times 1-2\times -1-3=\underline{1}$$

Since the function value changes from negative to positive when x changes from 0 to -1 there must be a value of x between 0 and -1 for which the value of $f(x)$ is zero. The function value changes by 4 units (-3 to 1) while x changes by 1 unit (0 to -1) and so the required value of x at which $f(x)=0$ is estimated at 3/4 of the interval from 0 to -1, that is, $\underline{-0.8}$ to 1 significant figure.

14.2 MAPPING DIAGRAMS

(a) Arrow diagrams

Diagrams like those of Fig. 14.1 are called *arrow diagrams.*

(b) Parallel axes

The domain and range can also be represented as points or intervals on two parallel number lines.

Fig. 14.2 is a diagram of the mapping

$$f: x \to \frac{2-x}{2+x}, x \in \{-1, 1, 2\},\ f(-1) = 3,\ f(1) = \frac{1}{3}, f(2) = 0$$

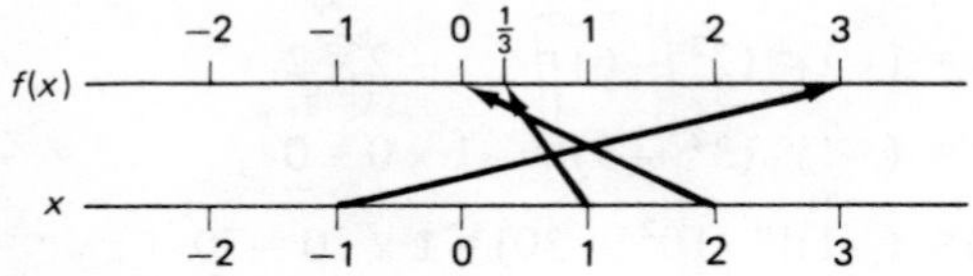

Fig 14.2

The mapping of Example 14.1(b) is illustrated in Fig. 14.3. The parallel axes may be drawn down the page instead of across, but this would take up more space on the printed page.

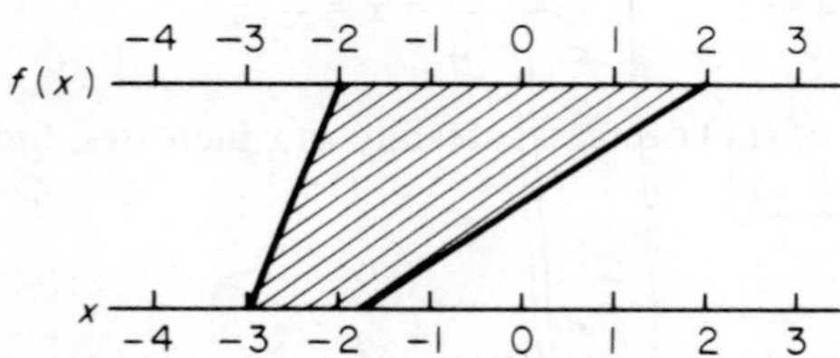

Fig 14.3

(c) Perpendicular axes

When the axes are drawn perpendicular to each other instead of parallel, each ordered pair of values $[x, f(x)]$ is represented by a point in the plane containing the axes. The scale need not be the same on both axes. Fig. 14.4 is a mapping diagram for Example 14.1(a).

When the domain of a function is an interval, a few points are plotted and joined by a smooth curve; the curve in the figure shows the value of the function $2x^2 - 3x + 4$ for values of x in the interval $-2 \leqslant x \leqslant 2$. It is called the Cartesian graph of the function, and graphs will be considered in the next chapter.

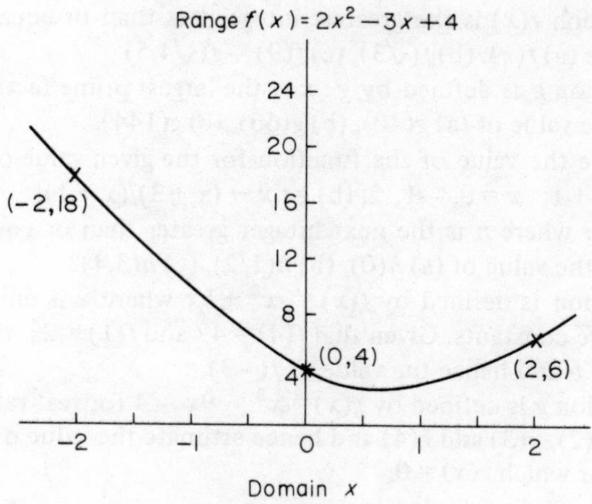

Fig 14.4

Exercise 14.1

(1) Fig. 14.5 shows a mapping f. Write down the element (i) $f(a)$, (ii) $f(b)$, (iii) $f(d)$.

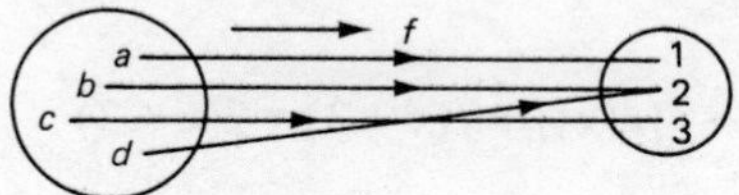

Fig 14.5

(2) Complete the statement for each of the mappings f, g, h

	Mapping	Domain	Range	Statement
(a)	f	1	4	$f: x \to$
		3	12	
		6	24	
(b)	g	4	1	$g: x \to$
		2	$\frac{1}{2}$	
		−1	$-\frac{1}{4}$	
(c)	h	−6	−4	$h: x \to$
		−3	−2	
		5	$3\frac{1}{3}$	

(3) A function f is defined by $f: x \to x^3$, where x is an integer in the interval $0 \leqslant x \leqslant 4$. List the corresponding set of elements in the range of the function.

(4) A function $f(x)$ is the greatest integer less than or equal to $x, x > 0$. Evaluate (a) $f(\pi)$, (b) $f(\sqrt{3})$, (c) $f(9) - f(\sqrt{4.5})$.
(5) A function g is defined by $g : x \rightarrow$ the largest prime factor of $x, x \in Z$. State the value of (a) $g(40)$, (b) $g(66)$, (c) $g(144)$.
(6) Calculate the value of the function for the given value of x. (a) $f(x) = 2x^3 - x + 1$, $x = 0, -1, 2$, (b) $g : x \rightarrow (x + 3)/(x + 1)$, $x = 6, 3, -2$.
(7) $h : x \rightarrow n$ where n is the next integer greater than or equal to $2x + 2$. What is the value of (a) $h(0)$, (b) $h(1/2)$, (c) $h(3.4)$?
(8) A function is defined by $f(x) = ax^2 + bx$ where x is an integer and a and b are constants. Given that $f(4) = 47$ and $f(1) = 2\frac{3}{4}$, find the values of a and b and hence the value of $f(-3)$.
(9) A function g is defined by $g(x) = x^3 - 9x - 4$ for real values of x. Calculate $g(2)$, $g(3)$ and $g(4)$ and hence estimate the value of x between 2 and 4 for which $g(x) = 0$.

14.3 COMBINING FUNCTIONS

There is only one operation for combining functions; it is a binary operation called *composition* of functions, and given the symbol $\circ$. The composite function $f \circ g$ is the function f of the function g so that $f \circ g : x \rightarrow f[g(x)]$.

For instance, if $f: x \rightarrow 2x$ and $g : x \rightarrow x + 3$ where x is a real number, then

$$f \circ g : x \rightarrow f[g(x)] = f(x + 3) = 2(x + 3)$$

and

$$g \circ f : x \rightarrow g[f(x)] = g(2x) = 2x + 3$$

The symbol $\circ$ is often omitted, and the function fg is taken to mean 'do g first, then f'. The method is shown in the next three examples.

Example 14.4
$f: x \rightarrow 2x + 1$ and $g : x \rightarrow x^2 + 3$ are functions defined on the set of real numbers. Find expressions for the composite function $f \circ f$ and $f \circ g$ and evaluate (a) $ff(2)$, (b) $fg(2)$, (c) $gf(3)$, (d) $gg(-1)$.

Solution

$$f \circ f : x \rightarrow f(2x + 1) = 2(2x + 1) + 1 = \underline{4x + 3}$$

$$f \circ g : x \rightarrow f(x^2 + 3) = 2(x^2 + 3) + 1 = \underline{2x^2 + 7}$$

(a) $ff(2) = 4 \times 2 + 3 = \underline{11}$ (b) $fg(2) = 2(2^2) + 7 = \underline{15}$

(c) $gf(3) = g(2 \times 3 + 1) = g(7) = 7^2 + 3 = \underline{52}$

(d) $gg(-1) = g(1 + 3) = g(4) = 4^2 + 3 = \underline{19}$

Example 14.5
$F: x \to x + 3$, $G: x \to 3/x$, $H: x \to 3x$, are functions defined on the domain $\{x : x \neq 0\}$. Find the value of (a) $FG(3)$, (b) $GF(3)$, (c) $FGH(4)$, (d) $HGF(6)$.
Zero is excluded from the domain because $G(0) = 3/0$ which cannot be evaluated.

Solution

(a) $FG(3) = F\frac{3}{3} = F(1) = 1 + 3 = \underline{4}$

(b) $GF(3) = G(3 + 3) = G(6) = \frac{3}{6} = \underline{\frac{1}{2}}$

(c) $FGH(4) = FG(3 \times 4) = FG(12) = F\left(\frac{3}{12}\right) = F\left(\frac{1}{4}\right) = \underline{3\frac{1}{4}}$

(d) $HGF(6) = HG(6 + 3) = HG(9) = H\left(\frac{3}{9}\right) = H\left(\frac{1}{3}\right) = \frac{1}{3} \times 3 = \underline{1}$

Example 14.6
Functions f and g are defined for real values of x by $f(x) = 3x + 2$ and $g(x) = mx + 1$. (a) Find the value of m for which $fg(x) = gf(x)$, (b) When m takes this value, for what value of x does $f(x) = g(2x)$?

Solution

(a) $fg(x) = f(mx + 1) = 3(mx + 1) + 2 = 3mx + 5$

$gf(x) = g(3x + 2) = m(3x + 2) + 1 = 3mx + 2m + 1$

These expressions are equal when $2m + 1 = 5$, or $\underline{m = 2}$.

(b) When $m = 2$, $g(x) = 2x + 1$ and $g(2x) = 2(2x) + 1 = 4x + 1$.

This is equal to $f(x)$ when $4x + 1 = 3x + 2$, or $\underline{x = 1}$.

14.4 INVERSE FUNCTIONS

The inverse of a function f is the function f^{-1} such that the composite functions $f \circ f^{-1}$ and $f^{-1} \circ f$ map each element of the domain to itself.
For instance, if

$f: x \to x + 1$, then $f^{-1} : x \to x - 1$

since

$$ff^{-1}(x) = f(x - 1) = x - 1 + 1 = x$$

and also, if

$$g : x \to 2x, \text{ then } g^{-1} : x \to \frac{x}{2}$$

since

$$g^{-1}g(x) = g^{-1}(2x) = \frac{2x}{2} = x$$

Not every function has an inverse. For example if $f(x) = x^2$, then $f(-x) = x^2$, and the inverse mapping has two different values in the range for each value in the domain. Thus in order for a function to possess an inverse function it must be a 1 : 1 mapping, each element in the domain being mapped to one and only one element in the range.

To find the inverse of a composite function $fg(x)$ the inverse operations are performed in the reverse order, that is, $g^{-1}f^{-1}(x)$.

Example 14.7

Find the inverse of the following function for the domain $x \in R, x \neq -2$. A flow diagram is used to show the method, but with practice the inverse function can usually be found directly.

Solution

(a) $f : x \to 3x - 2$

$$f : x \xrightarrow{\times 3} 3x \xrightarrow{-2} 3x - 2$$

$$\frac{x+2}{3} \xleftarrow{\div 3} x + 2 \xleftarrow{+2} x$$

The inverse is $f^{-1} : x \to \dfrac{x+2}{3}$.

(b) $f : x \to \dfrac{2x}{3}$

$$f : x \xrightarrow{\times 2} 2x \xrightarrow{\div 3} \frac{2x}{3}$$

$$\frac{3x}{2} \xleftarrow{\div 2} 3x \xleftarrow{\times 3} x$$

The inverse is $f^{-1}: x \rightarrow \dfrac{3x}{2}$.

(c) $f: x \rightarrow \dfrac{1}{2x+4}$ The value -2 is excluded since $f(-2)$ is $\dfrac{1}{0}$

$$f: x \xrightarrow{\times 2} 2x \xrightarrow{+4} 2x+4 \xrightarrow{\text{reciprocal}} \frac{1}{2x+4}$$

$$\frac{1}{2x} - 2 \xleftarrow{\div 2} \frac{1}{x} - 4 \xleftarrow{-4} \frac{1}{x} \xleftarrow{\text{reciprocal}} x$$

The inverse is $f^{-1}: x \rightarrow \dfrac{1}{2x} - 2$.

It is useful to check that $ff^{-1}: x \rightarrow x$, but checks will not be included in worked examples to save space.

When an inverse function exists, its domain and range are the same as the range and domain respectively of the original function.

Example 14.8

(a) If the function $f: x \rightarrow (x+1)^2$ is defined for $x \leqslant -1$, what is the value of $f^{-1}(4)$?

Solution The restriction on the domain makes f 1 : 1 so that the inverse function exists, and its domain is $\{x : x \geqslant 0\}$ since $(x+1)^2$ is non-negative

$$f: x \rightarrow (x+1)^2 \Rightarrow f^{-1}: x \rightarrow \sqrt{x} - 1$$

$$f^{-1}(4) = \sqrt{4} - 1 = -2 - 1 = \underline{-3} \qquad \text{since the range is } f^{-1}(x) \leqslant -1$$

(b) A function is defined by $f(x) = x^2 - 3, x \geqslant 0$. Find the inverse function f^{-1} and the value of $f^{-1}(-2)$ and $f^{-1}(1)$.

Solution The domain is restricted to ensure that the function has an inverse

$$f: x \rightarrow x^2 - 3 \Rightarrow f^{-1}: x \rightarrow \sqrt{(x+3)}, x \geqslant -3 \text{ and } f^{-1}(x) \geqslant 0$$

$$f^{-1}(-2) = \sqrt{(-2+3)} = \underline{1}; \quad f^{-1}(1) = \sqrt{(1+3)} = \underline{2}$$

Exercise 14.2

(1) $f: x \rightarrow 2x$, $g: x \rightarrow 3x - 1$, $h: x \rightarrow 4/x$, are functions defined on the domain $\{x : x \neq 0\}$. Find expressions for the composite functions (a) $fg(x)$, (b) $gh(x)$, (c) $hg(x)$, (d) $fh(x)$.

(2) Functions f and g are defined by $f: x \to x + 2$, $g: x \to ax - 1$, where x is a real number and a is a constant. Define the composite functions fg and gf.

(3) Find the inverse of the given function of x, $x \in \{\text{real numbers}\}$. (a) $2x$, (b) $x + 1$, (c) $3x/5$, (d) $(3x - 4)/2$, (e) $2x^3 + 4$.

(4) Find the inverse mapping of the following functions defined for $\{x : x > 3\}$. State whether each is a function and give its domain. (a) $f: x \to 4/(2x + 3)$, (b) $f: x \to 2/(x + 2)$, (c) $f: x \to \sqrt{x}$, (d) $f: x \to (3x - 1)^2$, (e) $f: x \to x^{-2}$.

(5) Given $f(x) = 2x - 1$, $g(x) = x/2$, $h(x) = 3(x + 1)$, find expressions for each of the following mappings in the form $x \to$. (a) f^{-1}, (b) gf^{-1}, (c) gh, (d) ff^{-1}, (e) $(fg)^{-1}$.

(6) If $f(x) = (3x + 1)/4$ and $gf(x) = x$, what is the function g?

14.5 THE FACTOR THEOREM

When a function $f(x)$ is divided by a linear function $x - a$, the remainder is equal to $f(a)$, the value of the function $f(x)$ when a is substituted for x. If $x - a$ is a factor of the function $f(x)$ then there is no remainder and $f(a) = 0$. This is known as the factor theorem and it is used to find factors of cubic and higher functions as a means of solving equations.

In general, if a polynomial function $f(x)$ is exactly divisible by $mx + c$, then $f(-c/m) = 0$.

Example 14.9
Show that $3x + 2$ is a factor of $3x^3 + 2x^2 - 24x - 16$.

Solution Let $f(x) = 3x^3 + 2x^2 - 24x - 16$

$$3x + 2 = 0 \Rightarrow x = -\tfrac{2}{3}$$

$$\begin{aligned} f(-\tfrac{2}{3}) &= 3(-\tfrac{2}{3})^3 + 2(-\tfrac{2}{3})^2 - 24(-\tfrac{2}{3}) - 16 \\ &= 3(-\tfrac{8}{27}) + 2(\tfrac{4}{9}) + 16 - 16 \\ &= -\tfrac{8}{9} + \tfrac{8}{9} + 16 - 16 \\ &= 0 \end{aligned}$$

Therefore <u>$3x + 2$ is a factor.</u>

Example 14.10
Show that $x - 2$ and $x + 1$ are both factors of the cubic function $f(x) = x^3 + x^2 - 4x - 4$ and hence factorise the expression.

Solution If $x-2$ is a factor then $f(2)=0$ and if $x+1$ is a factor $f(-1)=0$

$$f(2) \quad = 2^3 + 2^2 - 4(2) - 4 = 8 + 4 - 8 - 4 = 0$$

$$f(-1) = (-1)^3 + (-1)^2 - 4(-1) - 4 = -1 + 1 + 4 - 4 = 0$$

Therefore $\underline{x-2 \text{ and } x+1 \text{ are both factors of } f(x)}$.

Since a cubic function has three linear factors suppose the third factor is $x+a$. Then

$$x^3 + x^2 - 4x - 4 \equiv (x-2)(x+1)(x+a)$$

and since this is an identity the coefficients are the same on both sides. Considering the constant term, $-4 = (-2)(+1)(+a) = -2a \Rightarrow a = 2$. Hence $\underline{f(x) = (x-2)(x+1)(x+2)}$.

Example 14.11

Factorise the expression $x^4 - 5x^2 + 4$ and hence solve the equation $x^4 - 5x^2 + 4 = 0$.

Solution The factors of the constant term 4 are 1, 2, 4 and these are substituted in a systematic manner

$$f(x) \quad = x^4 - 5x^2 + 4$$

$$f(1) \quad = 1 - 5 + 4 = 0$$

$$f(-1) = 1 - 5 + 4 = 0$$

$$f(2) \quad = 16 - 5(4) + 4 = 0$$

$$f(-2) = 16 - 5(4) + 4 = 0$$

A polynomial of degree 4 cannot have more than 4 linear factors and therefore $x^4 - 5x^2 + 4 = (x-1)(x+1)(x-2)(x+2)$ and the solution set for $f(x) = 0$ is $\{\underline{1, -1, 2, -2}\}$.

Exercise 14.3

(1) Show that $(x-2)$ and $(x+1)$ are factors of the function $f(x)$ and find the third factor. (a) $f(x) = x^3 - 5x^2 + 2x + 8$, (b) $f(x) = x^3 + 2x^2 - 5x - 6$.

(2) Given that $2x^3 - 3x^2 - 2x + 3$ 9s exactly divisible by $x^2 - 1$, find the other factor and hence solve the equation $2x^3 - 3x^2 - 2x + 3 = 0$.

(3) What number must be subtracted from the expression $2x^3 - 3x^2 - 3x + 5$ to make $(x+1)$ and $(x-2)$ factors? Find the third factor of the new expression.

14.6 VARIATION

(a) Direct variation

If a mapping from x to y is defined by $y = Kx$ where K is a constant number, we say that y is directly proportional to x, or y varies directly as x. The symbol for proportional is $\propto$, and $y = Kx$ is the same mathematical statement as $y \propto x$. K is called the constant of proportion or the coefficient of variation, and its value is fixed by one pair of corresponding values of x and y [or an ordered pair (x, y)].

For example, the circumference of a circle is directly proportional to its diameter d, and the constant in this case is π (Section 13.1). $C \propto d$, and $C = \pi d$ are equivalent statements.

Example 14.12

The time of swing T of a pendulum varies as the square root of its length L. If T = 1 second when L = 25 cm calculate (a) the time of swing of a pendulum of length 36 cm, (b) the length of the pendulum when the time of swing is increased to 1.5 seconds, assuming that all other conditions remain constant.

Solution

$$T \propto \sqrt{L} \Rightarrow T = K\sqrt{L}$$

and substituting the given pair of values, $1 = K\sqrt{25} \Rightarrow K = 0.2$.

(a) When L = 36 cm, $T = 0.2\sqrt{36} = \underline{1.2 \text{ s}}$.

(b) When T = 1.5 s, $1.5 = 0.2\sqrt{L} \Rightarrow L = (1.5/0.2)^2 = \underline{56.2 \text{ cm}}$.

(b) Partial variation

When two variables x and y are connected by the general linear equation $y = ax + b$, y varies partly as x and is partly constant, and this is known as partial variation. Since there are two constants, a and b, two corresponding pairs of values of x and y are necessary in order to evaluate them.

Example 14.13

The speed v of a moving object is partly constant and varies partly as the time t. Given that v = 10 m/s when t = 0 and v = 50 m/s when t = 5 s, find the equation connecting v and t and hence the speed of the object after 3 seconds.

Solution Call the constants u and a, so that $v = u + at$, and substitute both pairs of values of v and t as in Section 8.4

$$10 = u + a \times 0 \Rightarrow u = 10$$

$$50 = 10 + a \times 5 \Rightarrow a = 8$$

Hence, the equation is $v = 10 + 8t$.

When $t = 3$s, $v = 10 + 8 \times 3 = 34$ m/s.

(c) Joint variation

When the value of a variable y is proportional to the product of variables p and x, y is said to vary jointly as p and x.

For example, in the formula for simple interest given in Section 5.6

$$I = \frac{PRT}{100}$$

I varies jointly as P, as R, and as T and the constant of proportion is 1/100. A change in any one of the variables P, R, or T causes a change in the value of the interest I.

Example 14.14

The kinetic energy E of an object varies jointly as its mass m and as the square of its speed v. If a mass of 20 kg has 1000 units of kinetic energy at a speed of 10 m/s calculate the constant of proportion, and hence the speed at which the kinetic energy of a 30 kg mass reaches 3375 units.

Solution Since the energy varies jointly as m and v^2

$$E \propto mv^2 \quad \text{or} \quad E = Kmv^2$$

Substituting the given values $E = 1000$, $m = 20$, $v = 10$

$$1000 = K \times 20 \times 100 \Rightarrow K = \tfrac{1}{2} \text{ and so } E = \tfrac{1}{2} mv^2$$

When $E = 3375$ and $m = 30$ kg

$$3375 = \frac{1}{2} \times 30 \times v^2 \Rightarrow v = \sqrt{\frac{3375}{15}} = 15$$

Therefore the speed is 15 m/s.

(d) Inverse variation

When one quantity varies inversely as another, as one increases the other decreases so that their product is constant

$$y \propto \frac{1}{x}, \; y = \frac{K}{x}, \text{ or } xy = K$$

where K is constant are all equivalent to the statement 'y varies inversely as x'.

Example 14.15
Assuming that the length of paper on a roll of fixed dimensions varies inversely as the thickness of the paper (a) calculate the constant of proportion if the roll holds 100 m of paper 0.75 mm thick, (b) calculate the thickness of paper on the roll when the length is 80 m.

Solution (a) If the length is L metres and the thickness T mm, $L \propto 1/T$

$$LT = K \Rightarrow K = 100 \times 0.75 = \underline{75}$$

(b) New length 80 m, $80T = 75 \Rightarrow T = 75/80 = 0.94$. The thickness is $\underline{0.94 \text{ mm}}$.

Example 14.16
The property R of a uniform wire varies directly as its length L and inversely as the square of its radius r. Given that $R = 6$ units when $L = 180$ and $r = 3 \times 10^{-2}$, calculate (a) the value of R when L is trebled and r doubled, (b) the value of L when $R = 3.5$ and $r = 2.4 \times 10^{-2}$.

Solution

$$R \propto \frac{L}{r^2} \quad \text{or} \quad R = K \times \frac{L}{r^2} \quad \text{where } K \text{ is a constant}$$

To find K, substitute the values $R = 6, L = 180, r = 3 \times 10^{-2}$. Transposing the equation for K

$$K = \frac{Rr^2}{L} = \frac{6 \times 9 \times 10^{-4}}{180} = 3 \times 10^{-5}$$

(a) $R = KL/r^2$; when $L_1 = 3L$, and $r_1 = 2r$

$$R_1 = K\frac{L_1}{{r_1}^2} = K\frac{3L}{4r^2} = \frac{3}{4}R = \underline{4.5 \text{ units}}$$

(b) Transposing for L

$$L = \frac{Rr^2}{K} = \frac{3.5 \times 2.4^2 \times 10^{-4}}{3 \times 10^{-5}} = \underline{67.2 \text{ units}}$$

Exercise 14.4

(1) If $y \propto x^2$ and $y = 54$ when $x = 3$, find (a) the value of y when $x = 5$, (b) the value of x when $y = 24$.

(2) If y is inversely proportional to x and when $x = 2.5$, $y = 20$, what is the value of x when $y = 4$?

(3) P is inversely proportional to s^2. If $P = 32$ when $s = 1/2$, find (a) the constant of proportion, (b) the value of P when $s = -3$.

(4) A quantity z varies directly as x and inversely as y^2. If $z = 3$ when $x = 12$ and $y = 2$, find the value of z when $x = 4$ and $y = 1/4$.

(5) The number of revolutions made per minute by a wheel varies as the speed of the vehicle and inversely as the diameter of the wheel. If a wheel of diameter 60 cm makes 100 revolutions per minute at a speed of 11.3 km/h, calculate the number of revolutions made per minute by a wheel of 50 cm diameter when the speed is 15 km/h.

(6) The area A of a triangle is directly proportional to the length of its base, b, and its height h. If $A = 90$ when $b = 18$ and $h = 10$, what is the height of a triangle with a base of length 30 and area 115?

(7) F is directly proportional to the square root of T, but varies inversely as the square root of M. Given that $F = 25$ when $T = 20$ and $M = 5$, calculate (a) the value of F when $T = 50$ and $M = 8$, (b) the value of M when $F = 80$ and $T = 10$.

(8) The distance s metres travelled by a particle is partly proportional to the time t seconds and partly proportional to t^2. Given that $s = 32$ metres when $t = 2$ seconds, and $s = 170$ metres when $t = 5$ seconds, calculate the value of t when $s = 112$ metres.

PROGRESS TEST 3

A

Where necessary take π as 22/7.
(1) In a rhombus PQRS (a) calculate QS when PR = 24 cm and QR = 20 cm. (b) If $\angle$ PSQ = 53.1°, what size is $\angle$ PRQ? (2) ABC is an equilateral triangle of side 12 cm and D and E are points on BC. Calculate the length (a) AD when BD = 6 cm, (b) AE when BE = 4 cm. (3) How many sides has a regular polygon if the exterior angles are 22.5°? (4) Calculate the area of (a) a circle of circumference 220 mm (b) a trapezium of height 12 cm and parallel sides of 8 cm and 14 cm, (c) the space between concentric circles of diameter 18 cm and 10 cm. (5) Find the area of a sector containing 42° in a circle of radius 35 mm. (6) Calculate the volume of (a) a sphere of diameter 7 cm, (b) a cone of base radius 24 cm and slant height 25 cm, (c) a metal rod of length 40 cm and diameter 2.1 cm. (7) If the area of a circle is 616 mm², what is the radius? (8) Construct a triangle XYZ having XY = 72 mm, XZ = 45 mm and $\angle$ ZXY = 60°. Draw the perpendicular from Z to XY. (a) Measure YZ. (b) Calculate the area XYZ. (9) The variable X is inversely proportional to Y^2 and $X = 5\frac{1}{16}$ when $Y = 2$. Calculate the value of Y when $X = 2\frac{1}{4}$. (10) Find the ratio of the volumes of two spheres when the diameters are 10 cm and 0.5 cm. (11) Show that $(x - 2)$ and $(x + 1)$ are both factors of $2x^3 - 3x^2 - 3x + 2$ and find the third factor. (12) (a) $f(x) = 2x + 4$, $g(x) = (x + 1)/2$. Find the value of (i) $fg(2)$, (ii) $f^{-1}(6)$, (iii) $fg^{-1}(1)$, (b) $f\colon x \to 3x + 2$, and $g\colon x \to 2x + k$. Find a value of k such that $fg(x) = gf(x)$.

B

(1) Calculate the adjacent angles of a parallelogram when one is four times the other. (2) The lines AC and BD bisect each other at right angles, AC is of length 100 mm and BD 120 mm. What is the length of AD? (3) The parallel sides of a trapezium are AB = 12 cm and DC = 24 cm, AD = 13 cm and BC is perpendicular to CD. Calculate (a) BC, (b) the area ABCD. (4) Calculate the total area of an equilateral triangle of side 12 cm and a semicircle drawn on one side of the triangle as diameter. (5) Which of these plane figures has only one line of symmetry: (a) parallelogram, (b) isosceles trapezium, (c) kite, (d) rhombus? (6) Calculate the area of the segment cut off by a chord subtending an angle of 60° at the centre of a circle of radius 10 cm. (7) Calculate the volume of concrete 0.1 m thick in a path 1.5 m wide surrounding a rectangular pool 7 m by 12 m. (8) Calculate (a) the surface area, (b) the volume of a cone of base diameter 7 cm and height 4 cm. (9) $P \propto Q/R^2$ and $P = 56$ when $Q = 16$ and $R = 2$. What is the value of R when $P = 14$ and $Q = 18$; (10) On a map drawn to a scale 1 : 20 000 a lake has an area 90 mm². Calculate the actual area. (11) A triangle ABC has BC = 80 mm and a point X on AB is such that BX = 40 mm and AX = 60 mm. A line through X, parallel to AC, meets BC at Y. (a) Calculate BY. (b) XY = 24 mm, calculate AC. (c) If AY and CX intersect at Z and the area of triangle XYZ is P mm², calculate in

terms of P the area of triangle ACZ. **(12)** Calculate the length of a cylinder when the volume is 2310 cm^3 and the radius of its cross section is 7 cm.

CHAPTER 15

CARTESIAN GRAPHS OF FUNCTIONS

15.1 CARTESIAN COORDINATES

An algebraic equation in the form $y = f(x)$, such as $y = x^2 - 2x + 3$, is called a *Cartesian* equation (after the French mathematician Descartes), and a mapping diagram referred to perpendicular axes is the Cartesian graph of the function $f(x)$. Since the value of y is calculated from a given value of x, x is the independent variable and y is the dependent variable.

The x-axis is drawn parallel to the top of the page and each point in the plane of the page can be represented by an ordered pair of values (x, y), called the *Cartesian coordinates* of the point.

The x-coordinate is called the abscissa and is the distance of the point from the y-axis; the y-coordinate is called the ordinate of the point and is its distance from the x-axis.

In Fig. 15.1 the point of intersection of the axes $(0, 0)$ is called the origin of coordinates. The point A is 3 units to the right of the y-axis and 2 units above the x-axis, and so it is written A$(3, 2)$. The point B$(-2, 4)$ is 2 units to the left of the y-axis and its abscissa is -2. Similarly the point C has coordinates $(-3, -2)$ and D is the point $(2, -4)$.

Example 15.1

(a) Draw Cartesian axes and plot the points A$(3, 2)$, B$(2, 5)$, C$(-1, 3)$, D$(-4, 2)$, E$(-4, -3)$ and F$(2, -4)$.

(b) What are the coordinates of the mid-point of BF and the mid-point of AD?

Solution (a) The scale must extend 5 units from the origin on the y-axis since the greatest ordinate is 5; the points are marked in Fig. 15.2.

(b) The point B is 5 units above the x-axis and the point F is 4 units below it; the mid-point is therefore $\frac{1}{2}$ unit above the axis and its coordinates are $(2, \frac{1}{2})$.

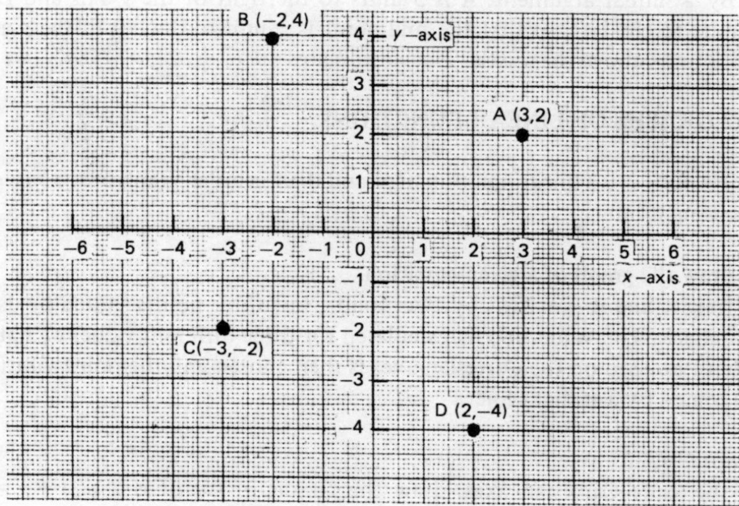

Fig 15.1

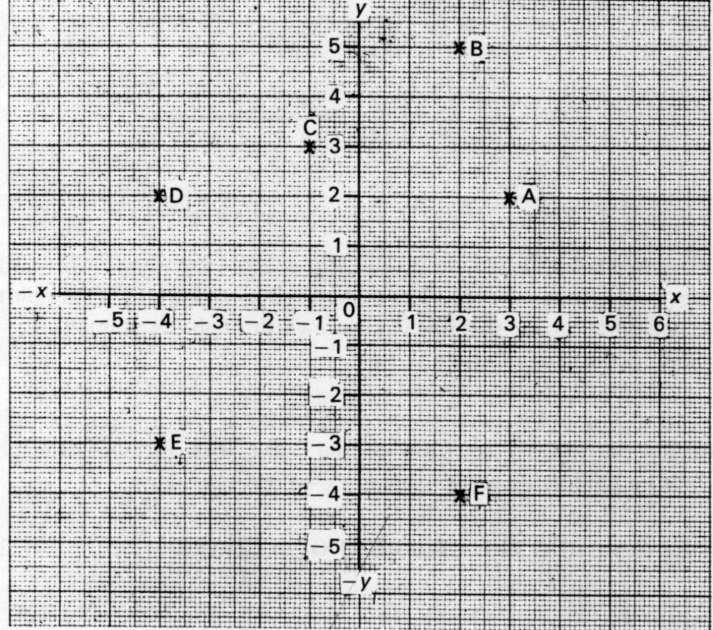

Fig 15.2

By a similar argument, A is 3 units to the right of the y-axis and D is 4 units to the left, so that the mid-point is $\frac{1}{2}$ unit to the left of the y-axis and 2 units above the x-axis. The coordinates are $\underline{(-\frac{1}{2}, 2)}$.

15.2 GRAPHS OF LINEAR FUNCTIONS

$f(x) = mx + c$ is a linear function of x, and the Cartesian graph of $y = mx + c$ is a straight line.

To draw the graph of a given equation

A straight line is specified by two points, but three points should be used as a check. The value of y is calculated for three different values of x, the scale to be used on the Cartesian axes is then chosen and the points plotted and joined by a straight line.

Example 15.2

On the same Cartesian axes, draw the graphs of the lines (1) $2y + x = 4$ (2) $y = 2x + 1$, marking the scale on the x-axis from -3 to $+3$.

Solution $2y + x = 4$ is rearranged in standard form as $y = -\frac{1}{2}x + 2$ and the values are arranged in a table.

(1) $y = -\frac{1}{2}x + 2$

x	-2	0	2
$-\frac{1}{2}x$	1	0	-1
$+2$	2	2	2
y	3	2	1

(2) $y = 2x + 1$

x	-2	0	2
$2x$	-4	0	4
$+1$	1	1	1
y	-3	1	5

The axes are shown in Fig. 15.3, with the x-axis extending from -3 to 3 and the y-axis from -3 to 5. The points $(-2, 3)$, $(0, 2)$, $(2, 1)$ are plotted and joined to form line (1); line (2) is drawn through the points $(-2, -3)$, $(0, 1)$ and $(2, 5)$.

Comment If a point with given coordinates is on a given curve, the coordinates satisfy the equation of the curve. For instance, the point $(6, -1)$ is on the line $2y + x = 4$ since $2 \times -1 + 6 = 4$.

15.3 DETERMINING THE GRADIENT OF A GRAPH

(a) Straight line

The gradient of a line is defined as the increase in the ordinate divided by the increase in the abscissa between two points on the line. Since this

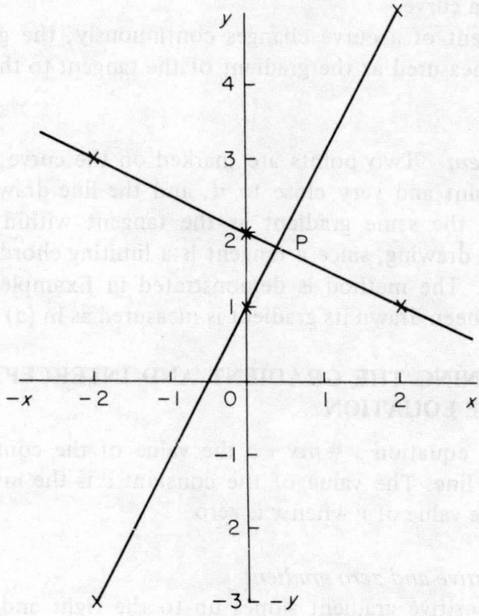

Fig 15.3

gradient is the same for any two points on the line, convenient values of x are chosen for easy division and the corresponding values of y are read from the graph.

In Fig. 15.4 the gradient is

$$\frac{7.5 - 4}{6 - 2} = \frac{3.5}{4} \text{ or } \frac{7}{8}$$

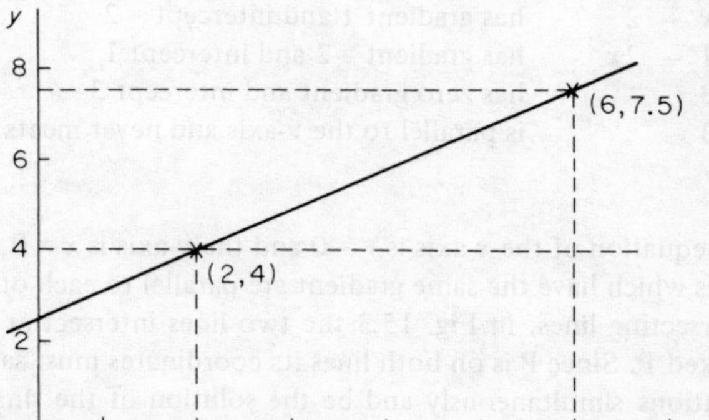

Fig 15.4

(b) Gradient of a curve

Since the gradient of a curve changes continuously, the gradient at any given point is measured as the gradient of the tangent to the curve at that point.

Drawing a tangent Two points are marked on the curve, one each side of the given point and very close to it, and the line drawn through the two points has the same gradient as the tangent within the limits of accuracy of the drawing, since a tangent is a limiting chord when the two points coincide. The method is demonstrated in Example 15.10. When the tangent has been drawn its gradient is measured as in (a) above.

15.4 DETERMINING THE GRADIENT AND INTERCEPT OF A LINE FROM THE EQUATION

In the standard equation $y = mx + c$ the value of the constant m is the gradient of the line. The value of the constant c is the intercept on the y-axis, that is the value of y when x is zero.

(i) Positive, negative and zero gradient

A line with a positive gradient slopes up to the right and a line with a negative gradient slopes down to the right, as shown in Fig. 15.5(a). Lines which are parallel to the x-axis have zero gradient.

(ii) Positive, negative and zero intercept

A line with a positive intercept meets the y-axis above the origin, a line with a negative intercept cuts the y-axis below the origin, and lines which pass through the origin of coordinates have zero intercept.

In Fig. 15.5(b)

$y = x - 2$	has gradient 1 and intercept -2
$y = 1 - 2x$	has gradient -2 and intercept 1
$y = 3$	has zero gradient and intercept 3
$x = 3$	is parallel to the y-axis and never meets it

Comment

(i) The equation of the x-axis is $y = 0$ and the y-axis is $x = 0$,

(ii) Lines which have the same gradient are parallel to each other,

(iii) Intersecting lines: in Fig. 15.3 the two lines intersect at the point marked P. Since P is on both lines its coordinates must satisfy both equations simultaneously and be the solution of the simultaneous equations $2y + x = 4$ and $y = 2x + 1$. This illustrates a graphical method of solving simultaneous equations which is discussed later in Section 15.6.

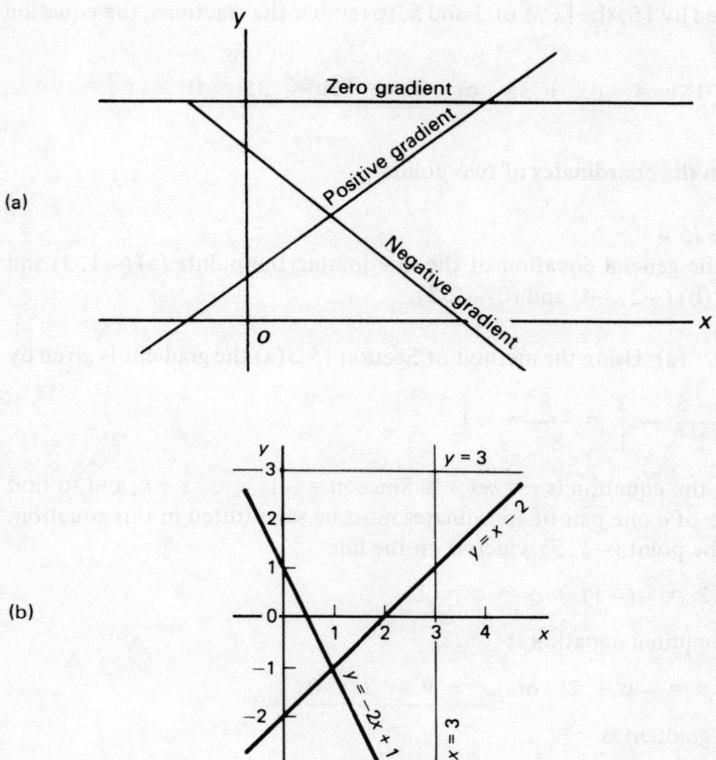

Fig 15.5

15.5 DETERMINING THE EQUATION OF A STRAIGHT LINE

(a) Given the gradient and intercept

The method is demonstrated in the following example.

Example 15.3

Determine the equation of the straight line having (a) intercept −2 and gradient −1/2, (b) intercept $2\frac{1}{3}$ and gradient −2/5.

Solution The standard equation is $y = mx + c$.

(a) When $c = -2$ and $m = -1/2$, $y = -\frac{1}{2}x - 2$. Multiplying each term by 2 removes the fraction and gives integral coefficients; the equation is

$$2y = -x - 4 \quad \text{or} \quad \underline{x + 2y + 4 = 0}$$

(b) Substituting $c = 2\frac{1}{3}$ and $m = -2/5$, $y = -\frac{2}{5}x + 2\frac{1}{3}$. Each term is multiplied by 15, the LCM of 3 and 5, to remove the fractions; the equation becomes

$$15y = -6x + 35 \quad \text{or} \quad \underline{6x + 15y - 35 = 0}$$

(b) Given the coordinates of two points

Example 15.4
Obtain the general equation of the line joining the points (a) $(-1, 3)$ and $(7, -5)$, (b) $(-2, -4)$ and $(0, -2/3)$.

Solution (a) Using the method of Section 15.3(a) the gradient is given by

$$\frac{-5 - 3}{7 - -1} = \frac{-8}{8} = -1$$

Suppose the equation is $y = mx + c$. Since $m = -1$, $y = -x + c$, and to find the value of c one pair of coordinates must be substituted in this equation. Taking the point $(-1, 3)$ which is on the line

$$3 = -(-1) + c \Rightarrow c = 2$$

and the required equation is

$$y = -x + 2 \quad \text{or} \quad \underline{x + y - 2 = 0}$$

(b) The gradient is

$$\frac{-2/3 - (-4)}{0 - (-2)} = \frac{3\frac{1}{3}}{2} = \frac{5}{3}$$

Since $x = 0$ when $y = -2/3$ [the point $(0, -2/3)$ is on the line] the intercept is $-2/3$, and the equation is

$$y = \tfrac{5}{3}x - \tfrac{2}{3} \quad \text{or} \quad \underline{5x - 3y - 2 = 0}$$

Exercise 15.1

(1) Draw Cartesian axes and mark the x-axis from 0 to 5 and the y-axis from 0 to 12, using a scale of 2 cm for 1 unit of x and 1 cm for 1 unit of y. Plot the points (1, 5), (3, 9), (4, 11) and join them by a straight line. From your graph find (a) the value of y when $x = 2$, (b) the value of x when $y = 8$, (c) the value of y at which the line cuts the y-axis, (d) the gradient of the line.

(2) Where necessary rewrite the equation in standard form and state (i) the gradient of the line, (ii) the intercept on the y-axis, (iii) the coordinates of the point at which the line cuts the x-axis.

(a) $y = 2x + 5$, (b) $y = 2 - 4x$, (c) $2y = 4x - 5$,
(d) $2y + 3x = 4$, (e) $5x - 7y = 8$, (f) $2x - y = 1/2$.

(3) State which of the lines in Question (2) are parallel to each other.

(4) Determine whether the given point is on the given line.
(a) $(1, -1)$ $y = 2x - 3$, (b) $(-2, 1)$ $2y = 3x + 8$,
(c) $(0, 0)$ $3y + 2x = 1$, (d) $(0.4, 1.2)$ $y - 3x = 0$.

(5) Find (i) the gradient, (ii) the equation with integral coefficients, of the line joining the given points. (a) $(2, 3)$ and $(-4, 5)$, (b) $(0, 0)$ and $(2, -4)$, (c) $(2, -3)$ and $(-1, -6)$, (d) $(2, 4)$ and $(5, -4)$.

(6) What is the equation of the line AB which has gradient $-3/4$ and passes through the point $(3, -1)$? Find the equation of a line parallel to AB which passes through the origin.

15.6 THE GRAPHICAL SOLUTION OF SIMULTANEOUS LINEAR EQUATIONS AND INEQUALITIES

Example 15.5

Use a graphical method to solve the simultaneous equations $y = 3x + 1$ and $y = x + 4$. Represent, by shading out, the regions

$$A = \{(x, y): y \leqslant 3x + 1\} \quad \text{and} \quad B = \{(x, y): y \geqslant x + 4\}.$$

Solution

(1) $y = 3x + 1$

x:	-2	0	2
y:	-5	1	7

(2) $y = x + 4$

x:	-2	0	2
y:	2	4	6

The graphs of the two lines are shown in Fig. 15.6

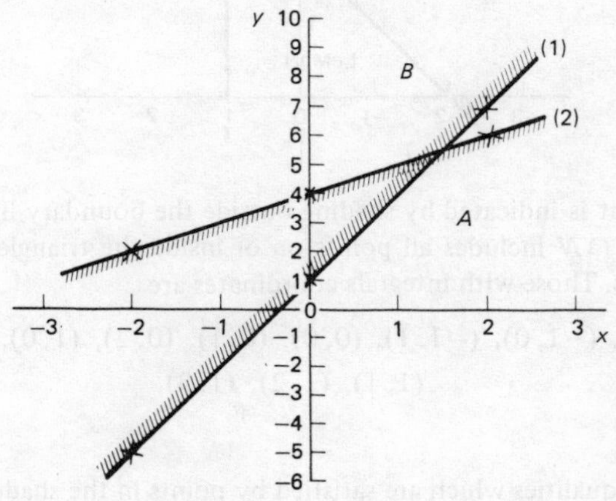

Fig 15.6

From the graph the point of intersection of the lines is (1.5, 5.5) and so the solution of the simultaneous equations is $x = 1.5, y = 5.5$.

The region A contains all the points on the line (1) or below it and the shading out is above the line.

The region B includes all points on line (2) and above it, and the shading is below the line. A boundary which is not included is sometimes shown by a broken line.

Example 15.6

$$L = \{(x,y): y \leqslant x + 2\}, \quad M = \{(x,y): y \geqslant 0\}, \quad N = \{(x,y): x \leqslant 1\}$$

On the same Cartesian axes draw the graphs of two straight lines and indicate, by shading out, the region representing the set $L \cap M \cap N$. List the ordered pairs (x, y) in this set such that x and y are integers.

Solution The region L includes all the points on or below the line $y = x + 2$, and this is one of the boundary lines to be drawn.

The region N includes points on the line $x = 1$ and to the left of it.

The boundary line for region M is the x-axis, $y = 0$.

To plot the graph of $y = x + 2$ values of y are calculated for three values of x. $x = 1 \Rightarrow y = 3$, $x = 0 \Rightarrow y = 2$, $x = -2 \Rightarrow y = 0$. $x = 1$ is parallel to the y-axis and passes through (1, 0) and the two lines are shown on the graph in Fig. 15.7.

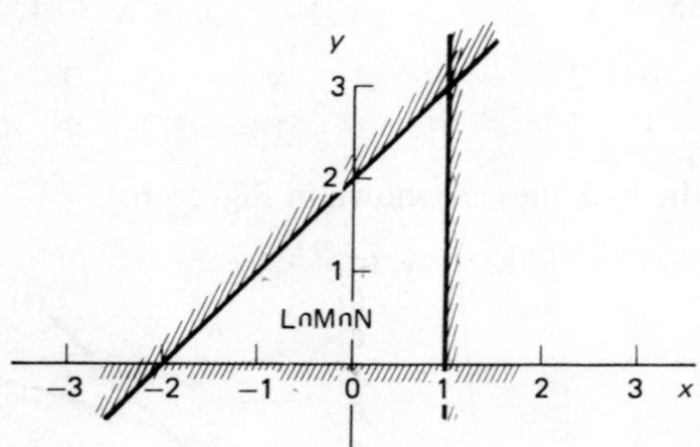

Fig 15.7

Shading out is indicated by shading outside the boundary lines, and the region $L \cap M \cap N$ includes all points on or inside the triangle formed by the three lines. Those with integrals coordinates are

$$(-2, 0),\ (-1, 0),\ (-1, 1),\ (0, 0),\ (0, 1),\ (0, 2),\ (1, 0),$$
$$(1, 1),\ (1, 2),\ (1, 3).$$

Example 15.7

State four inequalities which are satisfied by points in the shaded region of Fig. 15.8.

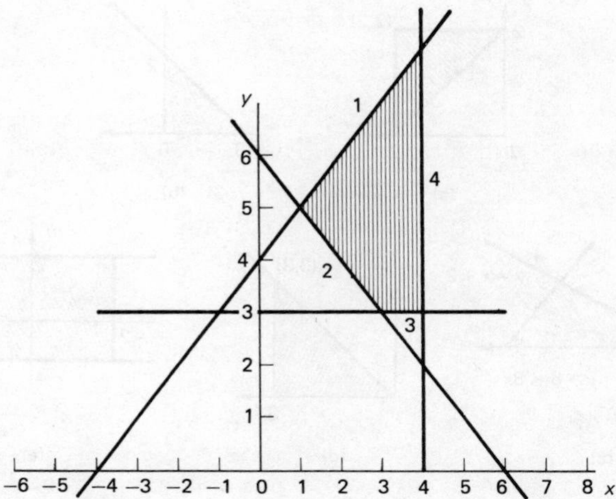

Fig 15.8

Solution The shaded region is a quadrilateral, and we must find the equations of the four boundary lines marked (1), (2), (3) and (4) in the diagram.

Line (1) passes through $(-4, 0)$ and $(0, 4)$ and using the method of Section 15.4(b) the gradient is $4 \div 4 = 1$ and the intercept is 4. The equation is $\underline{y = x + 4}$.

Line (2) passes through $(0, 6)$ and $(6, 0)$. Its gradient is $6 \div -6 = -1$ and the intercept is 6. The equation is $\underline{y = 6 - x}$.

Line (3) is parallel to the x-axis and has equation $\underline{y = 3}$.

Line (4) is parallel to the y-axis and its equation is $\underline{x = 4}$.

The shaded region is below (1), above (2), above (3), and to the left of (4) and if the boundary lines are included it is defined by the simultaneous inequalities

$$\underline{y \leqslant x + 4, \quad y \geqslant 6 - x, \quad y \geqslant 3, \quad x \leqslant 4}$$

Exercise 15.2

(1) Solve the simultaneous equations by a graphical method.
(a) $3y + x = 7$, $y - x = 5$, (b) $2y + 2x = 1$, $y + 4x = 4$.

(2) By drawing suitable straight lines on the same axes, find all the points with integral coordinates satisfying the simultaneous inequalities $0 < x < 5$, $0 < y < 4$, $x + 2y \geqslant 9$.

(3) Given that $x \geqslant 0, y \geqslant 0$, and $3x + 2y \leqslant 6$, find the maximum value of x.

(4) For each of the graphs sketched in Fig. 15.9 write down the inequalities which define the shaded region, including the boundary.

(5) Using the same Cartesian axes, draw the graphs of suitable boundaries

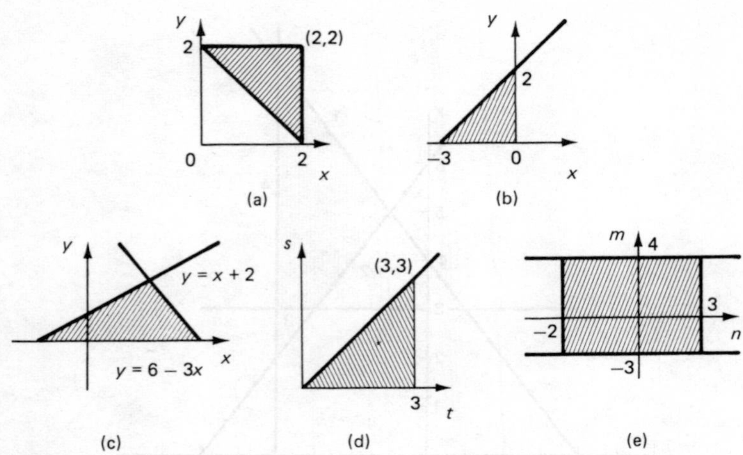

Fig 15.9

for the inequalities $y > x, y < 2x, x + y < 6$. Indicate by shading out the region defined by the inequalities, and if x and y are integers, state the set of ordered pairs included in this region.

(6) $A = \{(x, y): x + y \geqslant 1\}$ and $B = \{(x, y): y \leqslant x \leqslant 1\}$. Draw a graph and shade in the region defined by $A \cap B$ and state the coordinates of the points of intersection of the boundaries.

(7) Find the set $A \cap B$ when $A = \{(x, y): 2x - 3y = 3\}$ and $B = \{(x, y): x - y = 2\}$.

(8) Using the same axes and scales, draw the graphs of $x + y = 6$ and $x + 2y = 8$, and shade in the area for which $x + y < 6, x + 2y < 8, x > 0, y > 2$. List the points with integral coordinates which satisfy all these inequalities simultaneously.

15.7 THE GRAPHS OF QUADRATIC FUNCTIONS

The graph of a non-linear function is a curve, and more than three points must be plotted, preferably at least six. A table of values of x and y is constructed first, before the scales on the axes are chosen, and when the points have been plotted they are joined by a smooth curve.

The graph of a general quadratic function $f(x) = ax^2 + bx + c$ has a characteristic shape called a parabola.

Example 15.8

Draw the graphs of $y = (2x + 3)(x - 2)$ for values of x from -3 to 3. From the graph find (a) the values of x for which $(2x + 3)(x - 2) \geqslant 0$, (b) the solution of the equation $2x^2 - x = 11$.

Solution

x	-3	-2	-1	0	1	2	3
$2x+3$	-3	-1	1	3	5	7	9
$x-2$	-5	-4	-3	-2	-1	0	1
y	15	4	-3	-6	-5	0	9

Since the y values vary from 15 to -6, the x-axis is drawn nearer the bottom of the page, and the scales should be chosen so that the graph is as large as is convenient on the page.

The seven points are then plotted on the graph and joined by a smooth curve, using a hard pencil with a sharp point. Drawing a smooth curve free-hand takes a lot of practice, but a Flexicurve or a set of standard curves can be helpful. Fig. 15.10 represents a sketch of the graph.

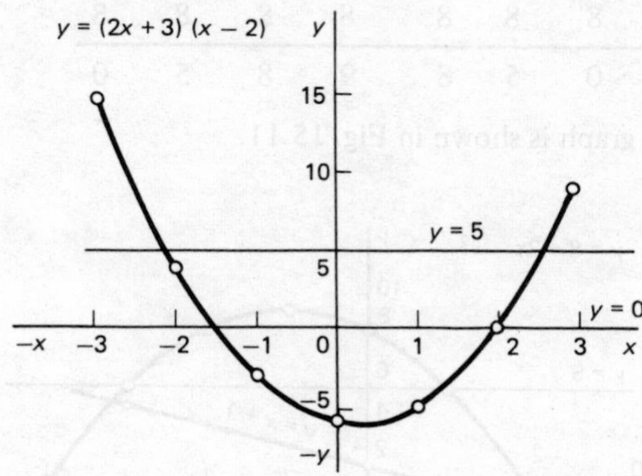

Fig 15.10

(a) The curve crosses the x-axis, $y = 0$ at two points, when $x = 2$ and when $x = -1.5$, and these are the solutions of the quadratic equation $(2x + 3)(x - 2) = 0$. The curve is above the x-axis for $x > 2$ and for $x < -1.5$, and therefore the solution of the inequality

$$(2x + 3)(x - 2) \geqslant 0 \quad \text{is} \quad \underline{\{x: x \leqslant -1.5, x \geqslant 2\}}$$

(b) To solve the equation $2x^2 - x = 11$ we must compare it with the equation of the curve we have drawn.

$$(2x + 3)(x - 2) = 2x^2 - x - 6$$

and we have drawn the graph of

$$y = 2x^2 - x - 6 \quad \text{or} \quad 2x^2 - x = y + 6$$

$$2x^2 - x = 11 \Rightarrow y + 6 = 11 \quad \text{or} \quad y = 5$$

The line $y = 5$ is drawn on the graph and the values of x at which it cuts the curve are found to be -2.15 and 2.6 and these are the solutions of the quadratic equation $2x^2 - x = 11$.

$$x = -2.15 \quad \text{or} \quad x = 2.6.$$

Example 15.9

Draw the graph of the function $f(x) = 8 + 2x - x^2$ for values of x from -3 to 4 and use the graph to solve (a) $8 + 2x - x^2 \geqslant 5$, (b) $7 + x - x^2 = 0$.

Solution Let $y = f(x)$ and construct a table of values of x and y

x	-3	-2	-1	0	1	2	3	4
$2x$	-6	-4	-2	0	2	4	6	8
$-x^2$	-9	-4	-1	0	-1	-4	-9	-16
$+8$	8	8	8	8	8	8	8	8
y	-7	0	5	8	9	8	5	0

A sketch of the graph is shown in Fig. 15.11.

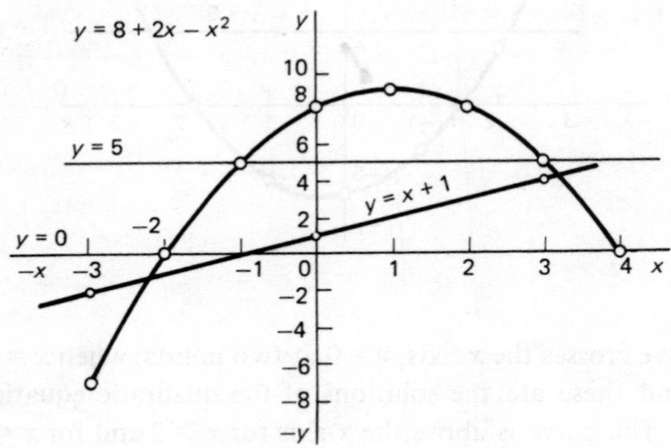

Fig 15.11

(a) We have drawn the graph of $y = 8 + 2x - x^2$, and $8 + 2x - x^2 \geqslant 5$ is satisfied by points on the curve for which $y \geqslant 5$.

The line $y = 5$ cuts the curve when $x = -1$ and 3 and points on the curve above this line have values of x between -1 and 3. The solution of $8 + 2x - x^2 \geqslant 5$ is $-1 \leqslant x \leqslant 3$.

(b) To solve the equation $7 + x - x^2 = 0$ it must be compared with the equation $y = 8 + 2x - x^2$

$$8 + 2x - x^2 = 7 + x - x^2 + x + 1, \text{ and } 7 + x - x^2 = 0 \Rightarrow y = x + 1$$

The straight line $y = x + 1$ is drawn on the same graph and the values of x

at which it cuts the curve $y = 8 + 2x - x^2$ are found to be -2.2 and 3.2.

Hence, the solution of $7 + x - x^2 = 0$ is $x = \underline{-2.2 \text{ or } 3.2}$.

Comment (i) In Fig. 15.10 the parabola has a lowest or minimum point, while in Fig. 15.11 it is the other way up and has a maximum point. The difference is in the coefficient of x^2, which was positive in the first example and negative in the second

(ii) When the graphs of two equations are drawn on the same axes, the values of x at which they intersect are the solutions of the simultaneous equations. For example, if the graphs of $y = x^2 + 3x - 2$ and $y = 3 + 2x - x^2$ are drawn they intersect in two points, and the values of x must satisfy the equation $x^2 + 3x - 2 = 3 + 2x - x^2$ or $2x^2 + x - 5 = 0$. The corresponding values of y can be read from the graph or calculated from one of the equations.

Exercise 15.3

(1) Draw the graph of the given function and use the graph to solve the equation $f(x) = 0$. (a) $f(x) = x^2 + 2x - 3$, (b) $f(x) = x^2 - 3x - 2$, (c) $f(x) = 8x^2 + 2x - 1$, (d) $f(x) = 2 - 3x - x^2$.

(2) Use a graphical method to find the minimum value of the function and the gradient of the curve when $x = 3$. (a) $3x^2 - 5x - 6$, (b) $x^2 - 4$.

(3) Draw the graph of $y = 2x^2 - 3x - 2$ for $-2 \leqslant x \leqslant 3$ and use your graph to solve (a) $2x^2 - 3x - 2 = 0$, (b) $2x^2 - 3x - 2 < 0$, (c) $2x^2 - 3x + 1 = 0$, (d) $2x^2 - 4x + 2 = 0$.

(4) Given $f(x) = 6 + 5x - 3x^2$, plot a graph of the function for the domain $\{x: -1 \leqslant x \leqslant 3\}$ and estimate (a) the greatest positive value of the function, (b) the roots of the equation $5 + 2x - 3x^2 = 0$, (c) the range of values of x for which $2 - 3x^2$ has a positive value in the given domain.

(5) On the same Cartesian axes draw the graphs of the sets $A = \{(x, y): y = x^2; -2 \leqslant x \leqslant 5\}$, $B = \{(x, y): y = 3x + 4\}$ and list the members of the set $A \cap B$.

15.8 THE GRAPHS OF CUBIC AND RECIPROCAL FUNCTIONS

(a) Cubic functions

The graph of a general cubic equation $y = ax^3 + bx^2 + cx + d$ has one maximum point and one minimum point. If the equation $y = 0$ has real solutions the curve cuts the x-axis at three points (not necessarily different) and the y-axis at one point.

In Fig. 15.12(a) the coefficient of x^3 is positive, the graph slopes up to the right and the maximum point is at the lower value of x. When the coefficient of x^3 is negative, the maximum point is at the higher value of x [Fig. 15.12(b)].

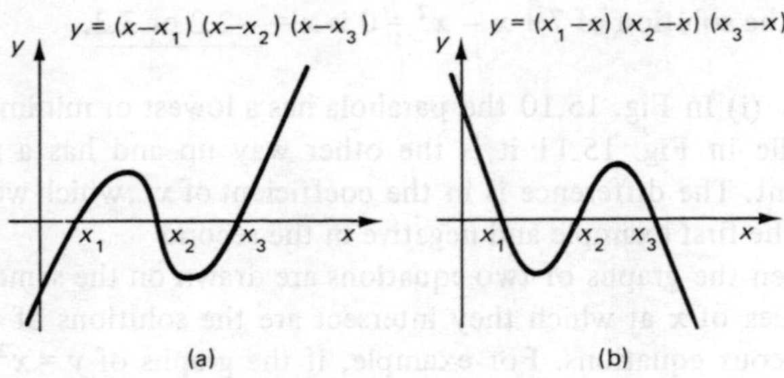

Fig 15.12

Example 15.10

Draw the graph of the function $y = 5x^2 - 6x - x^3$ for values of x from -1 to 4 and use it to solve the equation $5x^2 - 6x - x^3 = 5$. Find the gradient of the curve when x is (a) 0, (b) 2.

Solution For more difficult functions and for fractional values of x an electronic calculator can simplify the working.

x	-1	-0.5	0	0.5	1	1.5	2	2.5	3	4
$5x^2$	5	1.25	0	1.25	5	11.25	20	31.25	45	80
$-6x$	6	3	0	-3	-6	-9	-12	-15	-18	-24
$-x^3$	1	0.13	0	-0.13	-1	-3.38	-8	-15.63	-27	-64
y	12	4.38	0	-1.88	-2	-1.13	0	0.63	0	-8

The graph is shown in Fig. 15.13, and the three values of x at which the curve crosses the x-axis are the solutions of $5x^2 - 6x - x^3 = 0$. There is only one point on the curve for which $y = 5$, and the value of x at that point is the solution of $5x^2 - 6x - x^3 = 5$. From the graph the value is estimated as $\underline{-0.55}$.

To find the gradient of the curve tangents are drawn at the points (0, 0) and (2, 0) by the method described in Section 15.3.

(a) The tangent at the origin passes through the points $(-1, 6)$ and $(1, -6)$ and the gradient is $-12 \div 2 = \underline{-6}$.

(b) The tangent drawn at the point (2,0) passes through $(0, -4)$ and (3,2) and the gradient is $6 \div 3 = \underline{2}$.

(b) Reciprocal functions

If y is a reciprocal function of x it is inversely proportional to x [see Section 14.6(d)] .

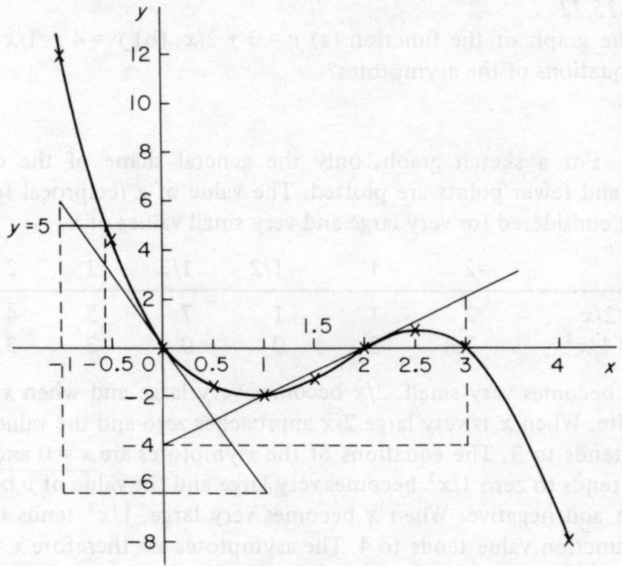

Fig 15.13

The general shape of the curves $y = a/x$ and $y = -a/x$, where a is a positive constant, is shown in Fig. 15.14.

As the value of x approaches zero, the value of y becomes very large but the curve never reaches the y-axis, which is called an *asymptote* to the curve. It is a tangent at infinity. Similarly, as y approaches zero the curve approaches the x-axis but never meets it.

The general reciprocal curve is called a rectangular hyperbola and it has two branches.

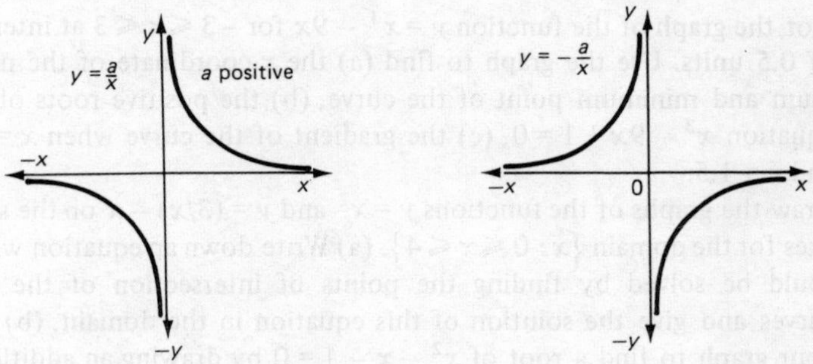

Fig 15.14

Example 15.11
Sketch the graph of the function (a) $y = 3 + 2/x$, (b) $y = 4 - 1/x^2$. What are the equations of the asymptotes?

Solution For a sketch graph, only the general shape of the curve is required and fewer points are plotted. The value of a reciprocal function should be considered for very large and very small values of x.

x	-2	-1	$-1/2$	$1/2$	1	2
$3 + 2/x$	2	1	-1	7	5	4
$4 - 1/x^2$	3.8	3	0	0	3	3.8

(a) As x becomes very small, $2/x$ becomes very large and when x is zero y is infinite. When x is very large $2/x$ approaches zero and the value of the function tends to 3. The equations of the asymptotes are $x = 0$ and $y = 3$.
(b) As x tends to zero $1/x^2$ becomes very large and the value of y becomes very large and negative. When x becomes very large, $1/x^2$ tends to zero, and the function value tends to 4. The asymptotes are therefore $x = 0$ and $y = 4$.

The graphs are shown in Fig. 15.15.

Exercise 15.4

(1) State which of the listed equations could represent each of the sketch graphs in Fig. 15.16. Draw sketch graphs of the remaining functions. (a) $y = 2x^2 + 5$, (b) $y = x + 5$, (c) $y = 1/x$, (d) $y = -1/x$, (e) $y = (x + 3)(x - 2)$, (f) $y = (x - 3)(x + 2)$, (g) $y = \frac{1}{3}(x + 3)(x - 2)$.

(2) Draw the graph of $y = 20x - 60/x^2$ for values of x from 2 to 6. From your graph determine (a) the value of y when $x = 4.6$, (b) the value of x when $y = 70$.

(3) Plot the graph of the function $y = x^3 - 9x$ for $-3 \leqslant x \leqslant 3$ at intervals of 0.5 units. Use the graph to find (a) the x-coordinate of the maximum and minimum point of the curve, (b) the positive roots of the equation $x^3 - 9x + 1 = 0$, (c) the gradient of the curve when $x = -1$ and $x = 1.5$.

(4) Draw the graphs of the functions $y = x^2$ and $y = (3/x) - x$ on the same axes for the domain $\{x : 0 \leqslant x \leqslant 4\}$. (a) Write down an equation which could be solved by finding the points of intersection of the two curves and give the solution of this equation in the domain. (b) Use your graph to find a root of $x^2 - x - 1 = 0$ by drawing an additional line. (c) Find, by drawing tangents, the gradient of each of the curves when $x = 2$.

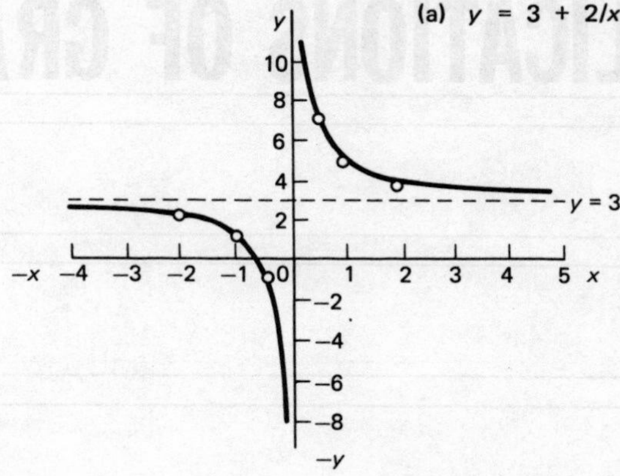

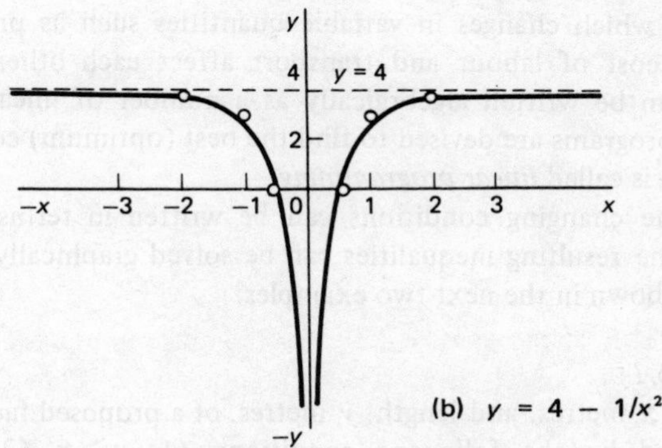

Fig 15.15

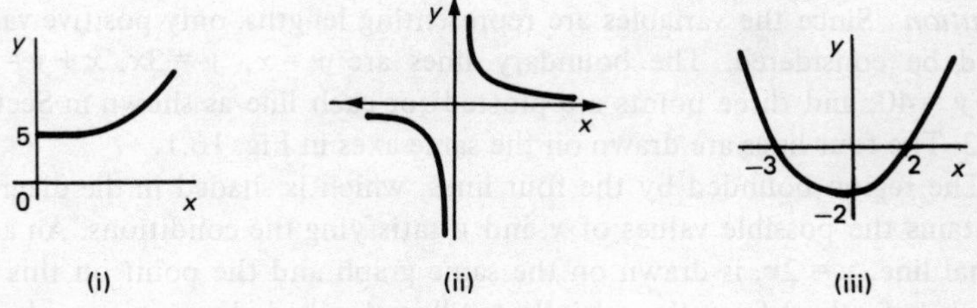

Fig 15.16

CHAPTER 16

APPLICATIONS OF GRAPHS

16.1 LINEAR PROGRAMMING

The solution of many problems in industry and commerce depends on the manner in which changes in variable quantities such as profit, value of stock and cost of labour and transport affect each other. When these variables can be written algebraically as a number of linear inequalities, computer programs are devised to find the best (optimum) conditions, and this process is called *linear programming.*

When the changing conditions can be written in terms of only two variables, the resulting inequalities can be solved graphically. The general method is shown in the next two examples.

Example 16.1
The width, x metres, and length, y metres, of a proposed factory building are specified by the following conditions: (1) $y > x$, (2) $y < 3x$, (3) $30 < x + y < 40$. Mark on a suitable diagram the region in which the possible values of x and y must lie and find the maximum ground area when $y = 2x$, assuming they take only integer values.

Solution Since the variables are representing lengths, only positive values need be considered. The boundary lines are $y = x$, $y = 3x$, $x + y = 30$, $x + y = 40$, and three points are plotted for each line as shown in Section 15.2. The four lines are drawn on the same axes in Fig. 16.1.

The region bounded by the four lines, which is shaded in the diagram, contains the possible values of x and y satisfying the conditions. An additional line, $y = 2x$, is drawn on the same graph and the point on this line which is furthest from the origin but still in the shaded region is marked P. At the point, P, $x = 13\frac{1}{3}$ and $y = 26\frac{2}{3}$. Therefore the greatest length is 26 metres. The maximum ground area is $26 \times 13 = \underline{338 \text{ m}^2}$.

The general method of approach is similar for all linear programming

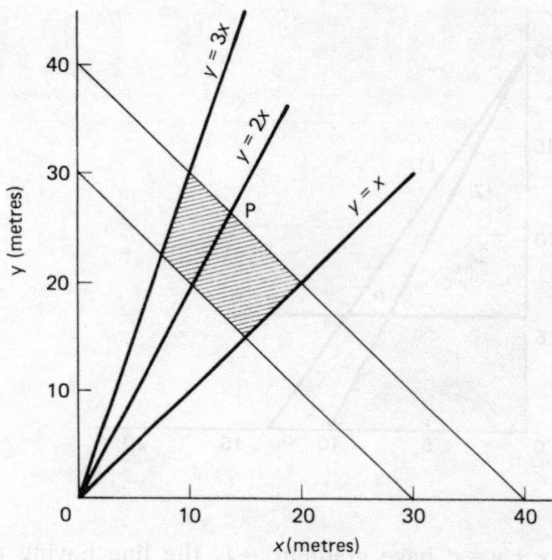

Fig 16.1

questions of this type and it can be summarised in four parts.

(i) Express the information in terms of two variables, usually x and y

(ii) From the given facts obtain two or more relations between the variables, and write down any inequalities given or implied such as $x \geqslant 0$

(iii) Draw the boundary lines on a suitable graph and indicate by shading in, or shading out, the region in which the possible solutions lie

(iv) Select the particular solution or solutions required in the question.

Example 16.2
Draw a suitable graph to show the ordered pairs (x, y) which satisfy simultaneously the inequalities (1) $3x + 2y \leqslant 40$, (2) $2x + y \leqslant 20$, (3) $y \leqslant 6$, (4) $x \geqslant 0$, (5) $y \geqslant 0$.

If x and y represent the numbers of two different kinds of machine sold by a firm in a certain week, calculate the maximum profit made in that week if the profit made on each machine was £40.

Solution Fig. 16.2 is a graph of the coordinate axes and the three boundary lines. The information given by line (1) was not necessary, since it is above line (2) for positive values of x and y. The possible solutions lie in the shaded region of the diagram, and we require the one giving the maximum value of $x + y$.

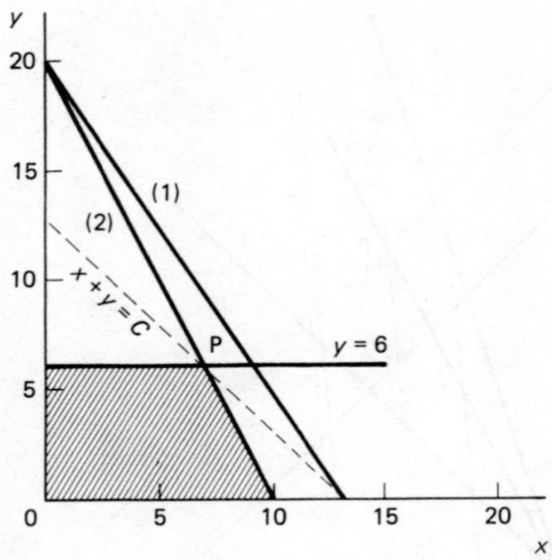

Fig 16.2

All the lines $x + y = c$ have gradient -1, the line having the greatest intercept c passes through the point P in the diagram at the limit of the shaded region. The coordinates of P obtained from the graph are (7, 6), and the maximum profit is therefore £40(7 + 6) = £520.

Exercise 16.1

(1) Draw the locus of $\{(x, y): x \geqslant 0, y \geqslant 1/2, x + y \leqslant 4, y \geqslant x - 2\}$. Subject to these conditions determine (a) the maximum value of $x + y$, (b) the minimum value of $x - y$, (c) the maximum value of $2x + y$.

(2) Represent graphically the region specified by the inequalities $x \geqslant 1$, $y \geqslant x - 3$, $y \leqslant x + 3$, and $y \leqslant 6 - x$. In this region find (a) the element with the greatest and the least value of x in the set $\{(x, y): y = x\}$, (b) the element with the least value of x in $\{(x, y : y = x + 3\}$.

(3) A holiday tour company hires two types of aircraft, Alphas which can take up to 120 passengers each, and Betas which can hold 300 each. On a certain day when there were 1200 holiday makers to be transported by air, only 3 Beta aircraft were available and 8 pilots were on duty. If x Alphas and y Betas were used, each making one journey and requiring one pilot, write down three inequalities in x and y and draw a suitable graph to show all the possible solutions. Which of these solutions is cheapest when a Beta costs half as much again as an Alpha for the same distance?

(4) A small television rental company buys two types of set, the Perry and the Windsor, which cost £250 and £180 respectively. At the end of each year Perry sets are sold for £180 and Windsors for £100. The company intends to spend a maximum of £9000 on new sets and to

receive a minimum of £4500 when it sells the sets at the end of the year. Subject to the further condition that they must buy at least twice as many Windsor sets as Perry sets, what is the greatest number of Perry sets the company can buy?

(5) There are two production systems, X and Y, available to a company. X produces 2000 articles a day and Y produces 4000; X needs 8 people to work the machines and Y needs 6 people. The total number of articles produced daily must be at least 18 000 and the total labour force not more than 48. If at least two of each system must be used, what combination of X and Y systems meets all these conditions?

16.2 OBTAINING A LINEAR LAW FROM EXPERIMENTAL DATA

When the linear equation connecting two variables is known only two or three points are required in order to plot the graph, but if the equation is not known, pairs of corresponding values may be obtained by measurement. At least five points should be plotted and in general more values give a more reliable result. They may not all lie on a straight line and so the 'best' straight line is drawn.

Example 16.3
In an experiment to verify Hooke's law for a spring, the following results were obtained

Force (N)	0	0.5	1.5	3.5	4.0	5.0
Extension (mm)	0	0.9	3.1	7.0	8.2	9.8

Plot a graph and from it determine the equation connecting force and extension for the spring. What force would cause an extension of (a) 4 mm, (b) 6 mm, (c) 11 mm?

Solution Force is the independent variable since known weights were added and the corresponding extension measured.

The graph is shown in Fig. 16.3 with values of the force as abscissa and extension as ordinate. The 'best' straight line is drawn through the points (0, 0) and (3.5, 7.0), two points are slightly above this line and two points are below it.

The gradient of the line is $7.0 \div 3.5 = 2.0$; the intercept is zero since it passes through the origin. The equation of the spring is therefore $\underline{E = 2F}$.

(a) When $E = 4$ mm, $\underline{F = 2 \text{ N}}$. [point P (2, 4) on the graph]

(b) When $E = 6$ mm, $\underline{F = 3 \text{ N}}$. [point Q(3, 6)]

(c) When $E = 11$ mm, $\underline{F = 5.5 \text{ N}}$. [point R(5.5, 11)]

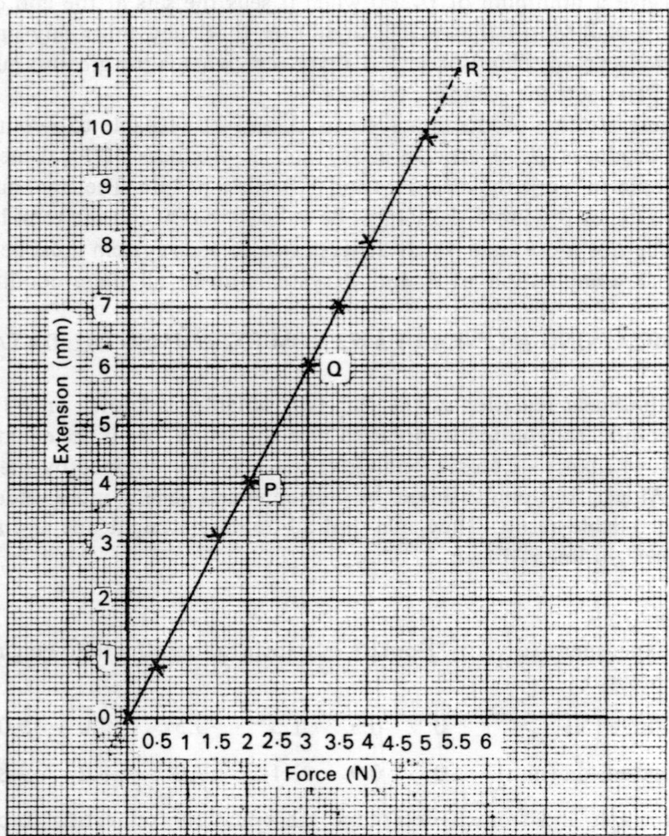

Fig 16.3

Comment Reading other values from a straight line graph is called linear *interpolation* and extending the line beyond the measured values is called *extrapolation*. This assumes that the line continues in the same direction and so extrapolated values are not always reliable.

Example 16.4

x	2.5	2.0	1.5	1.2	0.9
y	1.00	1.25	1.67	2.08	2.78

Verify by a graphical method that the above measurements are consistent with a relation of the form $y = K/x$, and find the value of the constant K.

Solution If y is inversely proportional to x the graph of y against x will *not* be a straight line (Fig. 15.14).

Let $z = 1/x$ so that $y = Kz$; plotting values of y against z will give a straight line having gradient K

z	0.40	0.50	0.67	0.83	1.11
y	1.00	1.25	1.67	2.08	2.78

The graph is shown in Fig. 16.4. It is a straight line passing through the origin and the gradient is $2.5 \div 1.0$ so that $y = 2.5z$. Hence the equation connecting y and x is $\underline{y = 2.5/x \text{ or } 2xy = 5}$.

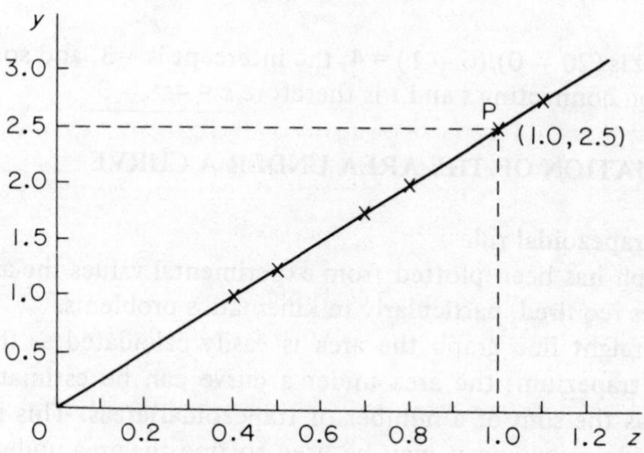

Fig 16.4

Example 16.5

t	1.2	1.4	1.6	1.7	1.8
s	3.9	8.0	13.4	16.7	20.3

It is thought that the variables s and t are connected by an equation of the form $s = at^3 + b$. Plot a suitable graph to verify that this is the relation and find the value of the constants a and b.

Solution Let $T = t^3$, then $s = aT + b$.

Comparing this with the standard equation of a straight line, $y = mx + c$, it can be seen that if values of s as ordinates are plotted against T as abscissae, a straight line should be obtained having gradient a and intercept b

T	1.73	2.74	4.10	4.91	5.83
s	3.9	8.0	13.4	16.7	20.3

The graph is shown in Fig. 16.5.

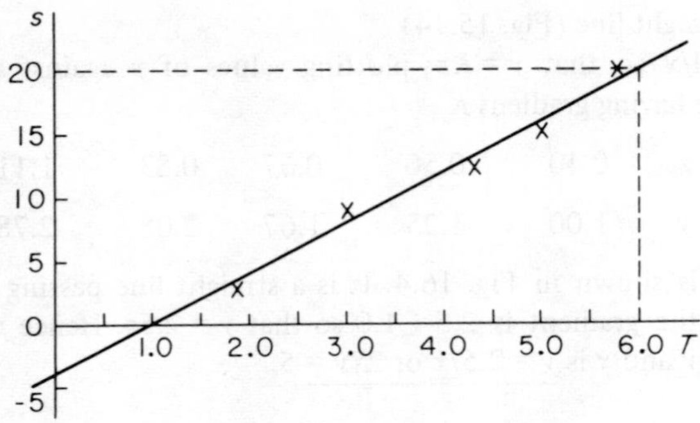

Fig 16.5

The gradient is $(20 - 0)/(6 - 1) = 4$, the intercept is -3, and so $s = 4T - 3$. The equation connecting s and t is therefore $\underline{s = 4t^3 - 3.}$

16.3 ESTIMATION OF THE AREA UNDER A CURVE

(a) By the trapezoidal rule

When a graph has been plotted from experimental values the area under it is sometimes required, particularly in kinematics problems.

For a straight line graph the area is easily calculated as the area of a triangle or trapezium; the area under a curve can be estimated by considering it as the sum of a number of trapezoidal areas. This is known as the *trapezoidal rule*, and it may be used to find the area under any curve. In Fig. 16.6 the shaded area under the graph of $y = f(x)$ is divided into a number of strips of equal width d by the ordinates y_0, y_1, y_2, etc. In this example there are six strips, but the greater the number of strips the closer is the approximation to the true area.

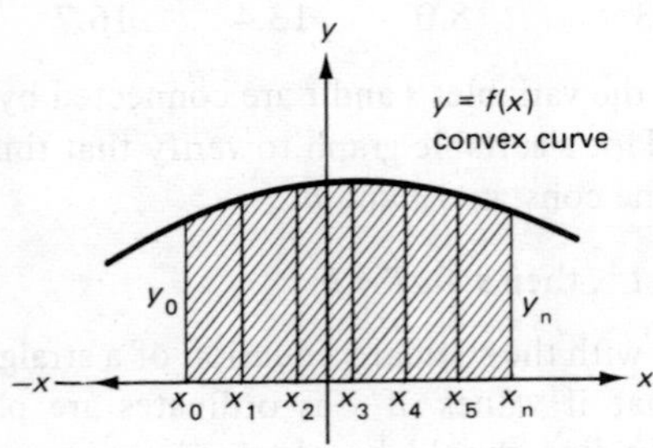

Fig 16.6

The formula is

$$A \approx \tfrac{1}{2}d(y_0 + y_1) + \tfrac{1}{2}d(y_1 + y_2) + \tfrac{1}{2}d(y_2 + y_3) + \ldots$$

or

$$A \approx \tfrac{1}{2}d\,[y_0 + y_6 + 2(y_1 + y_2 + y_3 + y_4 + y_5)]$$

since the first and last ordinates occur only once and the rest occur twice.

For a convex curve such as $y = f(x)$ in Fig. 16.7, the estimate is lower than the true area; the estimate for a concave curve such as $y = g(x)$ is slightly higher than the actual area.

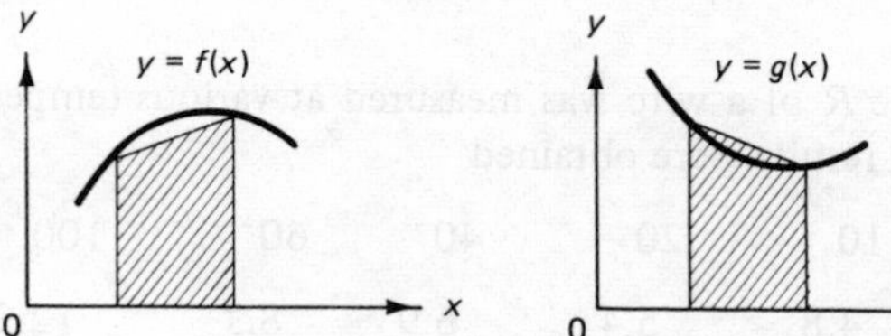

Fig 16.7

Example 16.6

The following pairs of values represent the coordinates of eleven points on a curve. Using all the given values, estimate the area enclosed by the curve and the coordinate axes when the unit on both axes is 4 cm

x	0.0	0.1	0.2	0.3	0.4	0.5	0.6	0.7	0.8	0.9	1.0
y	2.0	1.71	1.44	1.19	0.96	0.75	0.56	0.39	0.24	0.11	0.0

Solution In this example it is not necessary to plot the graph since the coordinates are given and the question does not require it.

The required area can be considered as ten strips each of width 0.1 units. By the trapezium rule

$$A \approx 0.05\ [2.0 + 0.0 + 2(1.71 + 1.44 + 1.19 + 0.96 + 0.75 + 0.56 + 0.39 + 0.24 + 0.11)]$$

$$= 0.05\ [2 + 2(7.35)]$$

$$= 0.835 \text{ square units}$$

Since the unit on both axes is 4 cm, each unit of area is 16 cm^2. The required area is therefore $16 \times 0.835 \approx$ 13.4 cm^2.

When the equation of a curve is given it is advisable to draw a sketch graph of the required area; the ordinates for selected points on the curve are then calculated and substituted in the formula.

(b) By counting squares

When a curve is drawn on squared or graph paper, the area under it can be estimated by counting the number of squares covered. The method is quite straightforward; complete squares are counted first, those parts that are under half are neglected and parts that are over half a square are counted as one unit. When the number of squares has been estimated it must be multiplied by the appropriate units.

For example, suppose the area under a velocity–time curve covered 50 unit squares. If one unit represented 5 m/s on the y-axis and 10 s on the x-axis, each unit square represented 5 m/s × 10 s or 50 metres, and the area would be estimated as 2500 metres.

Exercise 16.2

(1) The resistance R of a wire was measured at various temperatures and the following results were obtained

T	10	20	40	60	100
R	4.8	5.4	6.9	8.3	12.2

Making R the dependent variable, draw a graph and use it to find (a) the resistance of the wire when $T = 30$, (b) the value of T when the resistance was 7.5 units.

(2) Given that P is inversely proportional to V, draw a straight line graph from the following experimental results and use it to obtain a value for the constant of proportionality

V	4	6	9	11	12	15
P	96	63	42	35	32	25

(3) The following results were obtained in an experiment

r	1	2	3	4	5
s	0.25	1.0	2.3	4.0	6.2

By plotting values of s as ordinates against values of r^2, find the equation connecting s and r.

(4) A particle starts from rest at P, and its displacement s metres at a time t seconds is measured

t	1	2	3	4
s	25	40	45	40

It is known that s and t are connected by an equation of the form $s = bt^2$ or $s = at + bt^2$, where a and b are constants. (a) Draw a graph of s against t^2 to determine which is the correct relation and use the given values to determine the equation connecting s and t. (b) Calculate the time taken for the particle to return to P.

(5) Draw the graph of the function $x^2 + x - 12$ for $-5 \leqslant x \leqslant 4$. (a) Estimate the area enclosed by the curve and the x-axis, when the unit is 1 cm on each of the axes. (b) Using the same axes, draw the line $8y = 3x - 9$ and state the equation which can be solved by finding the points of intersection of this line and the curve $y = x^2 + x - 12$.

(6) The table gives corresponding values of two variables, x and y

x	0	0.1	0.2	0.3	0.4	0.5	0.6	0.7
y	1.00	0.98	0.92	0.82	0.70	0.54	0.36	0.17

Plot a graph using a scale of 2 cm for each 0.1 unit of x and 1 cm for each unit of y. Estimate the area between the curve and the x-axis for $0 \leqslant x \leqslant 0.7$, and from your graph obtain a value for the gradient of the curve when $x = 0.45$.

(7) The table shows values of x and y for the equation $y = x^3 - x^2 - 6x$. Complete the table and draw the graph using 2 cm for each unit of x and 1 cm for each unit of y

x	0	0.5	1.0	1.25	1.5	2.0	2.5	2.75	3.0
y		−3.1		−7.1	−7.9		−5.6	−3.3	

(a) Estimate the area between the curve and the x-axis by the trapezium rule, using six strips of width 0.5 units. (b) Draw the line $y = 2x - 6$ and estimate the positive roots of the equation $x^3 - x^2 - 8x + 6 = 0$.

16.4 GRAPHICAL KINEMATICS

Kinematics is the study of motion as a function of time. The speed of a moving object is the rate of change of distance with time and the average speed over a period of time is given by

$$\frac{\text{total distance covered}}{\text{total time elapsed}}$$

Displacement is distance measured in a given direction and if an object, such as a car for example, returns to its starting point the net displacement is zero but the actual distance travelled has a positive value.

Velocity is the rate of change of displacement with time, or speed in a given direction. The average velocity over a period is given by

$$\frac{\text{net displacement}}{\text{time elapsed}}$$

and if a car returns to its starting point the average velocity is zero, since the displacement is zero, but the average speed has a positive value.

Acceleration is rate of change of velocity, and so the average acceleration over a given period is

$$\frac{\text{change in velocity}}{\text{time elapsed}}$$

Negative acceleration, when the velocity is decreasing, is called retardation.

Displacement, velocity and acceleration are all vector quantities, but distance and speed are *scalars* since they have magnitude but not direction (Chapter 22).

Example 16.7

A hitch-hiker walks a distance of 3 km in 30 minutes and then takes a bus for a distance of 20 km which takes 40 minutes. He walks for a further hour, covering a distance of 5 km, and then stops for 18 minutes before getting a lift for 50 km which takes a further 47 minutes. Calculate his average speed for the journey.

Solution Average speed is

$$\frac{\text{total distance}}{\text{total time}} = \frac{3 + 20 + 5 + 50 \text{ km}}{30 + 40 + 60 + 18 + 47 \text{ minutes}}$$

$$= \frac{78}{195} \text{ km/min or } 24 \text{ km/h}$$

His average speed was 24 km/h.

(a) Time and distance graphs

If the velocity of a moving object is constant, the graph of its displacement from the origin, *s* metres, after a time *t* seconds is a straight line, and the velocity is represented by the gradient of the line. In Fig. 16.8(a)

$$\text{velocity} = \frac{\text{displacement}}{\text{time}} = \frac{25 \text{ m}}{10 \text{ s}} = 2.5 \text{ m/s}$$

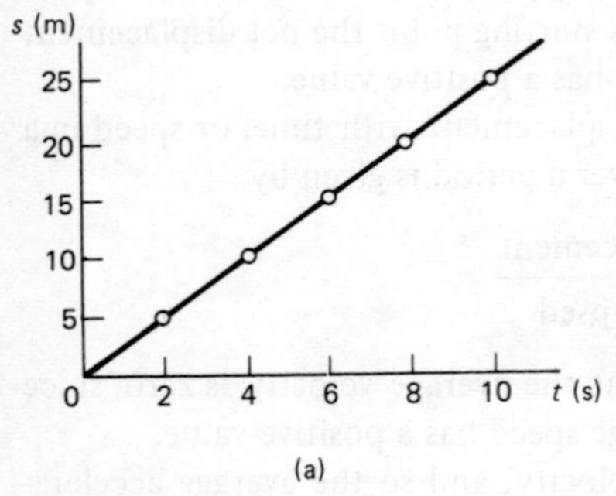

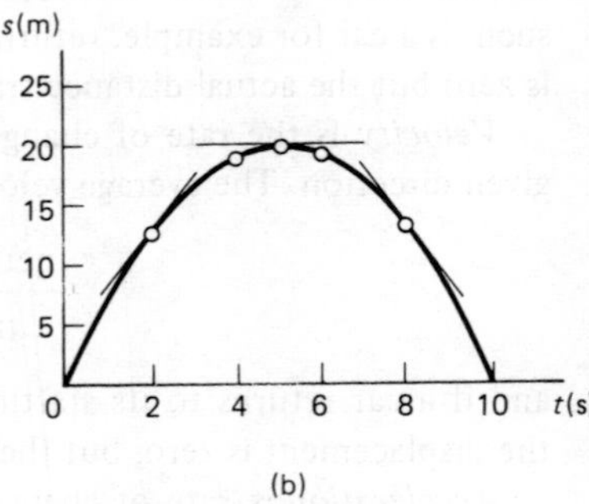

Fig 16.8

Fig. 16.8(b) represents the motion of a ball which is tossed straight up. Its velocity decreases until it reaches the highest point and it falls back to the starting point. The gradient of the curve changes continuously as the velocity changes, the velocity at any instant being given by the gradient of the curve at the particular value of t. For instance, when $t = 2$ s the height reached is about 12 m and the velocity about 6 m/s.

The greatest height of 20 m is reached after 5 s, and then the ball is momentarily at rest as it reverses direction. After 8 s the speed is again 6 m/s, but the velocity is negative since the ball is moving in the opposite direction and the displacement is decreasing.

(b) Time and velocity graphs

When the velocity of a moving object is plotted against the time elapsed the gradient of the graph at any point represents the acceleration and the area under the graph is a measure of the distance travelled.

For example in Fig. 16.9(a), the acceleration remains constant at 2 m/s per second, or 2 m/s^2, for 7 s and the distance travelled in that time is given by

$$A = \tfrac{1}{2} \times 7\text{ s} \times 14\text{ m/s} = 49\text{ m}$$

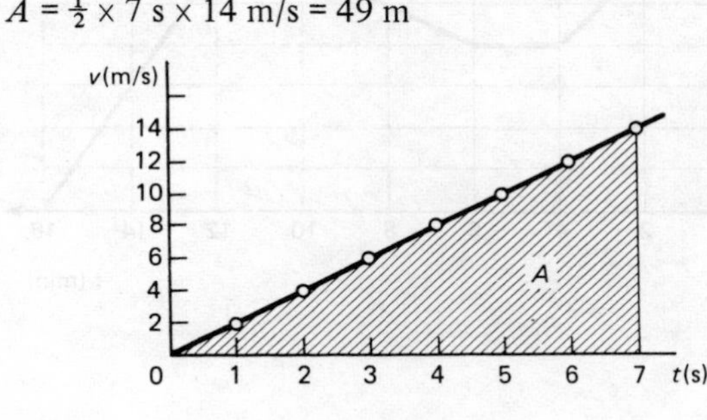

(a)

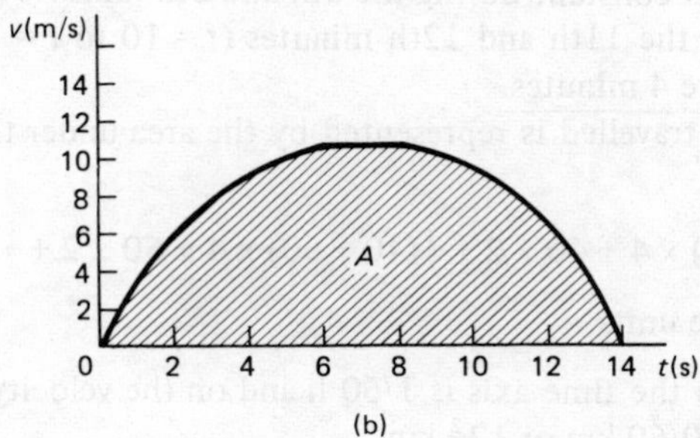

(b)

Fig 16.9

Fig. 16.9(b) represents the motion of an object starting from rest and travelling with variable acceleration for 6 s, maintaining a steady speed for 2 s and then slowing down, coming to rest after a further 6 s. The area under the graph represents the total distance travelled by the object in the 14 s, which could be estimated either by counting squares or by the trapezium rule.

The acceleration at a particular time could be estimated by drawing a tangent to the curve and measuring the gradient as described in Section 15.3.

Example 16.8

From the graph shown in Fig. 16.10 calculate (a) the time for which the object was travelling at a constant speed, (b) the total distance travelled during the 16 minutes, (c) the average speed in km/h.

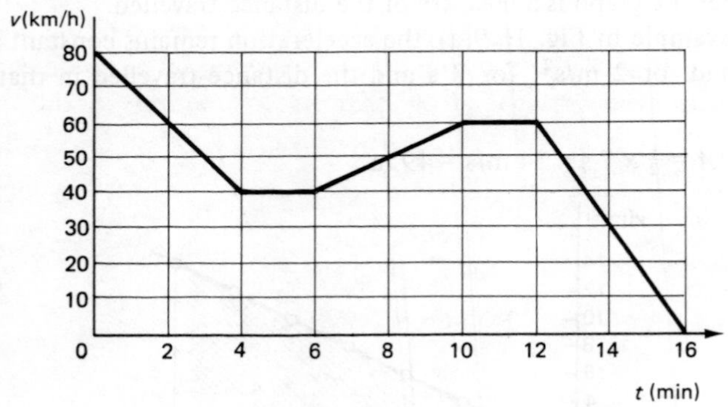

Fig 16.10

Solution

(a) The speed was constant during the 5th and 6th minutes ($t = 4$ to $t = 6$) and again during the 11th and 12th minutes ($t = 10$ to $t = 12$). The total time was therefore 4 minutes.

(b) The distance travelled is represented by the area under the graph from $t = 0$ to $t = 16$

$$A = \tfrac{1}{2}(80 + 40) \times 4 + 40 \times 2 + \tfrac{1}{2}(40 + 60) \times 4 + 60 \times 2 + \tfrac{1}{2}(60 + 0) \times 4$$

$$= 760 \text{ square units}$$

Since the unit on the time axis is 1/60 h and on the velocity axis 1 km/h, the distance is 760/60 km or $12\tfrac{2}{3}$ km.

(c) $$\text{Average speed} = \frac{\text{total distance}}{\text{total time}} = 12\tfrac{2}{3}\text{ km} \div \frac{16}{60}\text{ h}$$

$$= 47\tfrac{1}{2}\text{ km/h}$$

Example 16.9

A particle starts from rest and accelerates uniformly to reach a velocity v after 30 s, travels at that velocity for a further 3 minutes and then comes to rest after 30 s uniform retardation.

A second particle starts from the same position on a parallel track 1 minute after the first, accelerates uniformly to reach a velocity $2v$, and is then retarded uniformly for an equal time coming to rest at the same instant as the first particle. (a) Construct a time–velocity graph for the particles. (b) Given that the first particle travels 5 metres further than the second, calculate the value of v in metres/minute. (c) Calculate the distance each particle has travelled when they first have the same velocity.

Solution

(a) The graph consists of straight lines since the acceleration is uniform; the first particle has a travelling time of 4 minutes, the second particle has $1\frac{1}{2}$ minutes acceleration and $1\frac{1}{2}$ minutes retardation, as shown in Fig. 16.11.

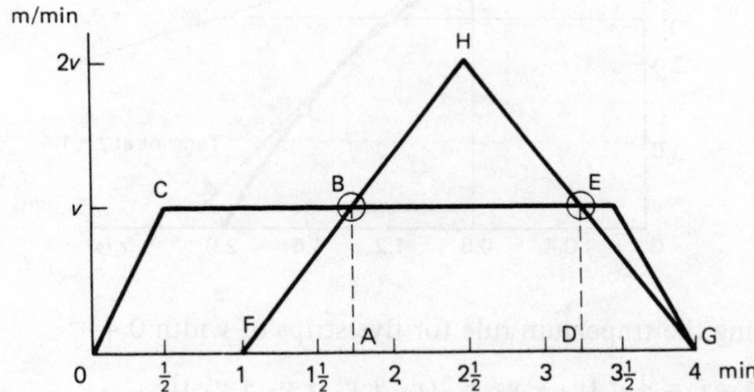

Fig 16.11

(b) Area under graph

1st particle $\quad \frac{1}{2}(3+4)\times v = 3\frac{1}{2}v$

2nd particle $\quad \frac{1}{2}\times 3\times 2v = 3v$

If the difference is 5 metres, $\frac{1}{2}v = 5 \Rightarrow v = 10$ m/minute.

(c) The particles have the same velocity when the graphs intersect, at the points marked B, E, and G in the diagram. The earliest is at B, $1\frac{3}{4}$ minutes after the start

1st particle $\quad$ Distance = Area OABC $= \frac{1}{2}(1\frac{3}{4}+1\frac{1}{4})\,10 = 15$ m

2nd particle $\quad$ Distance = Area FAB $= \frac{1}{2}\times\frac{3}{4}\times 10 = 3\frac{3}{4}$ m

Example 16.10
The table shows the velocity, v metres per second, of a particle after t seconds of a motion

t	0.0	0.4	0.8	1.2	1.6	2.0
v	4.0	3.84	3.36	2.56	1.44	0.0

Plot the points on a velocity-time graph and join them by a smooth curve. (a) Use the trapezium rule to estimate the area enclosed by the curve and the axes, and calculate the percentage error if the actual distance travelled by the particle was 5.33 metres. (b) Use your graph to determine the acceleration of the particle when $t = 0.4$ s and when $t = 1.4$ s.

Solution The graph is shown in Fig. 16.12.

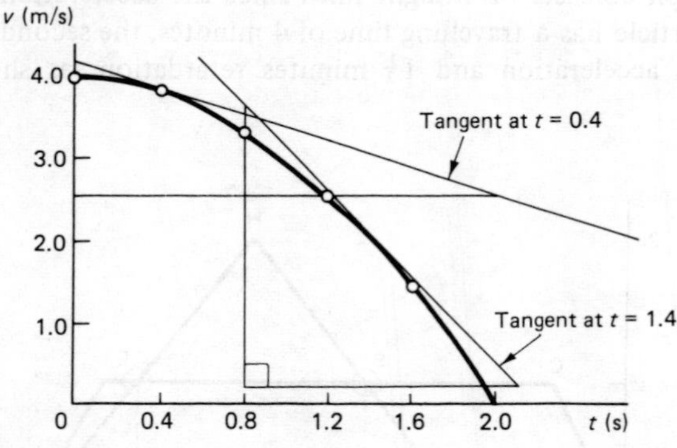

Fig 16.12

(a) Applying the trapezium rule for five strips of width 0.4

$$\text{Area} = \tfrac{1}{2} d\, [v_0 + v_5 + 2(v_1 + v_2 + v_3 + v_4)]$$
$$= 0.2\, [4.0 + 2(3.84 + 3.36 + 2.56 + 1.44)]$$
$$= 5.28 \text{ square units}$$

Hence the distance is estimated as 5.28 metres.

The error is $5.33 - 5.28 = 0.05$ m, and when expressed as a percentage of the true value this is $(0.05/5.33) \times 100\% \approx 1\%$.
(b) The gradient of the curve is found by drawing the tangent as described in Section 15.3 and completing the right angled triangle for two points on it, as shown in the diagram.
When $t = 0.4$

$$\text{the gradient} = \frac{4.2 - 2.55}{0 - 2.0} = -0.83$$

and the particle has a retardation of $\underline{0.83 \text{ m/s}^2}$ approximately.
When $t = 1.4$

$$\text{the gradient is } \frac{3.7 - 0.2}{0.8 - 2.08} = -2.7$$

and the retardation is approximately $\underline{2.7 \text{ m/s}^2}$.

Exercise 16.3

(1) A car is driven 50 km in 80 minutes. (a) What is the average speed? (b) At this speed how long would it take to travel 30 km?

(2) The average speed for a journey of 240 km was 60 km/h. (a) How long did the journey take? (b) If the average speed for the first 2 hours was 45 km/h calculate the average speed for the remaining time.

(3) The points (0, 0), (10, 15), (20, 10), (30, 30), (40, 30), (50, 0) are plotted in that order on a velocity–time graph with time in seconds as the abscissa and velocity in cm/s. The points are joined by straight lines. Use your graph to calculate the distance travelled (a) in the first 20 s, (b) in the 50 s of the motion. What was the acceleration in (c) the first 10 s, (d) the last 10 s?

(4) The table gives the velocity of a particle, v cm/s, at a time t s

t (s)	0	1	2	3	4	5	6	7	8	9	10
v (cm/s)	0	1	2.5	4.5	7	10.5	14.5	18	22	23	24

Plot a velocity–time graph and estimate (a) the time limits during which the acceleration was uniform, (b) the magnitude of this uniform acceleration, (c) the distance travelled in the first 10 s of the motion.

(5) Starting from rest a particle travels in a straight line. In the first 5 s it reaches a velocity of 15 cm/s, maintains this velocity for 40 s, accelerates to a velocity of 25 cm/s over the next 5 s and then comes to rest after a further 10 s. Assuming that the acceleration is uniform, construct a graph to illustrate the motion of the particle and use the graph to calculate (a) the acceleration during the first 5 s, (b) the distance travelled by the particle during the motion, (c) the average speed of the motion.

(6) A car on test starts from rest and maintains a constant acceleration for the first 10 s to reach 25 m/s. It travels at this speed for 30 s and then has a steady retardation, coming to rest in 15 s. (a) What was the total distance travelled? (b) What was the average speed for the test run? (c) What was the acceleration? (d) What was the retardation?

DIFFERENTIAL CALCULUS

The invention of the *calculus* was one of the greatest mathematical achievements of the seventeenth century and many famous mathematicians contributed, including Kepler, Fermat, Newton and Leibnitz, but the notation finally adopted was that of Leibnitz.

Differential calculus can be used to find the rate of change of variable quantities such as the speed and acceleration of a moving object, and the value at a particular instant can be calculated. Applications of calculus occur in most branches of science and engineering, but in elementary work it is used mainly for calculating the exact gradient of a given curve instead of making an estimate by drawing a tangent.

17.1 THE DIFFERENTIAL COEFFICIENT

The Greek letter δ (delta) is used to denote a very small change. If the equation of a curve is given by $y = f(x)$, δy represents a small change in the value of y corresponding to a small change (or increment) δx in the value of x.

The value of the ratio $\delta y/\delta x$ as δx becomes zero is written $\mathrm{d}y/\mathrm{d}x$ (pronounced 'dy by dx') and this is called the *differential coefficient*. It is *not* a fraction but a symbol meaning 'the *derivative* of y with respect to x' and it is a measure of the rate at which the value of y changes as x changes.

17.2 DIFFERENTIATION OF POWERS BY A FORMULA

While the derivative of any function of x could be calculated from first principles, this is not required in elementary work. Instead there is a simple formula which gives the differential coefficient for any power of x. If

$$y = x^n \text{ then } \frac{\mathrm{d}y}{\mathrm{d}x} = nx^{n-1}$$

This result holds for negative and fractional values of n, but only integral values are considered here. Thus if $y = x^3$ then $dy/dx = 3x^2$; if $y = x^{-2}$ then $dy/dx = -2x^{-3}$.

When a power of x is multiplied by a constant the derivative is multiplied by the same constant. A function containing a number of terms is differentiated term by term, as shown in the following Examples.

Example 17.1

(a) $y = 2x^5 \Rightarrow \dfrac{dy}{dx} = 2(5x^4) = \underline{10x^4}$

(b) $y = 3x^{-2} \Rightarrow \dfrac{dy}{dx} = 3(-2x^{-3}) = \underline{-6x^{-3}}$

(c) $y = 5x + 7 \Rightarrow \dfrac{dy}{dx} = 5(1x^0) + 0 = \underline{5}$

The derivative of any constant is zero.

Example 17.2

Find the derivative of y with respect to x

(a) $y = 3x^2 - 2x + 3 \Rightarrow \dfrac{dy}{dx} = 3(2x) - 2 = \underline{6x - 2}$

(b) $y = 7x + 3x^2 - 5x^3 \Rightarrow \dfrac{dy}{dx} = 7 + 6x - 5(3x^2) = \underline{7 + 6x - 15x^2}$

(c) $y = (x+3)(2x-4) = 2x^2 + 2x - 12 \Rightarrow \dfrac{dy}{dx} = 2(2x) + 2 = \underline{4x + 2}$

Example 17.3

Differentiate with respect to x

(a) $(2/x^2) - 3/x$: Written in index form this is $2x^{-2} - 3x^{-1}$.

The derivative is

$$2(-2x^{-3}) - 3(-1x^{-2}) = -4x^{-3} + 3x^{-2}$$

$$\text{or } \underline{\frac{-4}{x^3} + \frac{3}{x^2}}$$

The answer could be given in either form.

(b) $(x^4 - 2x^3 + 5)/x^2$: Division gives $x^2 - 2x + 5x^{-2}$.

The derivative is therefore $\underline{2x - 2 - 10x^{-3}}$.

Exercise 17.1

Find an expression for the differential coefficient of the following.

(1) $y = 3x$. (2) $y = 4x + 6$. (3) $y = x^2 - 3x$.

(4) $y = 4x^3 + x^2 + 3$. (5) $2y = 3$. (6) $y = (x + 2)^2$

Find the derivative of the following.

(7) $y = 6/x$. (8) $y = (2x)^5$. (9) $y = x^2(3x - 4)$.

Differentiate the following with respect to x.

(10) $(x + 1/x)^2$. (11) $(3/x^4) + 6x^2$. (12) $3x^3 - (4/x^2)$.

(13) $(x + 4)(x^2 - 3)$. (14) $x^n + mx$. (15) $3x^{n-2}$.

(16) If $f(x) = x^3 - 3x$, $g(x) = 3x^2 + 3x$ and their derived functions are f' and g', show that $f'(2) = g'(1)$.

17.3 A DERIVED FUNCTION AS A GRADIENT

(a) Gradient of a straight line

The gradient of the Cartesian graph of a straight line is the rate of change of y with x and is measured by the increase in the value of y divided by the corresponding increase in the value of x. It was shown in Section 15.4 that for the standard equation $y = mx + c$, m is the numerical value of the gradient.

Differentiating by the rule, $y = mx + c \Rightarrow \mathrm{d}y/\mathrm{d}x = m$, and so the derived function is also the gradient of the line, and measures the rate of change of y with x.

(b) Gradient of a curve

The gradient of a curve at a particular point is the gradient of the tangent at that point [Fig. 17.1(a)] and when the equation of the curve is known differentiation provides a method of calculating the gradient exactly.

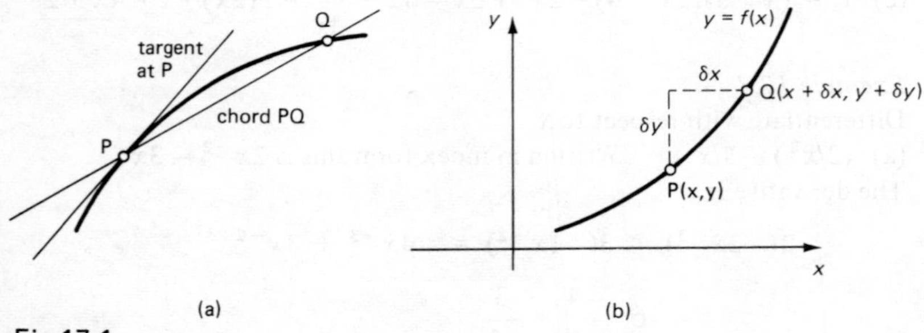

Fig 17.1

In Fig. 17.1(b) $P(x,y)$ and $Q(x + \delta x, y + \delta y)$ are two points close together on the graph of $y = f(x)$; the gradient of the chord PQ is $\delta y/\delta x$. As Q approaches P along the curve, δx approaches zero. When δx becomes zero, the chord PQ becomes the tangent at P and its gradient is given by

dy/dx, the differential coefficient. The gradient can therefore be calculated without plotting the graph and drawing a tangent.

Example 17.4
Calculate the gradient of the curve $y = 3x^3 - 2x^2 + 5x - 7$ at the points (a) (3, 71), (b) (−2, −49), (c) (0, −7).

Solution

$$y = 3x^3 - 2x^2 + 5x - 7 \Rightarrow \frac{dy}{dx} = 9x^2 - 4x + 5$$

(a) When $x = 3$, $dy/dx = 9 \times 9 - 4 \times 3 + 5 = 74$; gradient = $\underline{74}$.
(b) When $x = -2$, $dy/dx = 36 + 8 + 5 = 49$; gradient = $\underline{49}$.
(c) When $x = 0$, $dy/dx = 5$; gradient = $\underline{5}$.

Example 17.5
Calculate the coordinates of the point on the curve $y = 3x^2 + 2x - 5$ at which the gradient is (a) −2, (b) 0. At what point on the curve is the tangent parallel to the line $y = 8x + 6$?

Solution

$$y = 3x^2 + 2x - 5 \Rightarrow \frac{dy}{dx} = 6x + 2 = \text{gradient at } (x, y)$$

(a) When the gradient is −2, $6x + 2 = -2 \Rightarrow x = -2/3$. Substituting this value in the equation of the curve gives the corresponding value of y; $3(4/9) + 2(-2/3) - 5 = -5$. Hence the gradient is −2 at the point $\underline{(-2/3, -5)}$.
(b) $6x + 2 = 0 \Rightarrow x = -1/3, y = 3(1/9) + 2(-1/3) - 5 = -5\frac{1}{3}$.
The gradient is 0 at the point $\underline{(-\frac{1}{3}, -5\frac{1}{3})}$.

The line $y = 8x + 6$ has gradient 8, and so the tangent parallel to this line has gradient 8, and the gradient of the curve is 8

$$\frac{dy}{dx} = 8 \Rightarrow 6x + 2 = 8 \Rightarrow x = 1 \text{ and } y = 3 + 2 - 5 = 0$$

The required point is therefore $\underline{(1, 0)}$.

Exercise 17.2
(1) Find the gradient of each curve at the given point:
(a) $y = x^2 - 2$; (2, 2), (b) $y = 6x - x^2$; (−1, −7),
(c) $y = 2(x^2 - 1) + x(x + 1)$; (3, 28),
(d) $y = 3x^2 + 2x^3 + 6$; (−2, −10),

(e) $y = 2x + (3/x^3)$; $(-1, -5)$, (f) $y = x^2 + (1/x)$; $(-1, 0)$, (g) $xy = 9$; $(3, 3)$.

(2) Find the coordinates of the point on the curve $y = (x + 2)(2x - 3)$ at which the gradient is (a) 13, (b) -1, (c) 0.

(3) Find the coordinates of the point on the curve $y = 4x^2 - 7x + 2$ at which the tangent is parallel to the line $y = 9x$, and the equation of the tangent at that point (see Example 15.4).

(4) Given that the gradient of the curve $y = f(x)$ is -1 when $x = 1/2$, find the value of a when $f(x) = 6 + 3x - ax^2$.

(5) Find the equation of the tangent to the curve $y = (x - 2)(x + 3)$ at the point (a) $(-2, -4)$, (b) $(2, 0)$.

17.4 MAXIMUM AND MINIMUM POINTS ON A CURVE

The points on a Cartesian graph at which the tangents are parallel to the x-axis are called *stationary* or *turning* points, and the value of the gradient at such points is zero. Using differential calculus the coordinates of turning points can be found without plotting the graph, and the method is illustrated in the following Examples.

Example 17.6

Find the coordinates of the turning point of the given curve, stating whether it is a maximum or a minimum point. (a) $y = 2x - x^2$, (b) $y = 2x^2 + 3x - 4$.

Solution (a)

$$y = 2x - x^2 \Rightarrow \frac{dy}{dx} = 2 - 2x = 0 \quad \text{at the turning point}$$

$$2 - 2x = 0 \Rightarrow x = 1$$

and the corresponding value of y is $2 - 1 = 1$. Since the coefficient of x^2 is negative, the curve has a maximum at (1, 1).

(b)

$$y = 2x^2 + 3x - 4 \Rightarrow \frac{dy}{dx} = 4x + 3$$

At the turning point

$$\frac{dy}{dx} = 0 \Rightarrow 4x + 3 = 0 \text{ or } x = -\tfrac{3}{4}$$

The corresponding value of y is

$$2(\tfrac{9}{16}) + 3(-\tfrac{3}{4}) - 4 = -5\tfrac{1}{8}$$

The stationary point is $(-\frac{3}{4}, -5\frac{1}{8})$; it is a minimum point since the coefficient of x^2 is positive in the equation of the curve (see Section 15.7).

Example 17.7
Calculate the values of x at which the curve $y = 2x^3 - 5x^2 - 2x + 5$ has turning points and state whether they are maximum or minimum points.

Solution

$$y = 2x^3 - 5x^2 - 2x + 5 \Rightarrow \frac{dy}{dx} = 6x^2 - 10x - 2 \text{ or } 2(3x^2 - 5x - 1)$$

At the turning points $3x^2 - 5x - 1 = 0$, and solving this quadratic equation by the formula gives

$$x = \frac{5 \pm \sqrt{(25 + 12)}}{6}$$

and hence $x = -0.18$ or 1.85.

The coefficient of x^3 in the equation of the curve is +2, and so the maximum is at the lower value (see Section 15.8).

The curve has a maximum point at $x = -0.18$ and a minimum point at $x = 1.85$.

Comment The terms maximum and minimum are used in a special way here to describe the shape of the curve, and they are not necessarily the greatest and least values of the function in the given domain. For example, in Fig. 15.13 the value of the function at the maximum point is about 0.6, but the greatest value plotted is 12.

17.5 RATES OF CHANGE

Differentiation can be used for other variables besides x and y. For example, if $s = 3t^2$ then $ds/dt = 6t$, and measures the rate of change of s with t. Similarly, if $A = 4\pi r^2$ then $dA/dr = 8\pi r$.

Example 17.8
If $P = 4M^3 - (1/M)$, calculate the rate of change of P with M when $M = 2$.

Solution

$$P = 4M^3 - \frac{1}{M} \Rightarrow \frac{dP}{dM} = 12M^2 + \frac{1}{M^2} = 12 \times 4 + \frac{1}{4} \text{ when } M = 2$$

Hence the rate of change is $48\frac{1}{4}$.

Example 17.9
A metal cylinder of fixed height 80 mm is turned on a lathe so that the diameter is decreasing at a steady rate. Find an expression for the rate of change of the volume with the radius of the cylinder and calculate this rate of change when the radius is 20 mm.

Solution The volume of a cylinder of height h and radius r is given by

$$V = \pi r^2 h \quad \text{cubic units}$$

When h is constant the rate of change $\mathrm{d}V/\mathrm{d}r = \underline{2\pi hr}$ and when $r = 20$ mm this becomes

$$2 \times 3.142 \times 80 \times 20 \approx \underline{10\,000 \text{ mm}^3/\text{mm}}$$

Exercise 17.3

(1) Find the coordinates of the turning points of the following functions and state whether they are maximum or minimum points.
(a) $y = 4x^2 - 1$, (b) $y = 5 - 8x - 3x^2$, (c) $y = 2x^3 + 3x^2 - 1$, (d) $y = x^3 - 2x^2 + 7$, (e) $y = (3 - x^2)(x + 4)$.

(2) Calculate the coordinates of the turning points of the curve $y = x(x^2 - 27)$, and the coordinates of the points on the curve at which the gradient is -15.

(3) Find the value of x at which the gradients of the curves $y = 3x^2 + 9x + 2$ and $y = x^3 + 3x^2 - 2$ are the same.

(4) A quantity R is the sum of two parts, one being proportional to t and the other to t^2. If $R = 0$ when $t = 2$ and $R = 6$ when $t = 3$, find the equation connecting R and t and hence the minimum value of R.

(5) A rectangle having a perimeter of 30 metres has one side of length x metres. Calculate (a) the length of the adjacent side in terms of x, (b) the area A in terms of x, (c) the rate of change of A with x, (d) the maximum value of A.

(6) If the radius of a circle is increasing at a steady rate, find the rate of change of the area with the radius when the radius is 3.5 cm.

(7) Calculate the rate of increase with the diameter of the volume of a spherical bubble when the diameter is (a) 2 cm, (b) 3 cm.

(8) Gas flows out of a cylinder at a rate V cm^3/s at a time t s given by $V = 4t^2 - 20t + 30$. Find the value of t when the flow is at a minimum and the rate of flow at that time.

17.6 DIFFERENTIATION APPLIED TO KINEMATICS

(a) Definitions

If s is the displacement from a fixed origin in a given direction the velocity v measured in the same direction at a time t is given by $v = \mathrm{d}s/\mathrm{d}t$, the rate of change of displacement with time.

Acceleration is rate of change of velocity with time, $a = \mathrm{d}v/\mathrm{d}t$.

(b) Second derivative

If $v = \mathrm{d}s/\mathrm{d}t$ and $a = \mathrm{d}v/\mathrm{d}t$ then

$$a = \frac{\mathrm{d}}{\mathrm{d}t}\left(\frac{\mathrm{d}s}{\mathrm{d}t}\right), \text{ or } \frac{\mathrm{d}^2s}{\mathrm{d}t^2}$$ the second derivative of s with respect to t

When the equation connecting displacement with time is known, differentiation can be used to calculate the velocity and acceleration at any time.

Example 17.10

A stone is thrown vertically downwards from the top of a cliff so that after t s it is h m below the top, where $h = 20t + 5t^2$.

(a) Calculate how far it has fallen at the end of the first second and its velocity and acceleration at that time.

(b) If the height of the cliff is 105 m, how long does the stone take to reach sea level?

Solution (a)

$$h = 20t + 5t^2; \qquad t = 1 \Rightarrow h = 20 + 5 = \underline{25\text{ m}}$$

$$\text{velocity } \frac{\mathrm{d}h}{\mathrm{d}t} = 20 + 10t; \qquad t = 1 \Rightarrow \text{velocity} = \underline{30\text{ m/s}}$$

$$\text{acceleration } \frac{\mathrm{d}^2h}{\mathrm{d}t^2} = 10; \qquad t = 1 \Rightarrow \text{acceleration} = \underline{10\text{ m/s}^2}$$

(b) The stone reaches sea level when $h = 105$ m.

$$105 = 20t + 5t^2 \Rightarrow 5t^2 + 20t - 105 = 0$$

$$\text{or } t^2 + 4t - 21 = 0$$

This factorises as $(t + 7)(t - 3) = 0 \Rightarrow t = -7$ or $t = 3$. Only the positive value is required, therefore $t = 3$, and the stone reaches sea level after $\underline{3}$ s.

Example 17.11

A ball is thrown upwards and its height after t s is s m, where $s = 25t - 5t^2$.

Find expressions for its velocity and acceleration after t s and calculate (a) the height and velocity after 2 s, (b) the greatest height reached, (c) the distance travelled in the third second, (d) the acceleration when $t = 2.5$ s.

Solution

$$\text{Height } s = 25t - 5t^2 \text{ m}$$

$$\text{Velocity } v = \frac{ds}{dt} = 25 - 10t \text{ m/s}$$

$$\text{Acceleration } a = \frac{dv}{dt} = -10 \text{ m/s}^2$$

(a) When $t = 2$

$$s = 25 \times 2 - 5 \times 4 = \underline{30 \text{ m}}$$

$$v = 25 - 10 \times 2 = \underline{5 \text{ m/s}}$$

(b) At the greatest height, the ball comes to rest before falling back.

$$v = 0 \Rightarrow 25 - 10t = 0 \Rightarrow t = 2.5 \text{ s}$$

The greatest height is reached after 2.5 s, and at this time

$$s = 25 \times 2.5 - 5 \times 2.5^2 = \underline{31.25 \text{ m}}$$

(c) After 2 s, height $s = 30$ m. After 3 s, height $s = 75 - 45 = 30$ m. Thus the displacement in the third second is zero, but the distance travelled by the ball is $2(31.25 - 30) = \underline{2.5 \text{ m}}$.
(d) When $t = 2.5$ s the acceleration is $\underline{-10 \text{ m/s}^2}$ it is the same for all values of t.

Example 17.12
A point moves along a straight line PX so that its distance x m from P at a time t s is given by the equation $x = t^3 - 6t^2 + 9t$. Calculate (a) the times at which the point is at rest and the values of x at these times, (b) the acceleration when the velocity is zero and after 2 s of the motion.

Solution (a)

$$x = t^3 - 6t^2 + 9t \Rightarrow \frac{dx}{dt} = 3t^2 - 12t + 9 \text{ or } 3(t^2 - 4t + 3)$$

The point is at rest when the velocity dx/dt is zero.

$$t^2 - 4t + 3 = 0 \Rightarrow (t - 3)(t - 1) = 0 \Rightarrow t = 1 \text{ or } 3$$

The point is at rest at $\underline{1\text{ s}}$ and at $\underline{3\text{ s}}$ after leaving P.
When $t = 1$, $x = 1 - 6 + 9 = \underline{4}$. When $t = 3$, $x = 27 - 54 + 27 = \underline{0}$. The point has returned to P after $\underline{3}$ s.
(b)

$$\text{Acceleration } \frac{d^2x}{dt^2} = 6t - 12$$

When the velocity is zero

$$t = 1 \Rightarrow \text{acceleration} = \underline{-6\text{ m/s}^2}$$

and

$$t = 3 \Rightarrow \text{acceleration} = \underline{+6\text{ m/s}^2}$$

When $t = 2$, acceleration $= 12 - 12 = \underline{0}$.

Exercise 17.4

(1) A particle moves in a straight line so that at the end of t s its displacement s m from a fixed point on the line is given by $s = t^3 - 3t^2 + t$. Calculate the velocity and accleration when $t = 3$.

(2) A particle moves along a straight line so that its displacement s m after t s is given by the equation. Calculate for the given value of t (i) the velocity, (ii) the acceleration of the particle, (a) $s = 4t^2 + 2t - 8$; $t = 4$, (b) $s = 3t^3 - 4t^2$; $t = 2$.

(3) The velocity v cm/s of a body after t s is given by $v = 2t^3 - 3t + 6$. Calculate (a) the velocity after 2 s, (b) the acceleration after 4 s, (c) the average acceleration in the fourth second.

(4) A light spot moves in a straight line so that its displacement s cm from a fixed point after t s is given by the equation $s = 30t + 9t^2 - t^3$. Calculate (a) the velocity and acceleration after 3 s, (b) the time at which the spot is at rest, (c) the time at which the velocity is a maximum.

CHAPTER 18

INTEGRAL CALCULUS

18.1 INTEGRATION AS THE INVERSE OF DIFFERENTIATION

(a) Notation

It was Leibnitz who first used the elongated 's' (for 'summa') as the symbol for an integral. If y is a function of x

$$\int y \, dx$$

means 'the integral of y with respect to x'.

(b) Indefinite integration by a formula

If $y = x^3 + C$, where C is a constant, since the derivative of any constant is zero, $dy/dx = 3x^2$ for any value of C.

Integration is the inverse of differentiation, and so if $dy/dx = 3x^2$ then

$$y = \int 3x^2 \, dx = x^3 + C$$

where C represents a constant. An arbitrary constant C must be added to every indefinite integral. In general

$$\int x^n \, dx = \frac{x^{n+1}}{n+1} + C \qquad (n \neq -1)$$

Comment To integrate a power of x, increase the power by 1 and then divide by the new index. The value $n = -1$ is excluded since it would make the denominator zero.

(c) Integration of functions with several terms

Functions with more than one term can be integrated term by term and a single arbitrary constant is added.

Example 18.1

(a)

$$\int 2x^3\,dx = 2\left(\frac{x^{3+1}}{3+1}\right) = \frac{x^4}{2} + C$$

(b)

$$\int \frac{3}{x^2}\,dx = \int 3x^{-2}\,dx = 3\left(\frac{x^{-1}}{-1}\right) = -\frac{3}{x} + C$$

Example 18.2

Integrate the given function; (a) $y = 3x^2 - 4x^3 + 2x - 7$, (b) $x = t^3 + t - 2 - 2/t^2$.

Solution

(a)

$$y = 3x^2 - 4x^3 + 2x - 7$$

$$\int y\,dx = 3\left(\frac{x^3}{3}\right) - 4\left(\frac{x^4}{4}\right) + 2\left(\frac{x^2}{2}\right) - 7x + C$$

$$= x^3 - x^4 + x^2 - 7x + C$$

(b)

$$x = t^3 + t - 2 - \frac{2}{t^2}$$

$$\int x\,dt = \frac{t^4}{4} + \frac{t^2}{2} - 2t - 2\left(\frac{t^{-1}}{-1}\right) + C$$

$$= \frac{t^4}{4} + \frac{t^2}{2} - 2t + \frac{2}{t} + C$$

18.2 DEFINITE INTEGRATION

(a) Evaluating the constant of integration

All curves which are parallel to each other have equations differing only by a constant; a particular curve in the set is specified when one point on it is known so that the constant of integration can be calculated.

Example 18.3

The gradient of a curve is given by $dy/dx = 3x^2 + 2x + 4$. Find the equation of the curve, given that (a) $y = 0$ when $x = 2$, (b) $y = 2$ when $x = 1$.

Solution

$$y = \int \frac{dy}{dx}\,dx = \int (3x^2 + 2x + 4)dx$$

and the equation of the set of curves is $y = x^3 + x^2 + 4x + C$. (a) If the point (2, 0) is on the curve

$$0 = 2^3 + 2^2 + 8 + C \Rightarrow C = -20$$

and the equation is $\underline{y = x^3 + x^2 + 4x - 20}$.

(b) Substitution of the coordinates of the point (1, 2) in $y = x^3 + x^2 + 4x + C$ gives

$$2 = 1 + 1 + 4 + C \Rightarrow C = -4$$

The equation of the curve is $\underline{y = x^3 + x^2 + 4x - 4}$.

(b) Integration between limits

Instead of finding its value, the arbitrary constant can be eliminated by evaluating the integral of $f(x)$ for two different values of x and then subtracting. The *definite* integral is written $\int_a^b f(x)\,dx$ and a and b are called the limits of integration. The value of the integral when $x = a$ is subtracted from the value when $x = b$.

Example 18.4

Evaluate the definite integral (a) $\int_1^2 2x^3\,dx$, (b) $\int_1^3 3/x^2\,dx$. The method is to integrate first and then substitute the limits.

Solution

(a)

$$\int_1^2 2x^3\,dx = \left[\frac{2x^4}{4} + C\right]_1^2 = \left[\frac{x^4}{2} + C\right]_1^2$$

$$= \left(\frac{2^4}{2} + C\right) - \left(\frac{1^4}{2} + C\right) = \underline{7\tfrac{1}{2}}$$

Comment The value of C is the same at both limits, since the points are on the same curve, and so it is eliminated by subtraction. It is usual to omit it altogether, as in the next Example.

(b) $$\int_1^3 \frac{3}{x^2}\,dx = \left[\frac{-3}{x}\right]_1^3 = \left(\frac{-3}{3}\right) - \left(\frac{-3}{1}\right) = -1 + 3 = \underline{2}$$

Example 18.5

(a)

$$\int_1^2 (x+3)^2\,dx = \int_1^2 (x^2+6x+9)\,dx = \left[\frac{x^3}{3} + 3x^2 + 9x\right]_1^2$$

$$\left(\frac{2^3}{3} + 3\times 2^2 + 9\times 2\right) - \left(\frac{1}{3} + 3 + 9\right) = 32\tfrac{2}{3} - 12\tfrac{1}{3} = \underline{20\tfrac{1}{3}}$$

(b)

$$\int_{-1}^{3} (2x - x^3)\,dx = \left[x^2 - \frac{x^4}{4}\right]_{-1}^{3}$$

$$\left(3^2 - \frac{3^4}{4}\right) - \left[(-1)^2 - \frac{(-1)^4}{4}\right] = (9 - 20\tfrac{1}{4}) - (1 - \tfrac{1}{4}) = \underline{-12}$$

Exercise 18.1

(1) Integrate with respect to x (a) x^4, (b) $3x^5$, (c) $5x - 4x^6$, (d) $3/x^2$, (e) $(4/x^3) + 2x^2$, (f) $x(3x^2 - 2x + 1)$.

(2) Integrate $dy/dx = 3x^2 - 5x + 4$, given that $y = -2$ when $x = 0$.

(3) Find s in terms of t given that $ds/dt = 2t^2 - 3t + 4$ and $s = 3\frac{1}{6}$ when $t = 1$.

(4) Evaluate the definite integral (a) $\int_0^4 3x^3\,dx$, (b) $\int_{-1}^{3} (2x+1)^2\,dx$, (c) $\int_{-2}^{2} (3 - t^2)\,dt$, (d) $\int_1^4 x^2\,(x+1)\,dx$.

(5) The gradient of a curve is given by $dy/dx = 9x^2 - 1$. Find the equation of the curve if it passes through the point $(-1, 2)$ and find also the x-coordinates of its stationary points.

(6) If the x-axis is a tangent to the curve which has gradient $8x - 4$, find the equation of the curve in the form $y = f(x)$.

18.3 THE AREA UNDER A CURVE

In Section 16.3 the area under a curve was estimated by considering it as a number of trapezia, but when the width of each strip is very small it can be considered as a rectangle. In Fig. 18.1 the area of the strip at point P(x, y) is approximately $y\,\delta x$, and the area under the curve is the sum of a very large number of such areas. The limit of this sum, when δx becomes zero, is the definite integral. In Fig. 18.1 $A = \int_a^b y\,dx$.

Thus integral calculus provides a method of calculating areas accurately. Areas below the x-axis can be calculated but, since y is negative, they will have a negative sign which can be ignored when the magnitude of the area is required.

It is important to make a sketch of the curve, showing where it cuts the axes, as the area required may be in two parts.

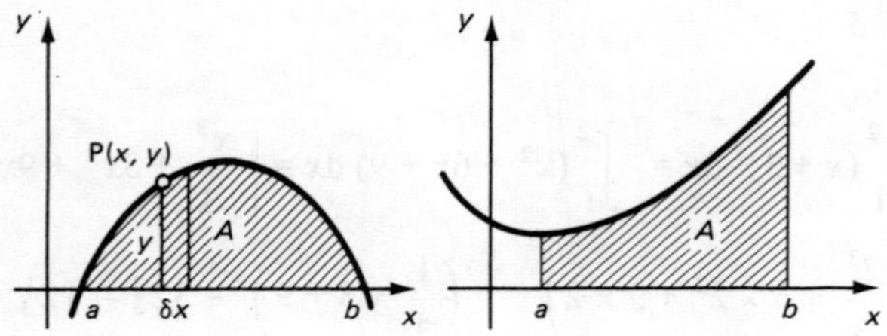

Fig 18.1

Example 18.6

Calculate the area under the curve $y = 4x^3 + 8x^2$ between the ordinates $x = -2$ and $x = 0$.

Solution A sketch of the curve is shown in Fig. 18.2(a). It has a turning point at the origin, and the required area is shaded

$$\int_{-2}^{0} (4x^3 + 8x^2)\,dx = \left[x^4 + \frac{8x^3}{3} \right]_{-2}^{0} = (0) - \left[(-2)^4 + \frac{8(-2)^3}{3} \right]$$

$$= -16 + 21\tfrac{1}{3}$$

The required area is $\underline{5\tfrac{1}{3}}$ square units.

Example 18.7

Calculate the area of the segment cut off from the curve $y = x^2 + 1$ by the line $y = 10$.

Solution The curve $y = x^2 + 1$ meets the line $y = 10$ when $x^2 + 1 = 10$, that is when $x = -3$ or $+3$, and these are the limits of integration. The area is shaded in Fig. 18.2(b); it is the difference between the area under the line $y = 10$ and the area under the curve between the ordinates $x = -3$ and $x = 3$

$$\int_{-3}^{+3} 10\,dx - \int_{-3}^{+3} (x^2 + 1)\,dx$$

$$= \Big[10x \Big]_{-3}^{3} - \left[\frac{x^3}{3} + x \right]_{-3}^{3}$$

$$= (30) - (-30) - (9 + 3) - (-9 - 3)$$

$$= 60 - 24$$

The required area is $\underline{36}$ square units.

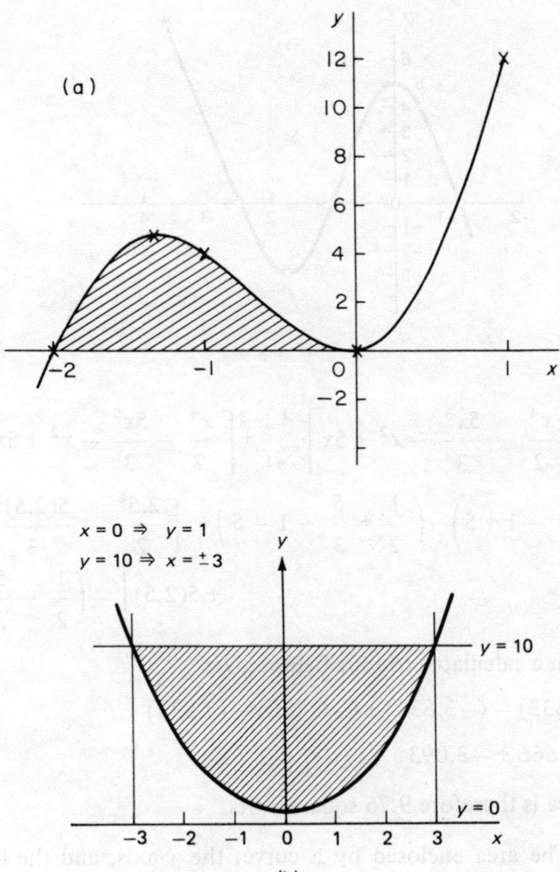

Fig 18.2

Example 18.8
Fig. 18.3 shows a sketch of the curve $y = 2x^3 - 5x^2 - 2x + 5$. Calculate the area enclosed between the curve and the x-axis.

Solution The area is in two parts, and since the part below the axis is negative integrating between the limits -1 and 2.5 would give the algebraic sum, and result in too small an answer

$$\int_{-1}^{1} (2x^3 - 5x^2 - 2x + 5)\, dx + \int_{1}^{2.5} (2x^3 - 5x^2 - 2x + 5)\, dx$$

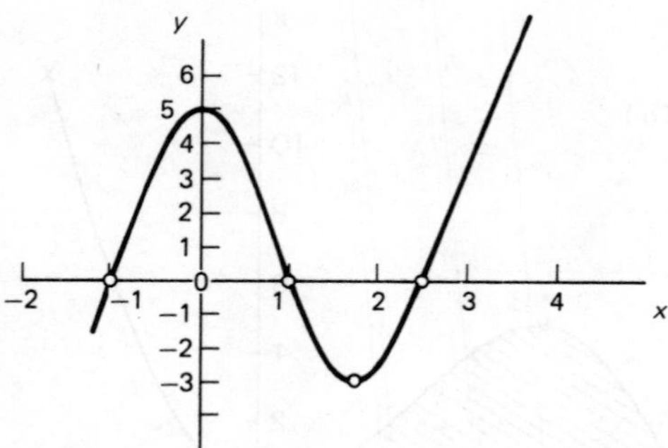

Fig 18.3

$$= \left[\frac{x^4}{2} - \frac{5x^3}{3} - x^2 + 5x\right]_{-1}^{1} + \left[\frac{x^4}{2} - \frac{5x^3}{3} - x^2 + 5x\right]_{1}^{2.5}$$

$$= \left(\frac{1}{2} - \frac{5}{3} - 1 + 5\right) - \left(\frac{1}{2} + \frac{5}{3} - 1 - 5\right) + \left[\frac{2.5^4}{2} - \frac{5(2.5)^3}{3} - (2.5)^2 + 5(2.5)\right] - \left(\frac{1}{2} - \frac{5}{3} - 1 + 5\right)$$

Working with a calculator to four figures gives

$$(2.833) - (-3.833) + (-0.260) - (2.833)$$

$$= 6.666 + -3.093$$

The total area is therefore 9.76 square units.

Comment The area enclosed by a curve, the y-axis, and the lines $y = a$ and $y = b$ is given by $\int_a^b x\,dy$. For example, the area enclosed by $y^2 = 4x$, $x = 0, y = 1$ and $y = 4$ is given by $\int_1^4 y^2/4\,dy$.

18.4 THE VOLUME OF A SOLID OF REVOLUTION

When the area under a curve is revolved once, through 360°, about the x-axis, it forms a three-dimensional shape called a solid of revolution.

The solid formed by revolving about the x-axis the small rectangle $y\,\delta x$, is a cylinder of radius y and length δx (Fig. 18.4). The volume of the cylinder is $\pi y^2\,\delta x$ (Section 13.4). The solid of revolution formed by revolving the area between the ordinates $x = a$ and $x = b$ is made up of a very large number of such cylinders, and in the limit, when δx becomes zero, the sum of the volumes becomes the definite integral

$$\text{Volume} = \int_a^b \pi y^2\,dx$$

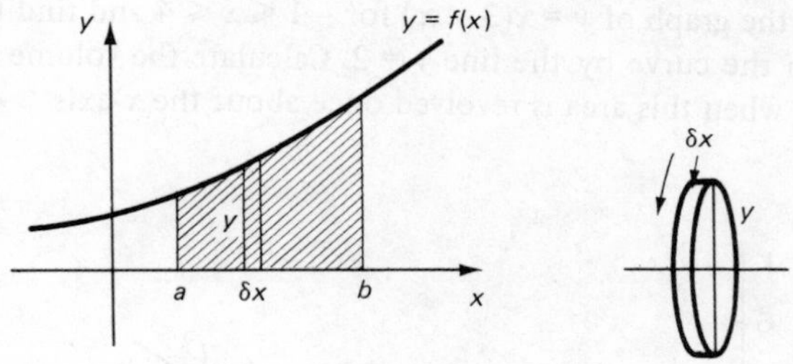

Fig 18.4

Example 18.9
Calculate the volume of the solid generated by rotating once about the x-axis the area enclosed by (a) $y = 2x$, $y = 0$, $x = 2$, (b) $y = 1/x^2$, $y = 0$, $x = 1$, $x = 4$. Leave π as a factor in the answer.

Solution The area to be rotated is shown in the sketch graph in Fig. 18.5.

(a)
$$\text{Volume} = \int_0^2 \pi y^2 \, dx = \int_0^2 \pi 4x^2 \, dx = \left[4\pi \frac{x^3}{3}\right]_0^2 = 4\pi \frac{8}{3}$$

The volume of the solid is $\underline{10\frac{2}{3}\pi}$ cubic units.

(b)
$$\int_1^4 \pi y^2 \, dx = \int_1^4 \pi \frac{1}{x^4} \, dx = -\pi \left[\frac{1}{3x^3}\right]_1^4 = -\pi \left(\frac{1}{192} - \frac{1}{3}\right)$$

The volume is $\underline{21\pi/64}$ cubic units.

Exercise 18.2

(1) Find the area enclosed by the curve $y = x^3$ and the lines $y = 0$, $x = 2$, $x = 4$.

(2) Sketch the graph of the function $y = x^3 - 4x^2 + 4x$ and calculate the area enclosed by the curve and the x-axis.

(3) Calculate the area enclosed by the x-axis and the curve (a) $y = 9 - x^2$, (b) $y = x(3 - x)$, (c) $y = (2x + 3)(x - 4)$, (d) $y = x^3 - 3x^2 + 2x$.

(4) Find the volume of the solid of revolution formed by rotating through 360° about the x-axis the area enclosed by the lines $y = 0$, $x = 1$, $x = 3$ and the curve (a) $y = 2x + 1$, (b) $y = 4 - x$, (c) $y = x^2$.

(5) Calculate the volume of the solid generated when the area described in question 3(a) is revolved once about the x-axis.

(6) Find the area enclosed by the x-axis, the curve $y = 1/x^2$ and the ordinates $x = 1$ and $x = 3$. What is the volume of the solid formed when this area is rotated through 360° about the x-axis?

(7) Sketch the graph of $y = x(3 - x)$ for $-1 \leqslant x \leqslant 4$ and find the area cut of from the curve by the line $y = 2$. Calculate the volume of the solid formed when this area is revolved once about the x-axis.

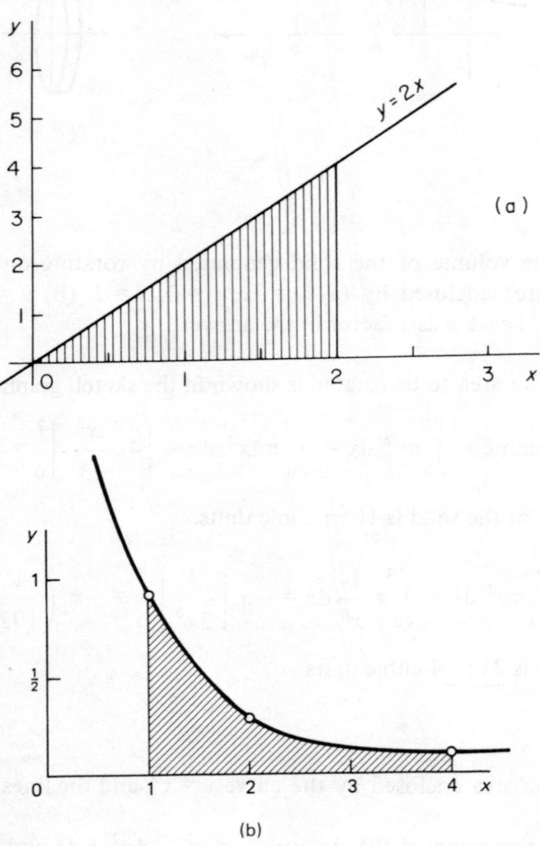

Fig 18.5

18.5 INTEGRAL CALCULUS APPLIED TO KINEMATICS

It was shown in Section 17.6 that if the displacement of a particle is given in terms of time, then its velocity and acceleration can be calculated by differentiation. Since integration is the inverse of differentiation, if we are told the acceleration of a body in terms of time, the velocity and displacement can be obtained by integration, but one pair of corresponding values must be known at each stage in order to evaluate the constant of integration.

Example 18.10

The velocity v of a particle at time t is given by $v = 1 + 2t + 6t^2$. Find (a) the average velocity, (b) the average acceleration, during the interval $1 \leqslant t \leqslant 3$, the units being metres and seconds.

Solution

(a) Average velocity = $\dfrac{\text{total displacement}}{\text{time elapsed}}$

$$\text{Displacement} = \int_1^3 v\,\mathrm{d}t = \int_1^3 (1 + 2t + 6t^2)\,\mathrm{d}t$$

$$\left[t + t^2 + 2t^3\right]_1^3 = (3 + 9 + 54) - (1 + 1 + 2) = 62\text{ m}$$

The average velocity during 2 s is 31 m/s.

(b) Average acceleration = $\dfrac{\text{change in velocity}}{\text{time elapsed}}$

When $t = 1$ s, $v = 1 + 2 + 6 = 9$ m/s

When $t = 3$ s, $v = 1 + 6 + 54 = 61$ m/s

The average acceleration during 2 s is 26 m/s^2.

Example 18.11

A particle starts from rest at O and moves along a straight line OA so that its acceleration after t seconds is $12t - 6t^2$ m/s^2. (a) Find when it returns to O and the velocity at that time. (b) What is the maximum displacement from O during this interval?

Solution

(a) Acceleration $\dfrac{\mathrm{d}v}{\mathrm{d}t} = 12t - 6t^2 \Rightarrow v = \int (12t - 6t^2)\,\mathrm{d}t$

$$= 6t^2 - 2t^3 + C$$

Using the initial condition, $v = 0$ when $t = 0 \Rightarrow C = 0$. Hence

$$v = 6t^2 - 2t^3$$

$$\text{Displacement } s = \int v\,\mathrm{d}t = \int (6t^2 - 2t^3)\mathrm{d}t = 2t^3 - \tfrac{1}{2}t^4 + C$$

The motion started at O; $s = 0$ when $t = 0 \Rightarrow C = 0$. Therefore the displacement is given by

$$s = 2t^3 - \tfrac{1}{2}t^4$$

The particle is at O when $s = 0$ or $t^3\ (2 - \frac{1}{2}t) = 0$

$$2 - \tfrac{1}{2}t = 0 \Rightarrow t = 4$$

and the particle returns to O after 4 s.

When $t = 4$, $v = 6 \times 4^2 - 2 \times 4^3 = 96 - 128 = -32$ m/s. The negative sign shows that the particle has reversed direction. (b) The displacement is a maximum when the velocity is zero

$$6t^2 - 2t^3 = 0 \Rightarrow t^2(6 - 2t) = 0 \Rightarrow t = 0 \text{ or } t = 3$$

When $t = 3$ $s = 2t^3 - \frac{1}{2}t^4 = 54 - 40.5 = 13.5$ m

Exercise 18.3

(1) Complete the following table; time t is the independent variable.

	Displacement s	Velocity v	Acceleration a	Conditions
(a)	....	$v = 4 + 6t$	....	$t = 0 \Rightarrow s = 0$
(b)	....	......	$a = 2t - 3t^2$	$t = 0 \Rightarrow s = 0$ $t = 0 \Rightarrow v = 6$
(c)	$s = 7t^2 - 5t^3$	......	....	

(2) The displacement, s m, of a body at t s is given by $ds/dt = 12 + 18t$, and $s = 188$ when $t = 4$. (a) Find an expression for s in terms of t. (b) Calculate the acceleration of the body.

(3) The acceleration of an object moving in a straight line from rest is 6 m/s^2. Calculate (a) its velocity after 4 s of the motion, (b) the distance it has travelled in 5 s.

(4) A particle starts from rest and its velocity v m/s after t s is given by $v = 6t - t^2$. Calculate (a) the displacement of the particle when it next comes to rest, (b) the displacement between $t = 2$ and $t = 3.5$.

(5) A particle starts from rest and travels in a straight line from P to Q, its velocity being given by $v = 11t^2 - 2t^3$ mm/s. Calculate (a) the distance travelled in the first two seconds of the motion, (b) the distance from P at which the particle comes to rest, (c) its maximum velocity between P and Q. (d) the acceleration of the particle after 3 s of the motion.

CHAPTER 19

STATISTICS

The subject of statistics includes the collection and processing of large amounts of numerical information (data). Vital statistics, for instance, are not the measurements of a beauty queen but the numbers of births, marriages and deaths in the population of a country. Social statistics are the information collected on behalf of the government, such as the number of people employed in various industries, and these are published in Statistical Abstracts by HMSO. Mathematical statistics is concerned with calculations based on samples from a population, and includes such quantities as, for example, the average weekly income of employees and the expected life of electrical batteries.

19.1 STATISTICAL DATA

(a) Discrete and continuous data

Information collected by counting is *discrete*, and usually takes integer values. Examples are the numbers of people employed in various industries and the production and sales figures for passenger cars. *Continuous* data are collected by measurement, and include such information as the height and weight of people in a group, and the time taken to assemble a certain piece of machinery on different occasions. The quantity which is counted or measured is called the *variable*.

(b) Crude and classified data

Crude data are individual values of a variable in no particular order of magnitude, written down as they occurred or were measured. When the numbers have been arranged in order and grouped into a small number of classes, they are called *classified* or *grouped* data.

(c) Populations and samples

In statistics, a population is not restricted to people but is like a universal set. If the variable being investigated is the life of electric light bulbs, the population could be all the bulbs produced by a particular machine; those produced on a certain day constitute a *sample* from this population.

In a *random* sample, every member of the population has an equal chance of being selected.

19.2 DIAGRAMMATIC REPRESENTATION OF DATA

Except for very simple tables, classified data can be appreciated more readily when presented in a diagram rather than lists of numbers. The most useful diagrams are *bar charts* and *pie charts*.

(a) Bar diagrams

A bar chart consists essentially of a number of parallel rectangular bars, the area of each bar being made proportional to the number of members in the class it represents. The bars may be drawn across the page or down, whichever is more convenient, and they are usually of equal width so that the length of a bar is proportional to the size of the group.

Example 19.1

In a certain month the value of the exports from the United Kingdom to selected countries, in millions of pounds sterling, was as follows: Australia 53, Canada 45, Iran 31, Italy 38, Japan 27, Nigeria 40, Greenland 1. Construct a bar diagram to represent the data.

Solution The biggest group is 53 and so a scale of 2 mm for each unit of £1 million is a convenient length of the diagram, and a width of 10 mm for each bar except that for Greenland. Since the value for Greenland is very much smaller than the rest, a narrower bar is used, but the area of 20 mm^2 per unit must be the same.

The printed diagram shown in Fig. 19.1 is not full size, and the bars are drawn across the page to make it easier to label them.

Comment Greater values could be represented on the same diagram by using a wider bar than the rest, provided that the area per unit remained the same. For example, a value of 200 units could be shown as a bar 2 cm wide and 20 cm long or 4 cm wide and 10 cm long, keeping the area of 20 mm^2 per unit.

To compare values for two different months, a double bar for each country could be drawn on the same diagram.

(b) 100% charts

This type of diagram is used when the number of values in each group is a percentage of a fixed total. The total amount is represented by the area of a circle or a rectangular bar.

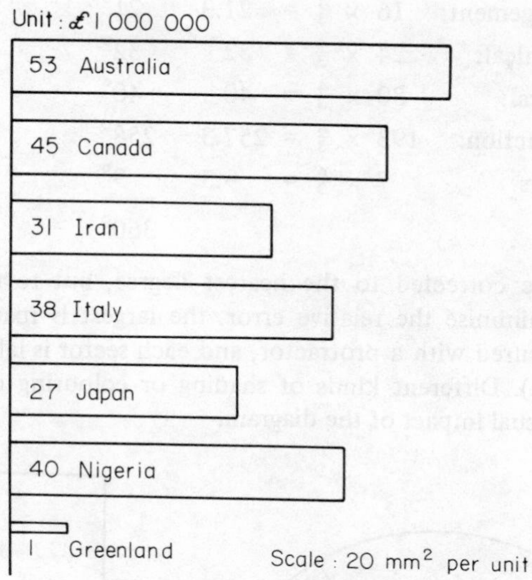

Fig 19.1 *exports from United Kingdom to selected countries in a month*

(i) Circular diagrams (pie charts)

The total number of values is represented by the area of a circle and the size of each class is proportional to the angle of the sector representing it (the slice of the pie).

Example 19.2

A student working during the vacation at a small factory in Sussex counted the number of staff in the various departments. Construct a pie chart to display this information.

Management and Sales	16	Clerical	30		
Technical	24	Production	193	Others	7

Solution The circle is drawn first, and its centre marked; of course the size depends on the space available and the purpose for which the diagram is required, but a radius of 3 or 4 cm is suitable for a notebook.

Next the angles of the sectors are calculated. The total number of staff is $16 + 24 + 30 + 193 + 7 = 270$ and this is represented by the whole circle, 360° (Section 11.2). Each person is therefore represented by an angle of $360^\circ/270$ or $4^\circ/3$. In this example the angles can be calculated easily without finding the percentages, but in general each 1% is equivalent to 3.6°.

Management:	$16 \times \frac{4}{3} =$	21.3	21°
Technical:	$24 \times \frac{4}{3} =$	32	32°
Clerical:	$30 \times \frac{4}{3} =$	40	40°
Production:	$193 \times \frac{4}{3} =$	257.3	258°
Others:	$7 \times \frac{4}{3} =$	9.3	9°
			360°

The angles are corrected to the nearest degree, but to make the total correct and minimise the relative error, the largest is made up to 258°. They are measured with a protractor, and each sector is labelled as shown in Fig. 19.2(a). Different kinds of shading or colouring can be used to increase the visual impact of the diagram.

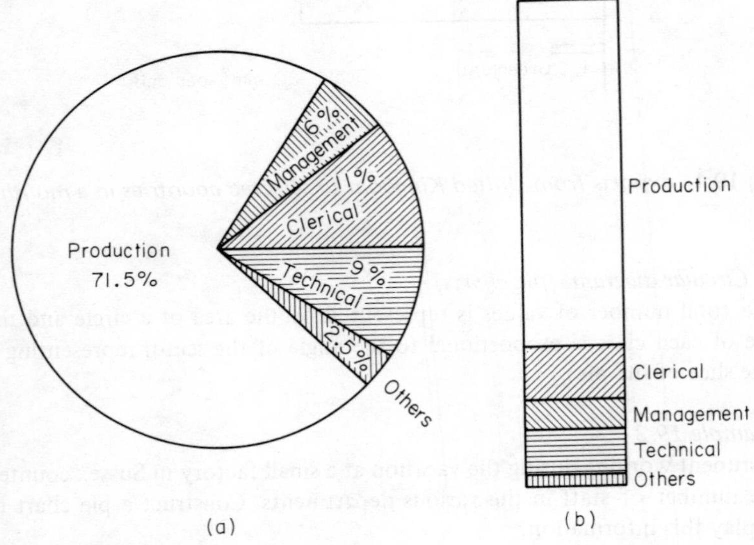

Fig 19.2 *distribution of staff in a factory*

(ii) Sectional bar charts

The number in each class is expressed as a percentage of the total and represented as a section of a rectangular bar. If a vertical bar is used, the sections can be labelled more easily.

Example 19.3

Construct a 100% bar diagram to represent the distribution of staff given in Example 19.2.

Solution Calculated to the nearest 0.5% of 270

Management 16 → 6%; Techanical 24 → 9%; Clerical 30 → 11%
Production 193 → 71.5%; Others 7 → 2.5%

The diagram is shown in Fig. 19.2(b).

(c) Comparison of distributions

Different totals can be represented by rectangles or circles of proportional areas. Since the area of a circle is proportional to the square of the radius (Section 13.2), it is not easy to assess the relative areas visually; for this reason 100% bars are more useful for comparing distributions for different totals.

For example, suppose the total amount spent of education was £5million for one county and £15 million for another county. Using bars of the same width, the lengths would be in the ratio 1 : 3 but using pie charts, the radii of the circles would be in the ratio 1 : $\sqrt{3}$ or 1 : 1.7.

Exercise 19.1

(1) For the pie diagram of Fig. 19.3, (a) if the profit was £180 000 find the cost of 'wages' and of 'other expenditure'. (b) If the profit is to be represented in a bar diagram by a bar of length 15 mm, calculate the length of the other bars (to the nearest mm).

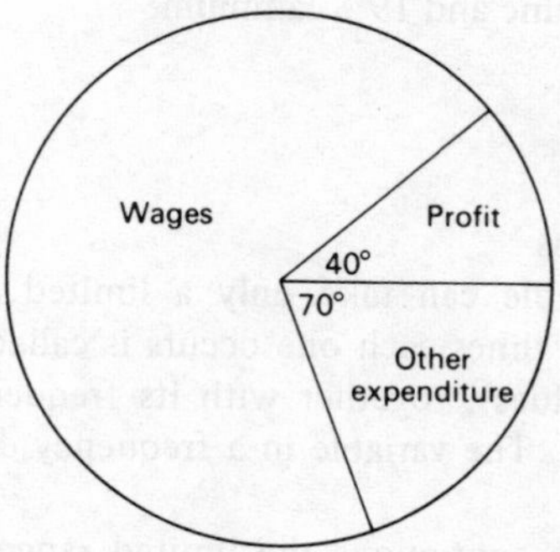

Fig 19.3

(2) Construct a pie chart and a percentage bar chart to represent each of the distributions (a) a sausage containing 45% meat, 40% cereal and 15% water, (b) a breakfast cereal consisting of 25% oats, 20% fruit, 30% wheat and the rest 'other cereals'. State the angles used in the pie charts.

(3) In Seahaven in a certain week, the average mass of refuse per household contained 35% paper, 9.8% metal, 7.9% cloth, 7.4% vegetable matter, 12.1% glass and 27.8% other material. Illustrate these statistics by means of a bar chart, and calculate the mass of glass and of metal if the total mass of refuse from a house was 12.3 kg.

(4) Use a suitable type of diagram to represent each of the following sets of data

(a) The number of students at a certain college: 982 female, 749 male.

(b) The estimated population (in millions) of the selected countries: United Kingdom 56.2, Belgium 9.9, Denmark 5.1, United States 211.

(c) United Kingdom Parliamentary elections (millions of votes):

Party	1964	1970	1974 (Oct)
Communist	0.08	0.05	0.02
Conservative	12.00	13.14	10.46
Labour	12.21	12.18	11.46
Liberal	3.09	2.17	5.35
Others	0.30	0.87	1.85

If the total electoral roll for the 1974 election was 40.36 million, what percentage voted?

(d) The percentage composition of a hard solder: 50% silver, 15% copper, 16% zinc and 19% cadmium.

19.3 HISTOGRAMS

Frequency distributions

When a discrete variable can take only a limited number of different values, the number of times each one occurs is called its frequency; each value (or group of values), together with its frequency, forms part of a *frequency distribution*. The variable in a frequency distribution is called a *variate*.

When the variate is continuous the limited range of values is divided into classes and the frequency of each class is counted. A frequency distribution for a continuous variate is represented graphically by a special kind of vertical bar chart called a *histogram*. There are no gaps between the bars of a histogram; the scale on the variate axis must be continuous, the upper boundary of one class coinciding with the lower boundary of the next. The bars are usually made equal in width, but the essential condition is that the *area* of each bar is proportional to the frequency of the class it represents, and the frequency scale must start at zero.

Fig. 19.4 is a histogram of the following distribution of times measured to the nearest 0.1 minute.

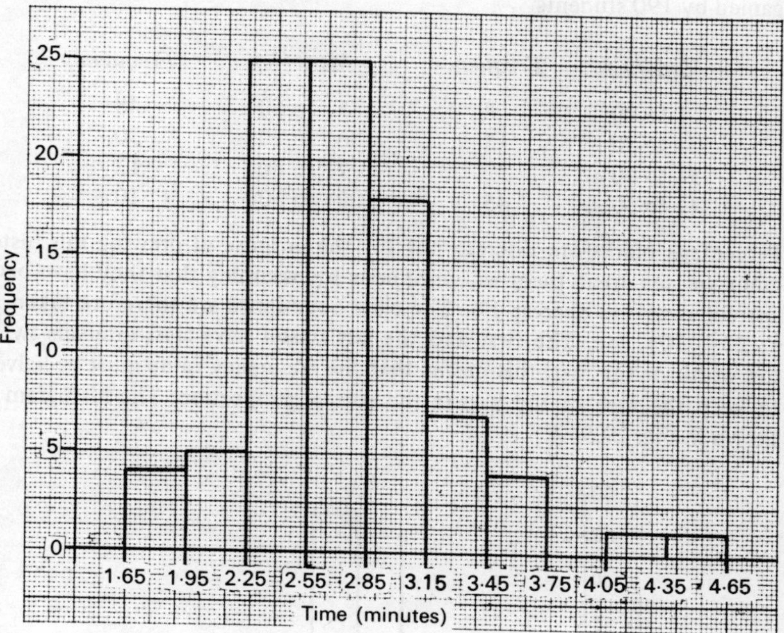

Fig 19.4

Time of Assembly of a Component on 90 Occasions

Time (min)	*Frequency*	*Time (min)*	*Frequency*
1.65 to 1.95	4	3.15 to 3.45	7
1.95 to 2.25	5	3.45 to 3.75	4
2.25 to 2.55	25	3.75 to 4.05	0
2.55 to 2.85	25	4.05 to 4.35	1
2.85 to 3.15	18	4.35 to 4.65	1

The first class contains the times measured as 1.7, 1.8 or 1.9 minutes and the class boundaries are placed at 1.65 and 1.95. This ensures that the time scale is continuous and also avoids any difficulty in deciding to which class a particular value belongs. In this example the data had been classified; in general a histogram should have about ten classes, but not less than five or more than fifteen.

Example 19.4
Construct a histogram to represent the following distribution of the marks gained by 190 students.

Mark (%)	*Frequency*	*Mark (%)*	*Frequency*
1 to 20	8	51 to 60	48
21 to 30	15	61 to 70	30
31 to 40	22	71 to 80	12
41 to 50	45	81 to 100	10

Percentage marks are discrete but for the purpose of drawing this histogram we can treat them as continuous by assuming that they have been corrected to the nearest 1%. The class boundaries are therefore taken at $\frac{1}{2}$, $20\frac{1}{2}$, $30\frac{1}{2}$, etc. The first and the last classes have class intervals twice the width of the rest; this means that the frequency value must be halved to keep the area of the bars in the correct proportion. The histogram is shown in Fig. 19.5.

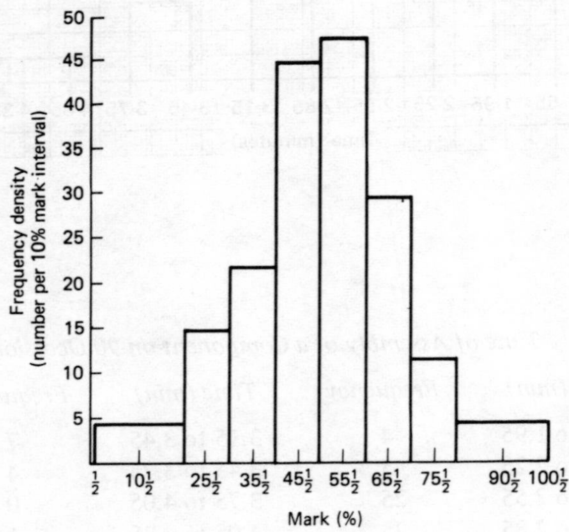

Fig 19.5 *histogram of percentage marks of 190 students*

Comment

(i) The scale on the vertical axis is labelled *frequency density* (number per 10% mark interval). This makes it clear that the frequency value has been adjusted because the interval widths are not all the same. In any case, it is becoming more common to label the vertical scale of any histogram as frequency density instead of frequency, to emphasise that it is the area of the histogram that is proportional to the total number of values of the variate

(ii) The midpoint of a class interval is called the class mark; for the first class it is $(1 + 20)/2 = 10\frac{1}{2}$ and for the next class it is $(21 + 30)/2$ or $25\frac{1}{2}$. The class marks are used in calculating the mean of a frequency distribution.

19.4 MEASURES OF LOCATION

Measures of location are average values, they locate values of a variate in a particular part of the number line. Three types are considered in this chapter, the *mean*, the *mode*, and the *median* of a frequency distribution.

(a) The mean

The mean, or arithmetic mean, is usually called the average; for crude data it is the sum of all the values divided by the number of values. For example, the mean of 2, 3, 5, 6 is $(2 + 3 + 5 + 6)/4 = 4$ and the mean of 1, 3, 4, 2, 8, 1, 4, 10, 11, 4 is $48/10 = 4.8$.

To calculate the mean of a frequency distribution for a discrete variate, each value is multiplied by its frequency and the sum of the products gives the total.

Example 19.5

Calculate the mean score

Score x:	1	2	3	4	5	6
Frequency f:	5	8	16	15	9	7

The Greek letter Σ, sigma, is used for sum, so that Σf means the sum of the frequencies or the total number of values. The total score is the sum of the products, Σfx.

The mean score is

$$\frac{\Sigma fx}{\Sigma f} = \frac{5 + 16 + 48 + 60 + 45 + 42}{5 + 8 + 16 + 15 + 9 + 7} = \frac{216}{60} = \underline{3.6}$$

Example 19.6

(a) Calculate the mean age of seven people whose ages are 21 years 4 months, 21 years 7 months, 20 years 8 months, 20 years 10 months, 21 years 4 months, 21 years 11 months, 21 years 3 months. (b) What was the age of an eighth person who joined the group if the mean age became 21 years 4 months?

Solution

(a) Since the values are all near to 21 years, the easiest method is to calculate the mean number of months over 21 years, which is called the working mean.

$$\text{The mean age is 21 years} + \frac{2 + 7 - 4 - 2 + 4 + 11 + 3}{7} \text{ months}$$

$$= \underline{\text{21 years 3 months}}$$

(b) If the average age of 8 people is 21 years 4 months, then the total number of months over 21 years is $4 \times 8 = 32$. For the seven people, the total number of months over 21 years was 21. Hence the eighth person contributed $32 - 21 = 11$ months and so was aged 21 years 11 months.

Comment

(i) When the mean of a set of numbers is given, the sum of the numbers is found by multiplying the mean by the number of values

(ii) For classified or continuous data, all the members of a particular class are assumed to have the same value, the class mark.

Example 19.7

Calculate the mean mark of the 190 students in Example 19.4.

Solution The table is rewritten to include the class marks and the *fx* products as shown below. Electronic calculators are invaluable for these calculations, but each product should be written down and checked before the addition.

Class mark x	*Frequency f*	*fx*	*Class mark x*	*Frequency f*	*fx*
10.5	8	84	55.5	48	2664
25.5	15	382.5	65.5	30	1965
35.5	22	781	75.5	12	906
45.5	45	2047.5	90.5	10	905

The total frequency $\Sigma f = 190$, $\Sigma fx = 9735$. The mean is $9735 \div 190 \approx 51.2$ and so the mean score was 51.2%.

Comment The error introduced by giving each member of a class the same value can be reduced by using a greater number of classes. In more advanced work correction factors are used to improve the accuracy.

(b) The mode

The mode of a distribution is the value with the highest frequency, and a distribution can have more than one mode. For a continuous variate, the *modal class* is the class with the highest frequency, and the position of the mode within the modal class can be estimated from a histogram of the data.

Example 19.8
From the histogram of Fig. 19.6 estimate the mode of the distribution of marks.

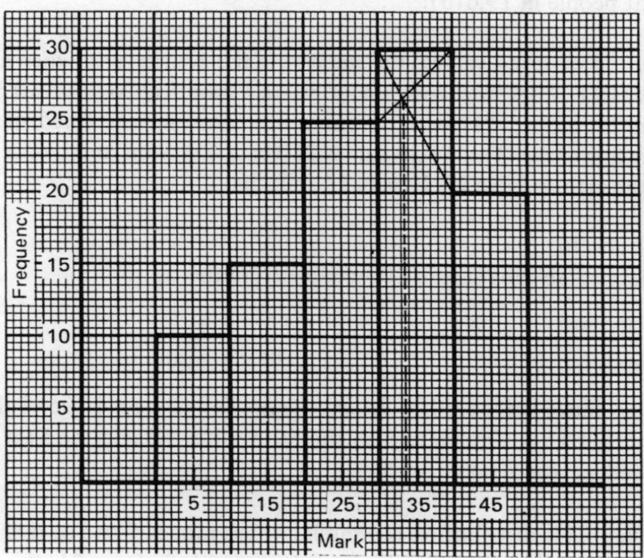

Fig 19.6 *histogram of marks of 100 students*

Solution The modal class has mark 35. To estimate the mode, diagonals are drawn as shown in the diagram, and the properties of similar triangles are used (Section 11.6). The position of the mode is about one-third of the interval, that is at $3\underline{3}$.

Comment The mean of the distribution could also be estimated from the histogram. Each class mark is multiplied by its class frequency and the sum of the products calculated. This total is then divided by 100, the number of students.

The mean mark is estimated as

$$\frac{50 + 225 + 625 + 1050 + 900}{100} = \underline{28.5}$$

(c) The median

The median of a set of numbers is the middle one when they are arranged in order of magnitude, or the mean of the two middle values. For example, the median of 2, 3, 7, 8, 10 is 7 and the median of 1, 2, 3, 4, 5, 6, is $(3 + 4)/2$ or $3\frac{1}{2}$.

Example 19.9

What was the median age of the seven people in Example 19.6(a) and of the eight people in 19.6(b)?

Solution For seven values, the median is the fourth, and when arranged in increasing order the ages are 20 years 8 months, 20 years 10 months, 21 years 2 months, 21 years 3 months, . . . and so the median age is 21 years 3 months. The next value is 21 years 4 months and so the median age of the eight people is 21 years $3\frac{1}{2}$ months.

Example 19.10

Find the median shoe size of 100 people from the distribution

Shoe size	3	4	5	6	7	8
Frequency	5	15	30	30	13	7

Solution Since there are 100 values, the median is the average of the 50th and 51st. $5 + 15 + 30 = 50$ and so the 50th value is 5 and the 51st value is 6; the median size is therefore $5\frac{1}{2}$.

Comment This distribution is bimodal, it has modes 5 and 6. If the mode had been estimated from a histogram, it would have been $5\frac{1}{2}$, the same as the median, but this is not generally true.

19.5 GRAPHS OF CUMULATIVE FREQUENCY

The *cumulative frequency* of a distribution is obtained by adding the frequency of each successive class to the total frequency of the lower classes. the cumulative frequency of each class is then plotted against the upper boundary of that class and the points are joined by a smooth curve called a cumulative frequency curve.

For a continuous variate the median value of a frequency distribution is estimated from a cumulative frequency curve, as shown in the next Example.

Example 19.11

Draw a cumulative frequency curve for the distribution of times shown in Fig. 19.4 and use the graph to estimate the median time.

Solution The frequency distribution table is rewritten to include the cumulative frequency of each class. Cumulative frequency must always extend from zero to the total number of values in the distribution.

Cumulative Frequency of Times of Assembly of a Component on 90 Occasions

Time (min)	*Frequency*	*Cumulative frequency*	*Time (min)*	*Frequency*	*Cumulative frequency*
1.65 (and under)	0	0	3.45	7	84
1.95	4	4	3.75	4	88
2.25	5	9	4.05	0	88
2.55	25	34	4.35	1	89
2.85	25	59	4.65	1	90
3.15	18	77			

The graph is shown in Fig. 19.7. The median value is the time corresponding to the mid-point of the frequency scale, that is a cumulative frequency of 45, and it is 2.7 minutes on the graph.

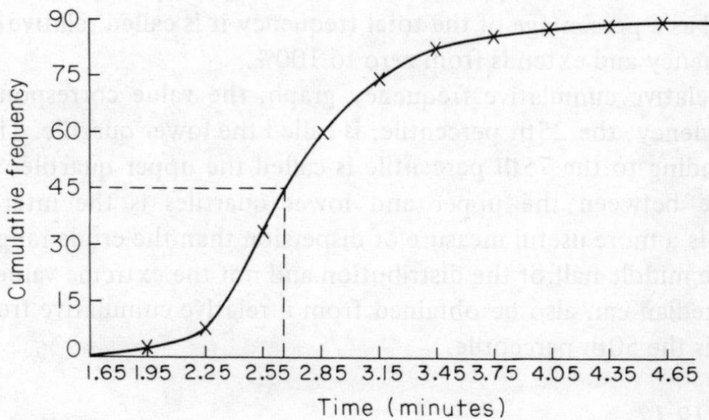

Fig 19.7 *cumulative frequency graph of times of assembly of a component*

19.6 MEASURES OF DISPERSION

The dispersion of a variate is the amount the values are spread out from the average. Only two measures of dispersion are required at this level, they are the *range* and the *interquartile range* of a frequency distribution.

(a) Range

The range of a distribution for a discrete variate is the difference between the greatest and the least values. For a continuous variate it is the difference

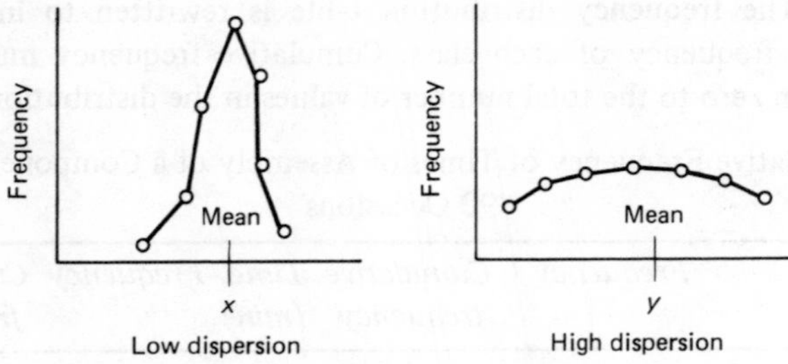

Fig 19.8

between the upper boundary of the highest class and the lower boundary of the lowest class, that is the greatest and least possible values.

For example, the range of the distribution of times in Example 19.11 is $4.65 - 1.65 = 3$ minutes. The range of shoe sizes in Example 19.10 is $8 - 3 = 5$.

(b) Interquartile range

When the cumulative frequency of each class in a frequency distribution is expressed as a percentage of the total frequency it is called relative cumulative frequency and extends from zero to 100%.

In a relative cumulative frequency graph, the value corresponding to 25% frequency, the 25th percentile, is called the lower quartile. The value corresponding to the 75th percentile is called the upper quartile, and the difference between the upper and lower quartiles is the interquartile range. It is a more useful measure of dispersion than the crude range, since it uses the middle half of the distribution and not the extreme values.

The median can also be obtained from a relative cumulative frequency curve: it is the 50th percentile.

Example 19.12

Four hundred children were given new shoes and the table shows the number of complete months before the shoes needed to be replaced. Draw a relative cumulative frequency graph of the distribution and from the curve estimate (a) the median time, (b) the interquartile range, (c) the number of children whose shoes lasted more than 9 months.

Time (mth)	*Frequency*	*Time (mth)*	*Frequency*	*Time (mth)*	*Frequency*
1	0	5	48	9	39
2	10	6	66	10	27
3	21	7	76	11	20
4	30	8	52	12	11

Solution First the cumulative frequency is calculated, and expressed as a percentage of 400.

Time (mth)	*Cumulative frequency*	*Relative frequency (%)*	*Time (mth)*	*Cumulative frequency*	*Relative frequency (%)*
1 (or less)	0	0	7 (or less)	251	$62\frac{3}{4}$
2	10	$2\frac{1}{2}$	8	303	$75\frac{3}{4}$
3	31	$7\frac{3}{4}$	9	342	$85\frac{1}{2}$
4	61	$15\frac{1}{4}$	10	369	$92\frac{1}{4}$
5	109	$27\frac{1}{4}$	11	389	$97\frac{1}{4}$
6	175	$43\frac{3}{4}$	12	400	100

The points $(1, 0)$, $(2, 2\frac{1}{2})$, etc., up to $(12, 100)$ are plotted and joined by a smooth curve, shown in Fig. 19.9.

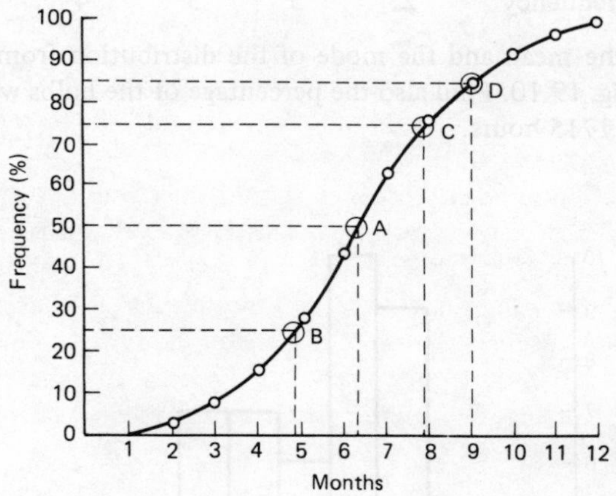

Fig 19.9 *relative cumulative frequency of the life of children's shoes*

(a) The 50th percentile is between 6 and 7 months and the median is estimated from the graph as 6.3 months (point A).
(b) The lower quartile, (point B) is at 4.8 months and the upper quartile (point C) is at 7.9 months. The interquartile range is therefore 3.1 months.
(c) The shoes of $85\frac{1}{2}\%$ of the children lasted 9 months or less (point D). Hence, $14\frac{1}{2}\%$ of the 400, that is 58 children wore their shoes for more than 9 months.

Another measure of dispersion which is sometimes useful is the *semi-interquartile range*; it is half the value of the interquartile range, as the name suggests.

Exercise 19.2

(1) Calculate the mean of the numbers (a) 2, 4, 1, 7; what 5th number would make the mean 6? (b) 12, 20, 20, 22, 21; what additional number would make the mean 17?

(2) The mean mark of n students was 40 but if another student's mark of 12 had been included, the mean would have changed to 36. What was the value of n?

(3) If the mean of three different positive integers is 7 and the median value is 4, what is the least possible difference between the lowest and highest numbers?

(4) Calculate the value of the mean and the boundaries of the modal class for the following distributions.

(a)

Class mark	6	9	12	15	18	21
Frequency	1	3	5	6	4	2

(b)

Class mark	2.5	7.5	12.5	17.5	22.5
Frequency	2	3	5	4	2

(5) Estimate the mean and the mode of the distribution from the histogram of Fig. 19.10. Find also the percentage of the bulbs which lasted more than 1715 hours.

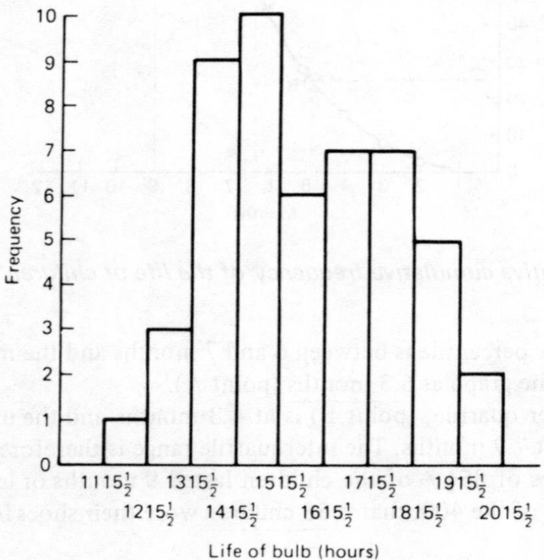

Fig 19.10 *frequency distribution of the life of electric light bulbs*

(6) The average mark obtained in a physics examination was 48% for one group of 30 students and 35% for another group of 24 students. Calculate the average mark for the whole group of 54.

(7) The following table shows the mass of packets of indigestion powder in a sample box. State the boundaries of the modal class and estimate the mean and the mode of the distribution.

Mass (g)	2.0	2.1	2.2	2.3	2.4	2.5	2.6
Number of packets	1	3	32	21	7	1	1

(8) The time needed to fasten four wheels to a model car on 24 occasions, measured to the nearest second, was 8, 6, 7, 6, 9, 5, 7, 7, 8, 10, 5, 9, 7, 6, 9, 6, 8, 7, 6, 10, 9, 6, 5, 11. Estimate the mean, median, and mode of these times.

(9) Find (a) the median weight, (b) the interquartile range from the following table of the weight of tea bags used in a small factory canteen.

Weight (g)	2.48 to 2.50	2.50 to 2.52	2.52 to 2.54
Number of bags	5	14	19
Weight (g)	2.54 to 2.56	2.56 to 2.58	2.58 to 2.60
Number of bags	21	9	2

(10) Draw a graph of the relative cumulative frequency of average weekly income of selected households in a small town.

Under £30	0%	£80 to £90	11.6%
£30 to £50	4.6%	£90 to £100	26.4%
£50 to £70	12.5%	£100 to £120	21.5%
£70 to £80	11.0%	£120 to £140	12.4%

From your graph estimate (a) the lower quartile, (b) the interquartile range, (c) the percentage of the households having a weekly income of not more than £95.

(11) The following table shows the distribution of marks obtained by candidates in an examination.

Class mark %	10	20	30	40	50	60	70	80
Frequency	85	120	200	382	372	200	90	21

Draw a graph of the relative cumulative frequency and from the curve find (a) the median mark, (b) the number of candidates who obtained less than 45%, (c) the pass mark, if 60% of the candidates passed the examination, (d) the interquartile range.

CHAPTER 20

PROBABILITY

20.1 OUTCOMES AND EVENTS

A possible result of an experiment or game is called an *outcome* and each combination of outcomes is an *event*. For instance, if an experiment consists of throwing two dice, the possible outcomes are all the ordered pairs such as (2,3) (6,6) (5,1) and there are 36 altogether. One event, A, might be showing the same number on both dice and event B might be a total score of 9.

The probability that an event will occur is defined as the number of outcomes favourable to the event divided by the number of possible outcomes, and the probability that event A will occur is written $p(A)$

$$p(A) = \frac{6}{36} \text{ or } \frac{1}{6}$$

since there are only six outcomes with both dice showing the same number (1, 1) (2, 2) (3, 3) (4, 4) (5, 5) (6, 6). This assumes that the dice are unbiased so that each number has the same chance.

Similarly

$$p(B) = \frac{4}{36} \text{ or } \frac{1}{9}$$

since a total score of 9 could occur in four ways (3, 6) (6, 3) (4, 5) (5, 4).

Probabilities may be expressed as common or decimal fractions or as percentages, and must be between 1 and 0. A probability of 1 means that an event is certain to occur.

Example 20.1

Fig. 20.1 represents the distribution of scores when two fair dice were thrown 160 times. Find the probability that for any one throw (a) the score was not more than 5,(b) the score was more than 5 but less than 10.

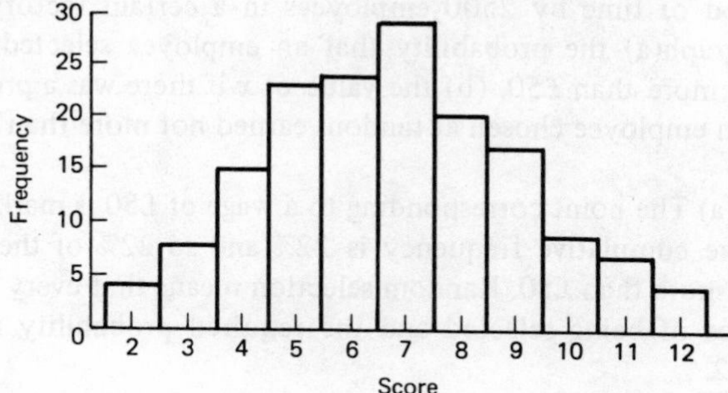

Fig 20.1 ***histogram of distribution of scores in 160 throws of two dice***

Solution The histogram has been drawn as though the scores were corrected to the nearest integer, although they must be whole numbers.

The total area represents the total frequency of 160 throws (Section 19.3).

(a) The area of the first four bars represents the number of scores which were 5 or less, 4 + 8 + 15 + 23 = 50. Hence, the probability that the score was not more than 5 is 50/160 = 5/16.

(b) A score of more than 5 but less than 10 is 6, 7, 8 or 9 and the area representing it is 24 + 29 + 20 + 17 = 90. The probability is therefore 90/160 = 9/16.

Example 20.2

Fig. 20.2 is a relative cumulative frequency curve of the wages earned in a

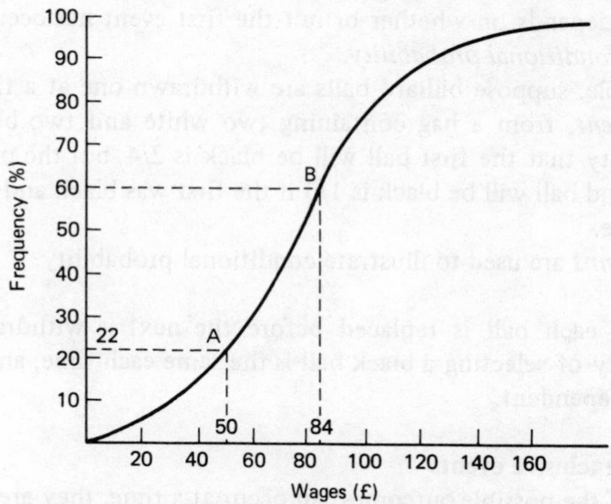

Fig 20.2 ***graph of relative cumulative frequency of wages earned by 2500 people***

given period of time by 2500 employees in a certain factory. Estimate from the graph(a) the probability that an employee selected at random earned not more than £50, (b) the value of x if there was a probability of 60% that an employee chosen at random earned not more than £x.

Solution (a) The point corresponding to a wage of £50 is marked A in the diagram; the cumulative frequency is 22% and so 22% of the employees earned not more than £50. Random selection means that every one has the same chance of being selected and the required probability is therefore 22% or 0.22.

(b) 60% of the employees earned not more than £x and this corresponds to point B on the graph. The wage associated with it is estimated as £84 and so $x = 84$.

20.2 INDEPENDENT EVENTS AND CONDITIONAL PROBABILITY

(a) Independent events

When the result of one experiment has no effect on the result of a second experiment, the two are said to be independent. For example, if a coin is tossed many times the probability that it will show a head is the same every time, 1/2, since there are only two possible outcomes. (The chance that it will land on edge is so small that it can be neglected.)

Venn diagrams can be used to illustrate independent events.

(b) Conditional probability

When two events are not independent, the probability that the second event occurs depends on whether or not the first event has occurred, and this is called *conditional probability*.

For example, suppose billiard balls are withdrawn one at a time, *without replacement*, from a bag containing two white and two black balls. The probability that the first ball will be black is 2/4, but the probability that the second ball will be black is 1/3 if the first was black and 2/3 if the first was white.

Tree diagrams are used to illustrate conditional probability.

Comment If each ball is replaced before the next is withdrawn, then the probability of selecting a black ball is the same each time, and the two events are independent.

(c) Mutually exclusive events

If only one of the possible outcomes can occur at a time, they are *mutually exclusive*. The events ‘drawing a white ball’ and ‘drawing a black ball’ are mutually exclusive, and so are ‘showing a head’ and ‘showing a tail’ on tossing a coin.

20.3 THE USE OF VENN DIAGRAMS TO REPRESENT INDEPENDENT EVENTS

If the possible outcomes of an experiment are considered as the elements of a universal set and various events as subsets, set notation and the operations union and intersection of sets can be used in calculating probabilities.

Suppose the universal set contains the 52 possible outcomes when a card is chosen at random from a well-shuffled pack. If A is the event 'selecting a diamond' and B is the event 'selecting a king', set A can be regarded as a subset with 13 elements and set B a subset with 4 elements. The set $A \cap B$ has only one element since there is only one king of diamonds in a pack; the set $A \cup B$ has 16 elements. The sets are shown in the Venn diagram of Fig. 20.3(a).

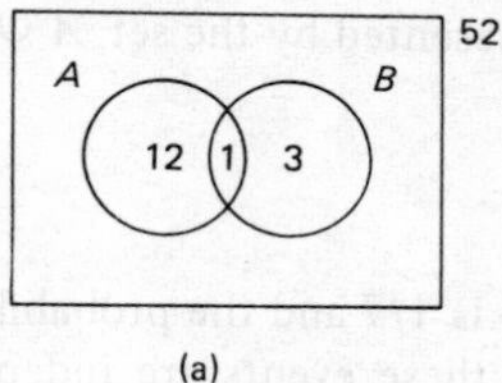

(a)

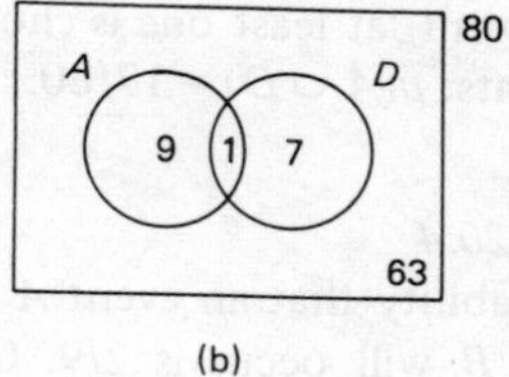

(b)

Fig 20.3

To find the probability of a particular event, the number of elements in the subset representing it is divided by the number in the universal set. Thus

$$p(A) = \frac{13}{52},\ p(B) = \frac{4}{52},\ p(A \cap B) = \frac{1}{52}$$

The sum and product rules of probability

The addition formula given in Section 9.4(c) for the number of elements $n(A) + n(B) = n(A \cup B) + n(A \cap B)$, can be adapted to give probabilities by dividing each term by $n(\mathscr{E})$. Since $p(A) = n(A)/n(\mathscr{E})$ and $p(B) = n(B)/n(\mathscr{E})$

$$\boldsymbol{p(A) + p(B) = p(A \cup B) + p(A \cap B)}$$

$$\boldsymbol{p(A) \times p(B) = p(A \cap B)}$$

$p(A \cup B)$ is the probability that at least one of the events will occur. $p(A \cap B)$ is the probability that both the events A and B will occur. If the events are mutually exclusive $p(A \cap B) = 0$.

Example 20.3

Angela is one of eight female players in a tennis club and David is one of the ten male players. If one mixed doubles pair is chosen by drawing lots, calculate the probability (a) that Angela is in the team but David is not, (b) that at least one of them is chosen. Draw a Venn diagram to show the these probabilities.

Solution Let the event A be 'Angela is chosen' and the event D 'David is chosen' and since the team was chosen by drawing lots the events A and D are independent.

$n(\mathscr{E}) = 80$, because any one of the eight women could partner any of the ten men, $n(A) = 10$, $n(D) = 8$. Only one team contains both Angela and David, so $n(A \cap D) = 1$. The Venn diagram is shown in Fig. 20.3(b).

(a) The event 'Angela is chosen but David is not' is represented by the set $A \cap D'$ and it has 9 elements: $p(A \cap D') = \underline{9/80}$.

(b) The event 'at least one is chosen' is represented by the set, $A \cup D$ with 17 elements: $p(A \cup D) = \underline{17/80}$.

Example 20.4

The probability that an event A will occur is 1/7 and the probability that an event B will occur is 2/9. Given that these events are independent, calculate the probability (a) that both A and B occur, (b) that neither occurs, (c) that at least one of them occurs, (d) that exactly one occurs.

Solution

(a) $p(A) = 1/7$, $p(B) = 2/9$ and since the events are independent

$$p(A \cap B) = p(A) \times p(B) = \frac{1}{7} \times \frac{2}{9} = \frac{2}{63}$$

The probability that both events occur is $\underline{2/63}$.

(b) The probability that A will not occur $p(A')$ is $1 - 1/7 = 6/7$ and similarly, $p(B') = 1 - 2/9 = 7/9$.

The probability that neither will occur is $p(A' \cap B')$, and since the events are independent this is

$$p(A') \times p(B') = \frac{6}{7} \times \frac{7}{9} = \underline{\frac{2}{3}}$$

(c) The probability that at least one occurs is $p(A \cup B)$ and from the addition formula $p(A) + p(B) = p(A \cup B) + p(A \cap B)$

$$p(A \cup B) = \frac{1}{7} + \frac{2}{9} - \frac{2}{63} = \frac{21}{63} = \underline{\frac{1}{3}}$$

(d) The probability that exactly one occurs is

$$p(A \cup B) - p(A \cap B)$$

or $$\frac{1}{3} - \frac{2}{63} = \frac{19}{63}$$

Exercise 20.1

(1) What is the probability that a single throw of a true die will result in a score of 4 or more?

(2) A card is chosen at random from a well shuffled pack of 52 playing cards. What is the probability that it is (a) a heart, (b) a seven, (c) the 4 of diamonds, (d) the ace or king of clubs?

(3) The table shows the marks obtained by a class for a test question.

Mark	1	2	3	4	5	6	7
Number of students	2	2	5	6	8	5	2

Calculate (a) the mean mark, (b) the probability of selecting at random a student who obtained a mark (i) greater than the mean, (ii) 4 or 5.

(4) Three unbiased coins are tossed into the air together. What is the probability of obtaining three 'heads'? If the coins are tossed again, what is the probability of obtaining three 'heads' at both the first and second throws?

(5) Forty equal discs numbered 1 to 40 are placed in a bag. If one disc is chosen at random what is the probability that the number on it contains a 7?

(6) A bag contains 14 discs, x of which are blue. What is the probability that a disc picked at random will be blue?

(7) A random selector contains 40 balls of different colours. If the probability of selecting a red ball is 0.4 how many red balls are in the selector?

(8) A multiple choice test provides five answers to each question but only one is correct. What is the probability of obtaining correct answers to the first four questions by sheer chance?

(9) If the probabilities of winning two independent events are 3/4 and 2/9 what is the probability of winning (a) both events, (b) only one event?

(10) When a fair die is thrown twice what is the probability of scoring (a) first 1 and then 2, (b) a double 6, (c) 11?

(11) The probability that a student in a college on a certain day will be carrying a set of four-figure tables is 2/7. (a) If three students are selected at random, what is the probability that they are all carry-

ing tables? (b) Calculate the probability that of four students picked at random two will have tables and two will not.

(12) In a set of 100 students, 37 studied mathematics, 32 physics and 31 chemistry. 9 students took both mathematics and physics, 12 took mathematics and chemistry, 11 physics and chemistry, and 28 none of these subjects. Assuming that the selection is random and with replacement, so that each choice is made from 100 students calculate (a) the probability that the first student chosen was studying all three subjects, (b) the probability that he took mathematics only, (c) the probability that of the first two chosen, one was studying other subjects.

20.4 TREE DIAGRAMS

Conditional probabilities are represented on a different type of diagram called a *tree diagram*. Every outcome of an experiment is represented by a different branch and when following a path from one branch to another the probabilities are multiplied together. This is the product rule for conditional probabilities. When branches come from one point the probabilities must add to 1. The method of drawing tree diagrams is shown in the following examples.

Example 20.5

Two balls are taken without replacement from a bag containing twelve similar balls, 4 red, 2 yellow and 6 blue. Represent the possible outcomes on a tree diagram and find the probability of selecting (a) two blue balls, (b) one red and one yellow ball.

Solution Since there are 12 balls all the same size, the probability of selecting at the first draw a red ball is 4/12, a yellow ball 2/12 and a blue ball 6/12.

The tree diagram is shown in Fig. 20.4; following the red branch, of the 11 balls left in the bag for the second draw only 3 are red. Hence the probability of selecting a red ball the second time is 3/11.

(a) The probability of selecting a blue ball twice is $(6/12) \times (5/11) = \underline{5/22}$.

(b) The probability that one ball is red and the other yellow is $(2/33) + (2/33) = \underline{4/33}$. It is in two parts since the result can be obtained in two ways.

Example 20.6

A game consists of tossing a coin. If a 'head' shows, a fair die is thrown once and the score noted; if a 'tail' shows, a card is selected from a well shuffled pack of 52 playing cards. Hearts score 1, diamonds 2, clubs 3, spades 4.

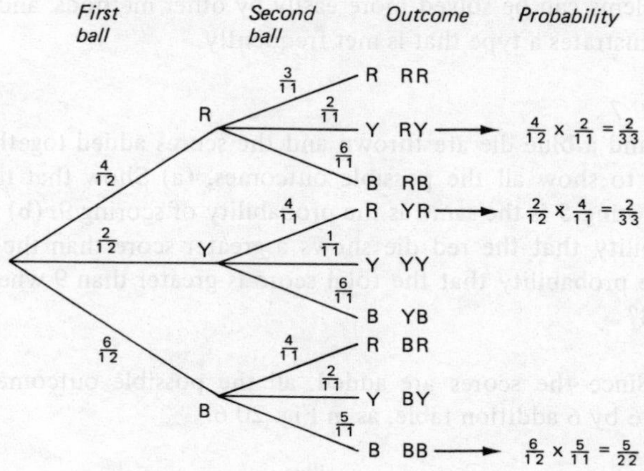

Fig 20.4

Draw a tree diagram to show all the possible outcomes and find the probability of scoring 3 or less.

Solution A score of 3 or less can be obtained in six ways and they are shown in the diagram of Fig. 20.5.

The probability of scoring 3 or less is given by the sum

$$\frac{1}{12} + \frac{1}{12} + \frac{1}{12} + \frac{1}{8} + \frac{1}{8} + \frac{1}{8} = \frac{5}{8}$$

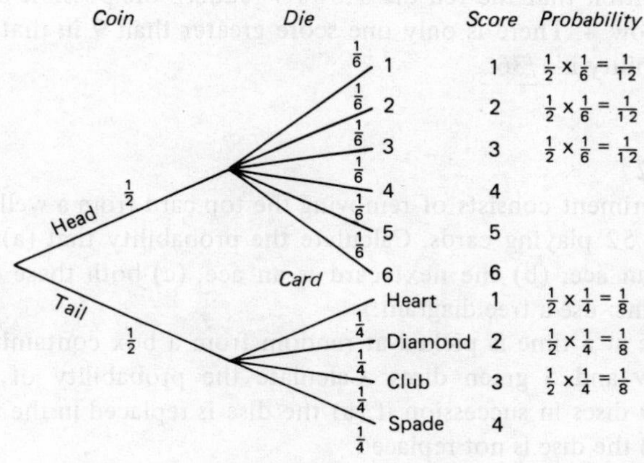

Fig 20.5

Some problems can be solved more easily by other methods, and the next Example illustrates a type that is met frequently.

Example 20.7
A red die and a blue die are thrown and the scores added together. Draw up a table to show all the possible outcomes. (a) Show that the probability of scoring 5 is the same as the probability of scoring 9. (b) Calculate the probability that the red die shows a greater score than the blue. (c) What is the probability that the total score is greater than 9 when the red die shows 4?

Solution Since the scores are added, all the possible outcomes can be shown in a 6 by 6 addition table, as in Fig. 20.6.

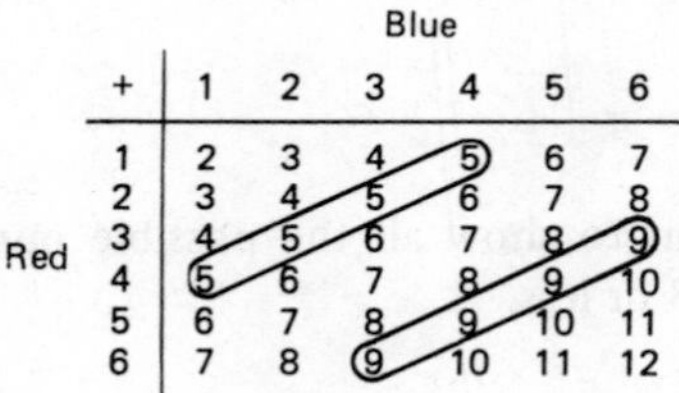

+	1	2	3	4	5	6
1	2	3	4	5	6	7
2	3	4	5	6	7	8
3	4	5	6	7	8	9
4	5	6	7	8	9	10
5	6	7	8	9	10	11
6	7	8	9	10	11	12

Fig 20.6

(a) There are four 5s in the table and four 9s, showing that the probability of scoring 5 is the same as scoring 9, 4/36.
(b) The red die shows a greater score than the blue in the bottom left part of the table, below the diagonal, and it includes 15 of the 36 outcomes. The probability is 15/36 or 5/12.
(c) The condition that the red die shows 4 reduces the possible outcomes to the 6 in row 4. There is only one score greater than 9 in that row and so the probability is 1/36.

Exercise 20.2
(1) An experiment consists of removing the top card from a well shuffled pack of 52 playing cards. Calculate the probability that (a) the first card is an ace, (b) the next card is an ace, (c) both these cards are aces. (Hint: use a tree diagram.)
(2) One disc at a time is picked at random from a box containing 5 red, 5 yellow and 5 green discs. Calculate the probability of selecting 3 yellow discs in succession if (a) the disc is replaced in the box each time, (b) the disc is not replaced.
(3) A, B and C are three bags containing coloured balls as indicated in Fig. 20.7 and balls drawn at random are not replaced.

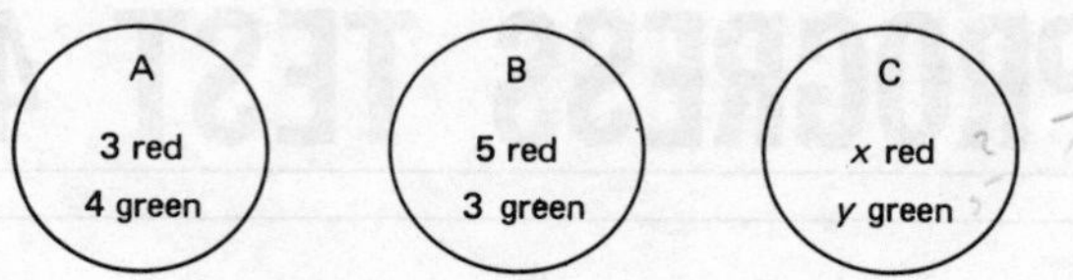

Fig 20.7

(a) A ball is drawn from A. What is the probability that it is red?
(b) A ball is drawn from B. Find the probability that it is green.
(c) Now a second ball is drawn from A. What is the probability that it is red?
(d) Two more balls are removed from B. Calculate the probability that all three balls removed are green.
(e) If the probability of drawing a red ball from C is 2/7 find a relation connecting x and y.

(4) A two-digit number is formed by choosing two digits at random, without replacement, from the set $\{1, 2, 3, 4, 5, 6,\}$. Calculate the probability that the number (a) is even, (b) is divisible by 5, (c) contains the digit 2, (d) is greater than 50.

(5) Two bags contain similar discs, bag A has 3 red and 5 black; bag B has 4 red and 3 black discs. If one disc is removed from A and put into B and *then* one is removed from B and put into A, calculate the probability (a) that the two discs are red, (b) that one is red and the other is black, (c) that at the end bag B contains 4 red and 3 black discs.

(6) For a side show at a school fair, a revolving barrel contained four red balls, four white, four blue and some black balls. The customer paid one penny to take a ball and was paid 1 p for a red ball, 2 p for a white ball and 3 p for a blue ball but nothing for a black ball, and the ball was replaced each time. What is the least number of black balls necessary in order that the stallholder should make a profit?

PROGRESS TEST 4

A

(1) Find the Cartesian equation of the line (a) with gradient -4, passing through the point (i) $(1, 4)$, (ii) $(-2, -1/2)$, (b) joining $(1, -5)$ and $(-3\frac{1}{2}, 4)$, (c) with gradient -3, which cuts (i) the x-axis at $x = 4$, (ii) the y-axis at $y = -1$, (d) through the origin, parallel to the line $2y = 5 - 6x$. **(2)** Draw the graph and measure the angle made with the x-axis by the line (a) $y = 0.4x + 3$, (b) $y = 3.5 - 2x$. **(3)** Find the point of intersection of the graphs of $2y + 3x = 2$ and $2x - 6y = 5$. **(4)** What is the equation of a curve which cuts the axes at $(-4, 0)$, $(3, 0)$ and $(0, -2)$? **(5)** Write down the equation of a straight line which must be drawn to intersect the graph of $y = x^2 - 4x - 12$ in order to solve (a) $x^2 - 4x - 12 = 0$, (b) $x^2 - 4x - 9 = 0$, (c) $x^2 - 3x + 4 = 0$. **(6)** State an equation for which the solution is given by the intersection of the graphs of (a) $y = x^2 + 1$ and $y = 3 + (1/x)$, (b) $2y = x + 4$ and $y = x^3 + (1/x^2)$. **(7)** A rectangle x cm by y cm has a perimeter of 12 cm to 20 cm inclusive and the length is twice the width. If x and y are integers calculate the greatest and the least possible value for the area of the rectangle. **(8)** Find the greatest and the least value of x in the set of integers satisfying simultaneously $y - x \leqslant 4, y \geqslant 3, x \leqslant 5$. **(9)** Calculate the average speed of a car travelling (a) 240 km in 3 hours 20 minutes, (b) 243 km in 2 hours 42 minutes. **(10)** A motorist travelled x km in y hours. Find an expression for the time the journey would take if he travelled at twice the speed but stopped on the way for 75 minutes. **(11)** The velocity v cm/s of a body t seconds after starting from rest is given by $v = ut - 2t^2$. If the body comes to rest after 5 seconds, find the value of the constant u. Using this value, plot the mapping $t \rightarrow v$ and find (a) the maximum velocity, (b) the distance travelled in the first four seconds of the motion. **(12)** If the mean of eight numbers if 13 and the mean of four of them is 9, calculate the mean of the remaining four. **(13)** Find the fifth number which must be included with 7, 4, 8, 13 to make the average 12. What is the median value of the five numbers? **(14)** The table shows the number of questions successfully completed in a test by 70 students. One mark was awarded for each correct answer.

Number of marks	0–3	4–7	8–11	12–15	16–19	20–23	24–27
Number of students	7	8	15	19	11	6	4

Estimate the mean and the mode of the distribution.
(15) Construct a cumulative frequency graph for the data in Question 14 and use it to estimate (a) the median mark, (b) the semi-interquartile range, (c) the pass mark if 40% of the students passed, (d) the probability that a student picked at random obtained more than 50%. **(16)** A bag contains six similar discs marked 1 to 6 and an experiment consists of taking two discs in succession. (a) If the first disc is replaced, (b) if it is *not* replaced, calculate the probability of obtaining (i) two even numbers, (ii) numbers adding to make 7.

B

(1) Differentiate with respect to x (a) $(x + 2)^2$, (b) $x^4 + (6/x^2)$, (c) $\sqrt{(6x^3)}$. (2) Calculate the gradient of the curve $y = x - 2x^2$ at (a) (0, 0), (b) (−1, −3). (3) Find the equation of the tangent at (2, −4) to the curve $y = x^3 - 3x^2$. (4) A particle moves in a straight line OP so that its distance from O, s m, after t s is given by $s = 3t^2 - t^3$. Calculate (a) the velocity and acceleration when $t = 3$, (b) the maximum distance from O. (5) Evaluate (a) $\int (2/\sqrt{x})\, dx$, (b) $\int \sqrt[3]{x}\, dx$, (c) $\int (3x - 2)^2\, dx$, (d) $\int_{-1}^{3} (3x + 1)\, dx$. (6) Calculate (a) the area enclosed by the curve $y = 4 - x^2$ and the x-axis, (b) the volume of the solid formed when this area is revolved once about the x-axis. (7) Find an expression for the rate at which the area A of a circle increases with the radius r. (8) A body starts from rest and its velocity after t seconds is given by $v = t(3 + 8t)$ cm/s. Calculate (a) its acceleration, (b) the distance travelled, when $t = 4$.

CHAPTER 21

TRIGONOMETRY

In geometry the angle between two lines is measured in degrees, but in *trigonometry* the size of an acute angle is described by the ratio of two of the sides of a right angled triangle containing the angle. The main advantage of using trigonometrical ratios is that distances can be calculated which can not be measured directly, in astronomy, surveying, navigation and travel problems. Another advantage is that angles can be introduced into equations without any change of units because ratios are numbers, and this is important in mathematics and physics.

21.1 SINE, COSINE AND TANGENT OF ACUTE ANGLES

Three trigonometrical ratios are considered in this book and they are called the sine, the cosine and the tangent of an angle. They are sometimes called trigonometrical functions since for each angle there is exactly one value for each of the trigonometrical ratios.

In Fig. 21.1 (a) the triangles OAB, OA_1B_1 and OA_2B_2 are similar since they have the same angles (Section 11.5), and therefore the ratios of corresponding sides are the same and depend only on the size of the angle x. Fig. 21.1 (b) shows a right angled triangle containing the angle x with sides marked hypotenuse, opposite and adjacent.

Definitions

$$\text{sine } x = \frac{AB}{OA} = \frac{A_1B_1}{OA_1} = \frac{A_2B_2}{OA_2} = \frac{\text{opposite}}{\text{hypotenuse}}$$

$$\text{cosine } x = \frac{OB}{OA} = \frac{OB_1}{OA_1} = \frac{OB_2}{OA_2} = \frac{\text{adjacent}}{\text{hypotenuse}}$$

$$\text{tangent } x = \frac{AB}{OB} = \frac{A_1B_1}{OB_1} = \frac{A_2B_2}{OB_2} = \frac{\text{opposite}}{\text{adjacent}}$$

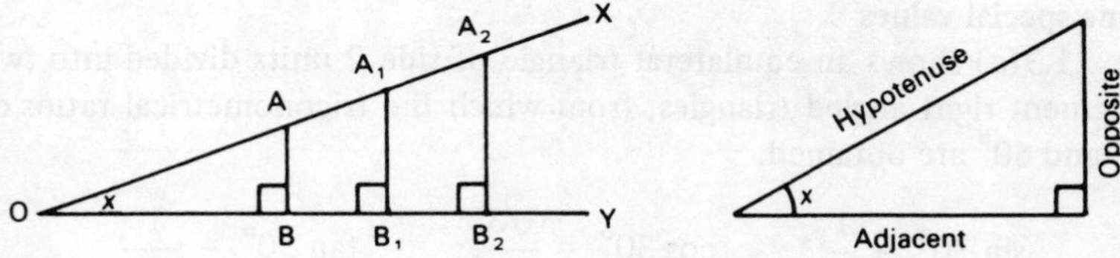

Fig 21.1

The usual abbreviations are sin, cos and tan.

Using the initial letters only the definitions may be remembered as S-O-H, C-A-H, T-O-A, and there are several mnemonics in use.

21.2 PYTHAGORAS' THEOREM IN TRIGONOMETRY

When one of the trigonometrical ratios is known for a particular angle, the others can be found from a right angled triangle containing the angle. For example in Fig. 21.2(a), $\sin x^\circ = 3/5$. From Pythagoras' theorem the third side of the triangle is of length 4 units, and therefore

$$\cos x^\circ = \tfrac{4}{5} \quad \text{and} \quad \tan x^\circ = \tfrac{3}{4}$$

Similarly, for the 5 : 12 : 13 triangle of Fig. 21.2(b)

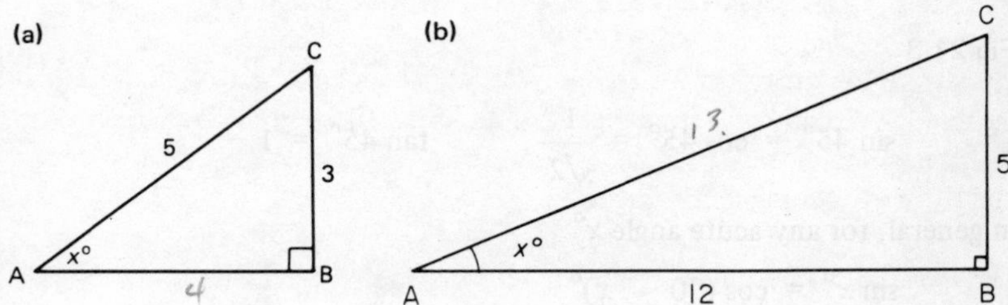

Fig 21.2

$$\tan x^\circ = \tfrac{5}{12} \Rightarrow \sin x^\circ = \tfrac{5}{13} \quad \text{and} \quad \cos x^\circ = \tfrac{12}{13}$$

In general, for any angle x

$$(\sin x)^2 + (\cos x)^2 \equiv 1$$

and also

$$\frac{\sin x}{\cos x} \equiv \tan x$$

These are trigonometrical identities, they are true for every angle x, and they can be verified for the triangles of Fig. 21.2.

Some special values

Fig. 21.3 (a) shows an equilateral triangle of side 2 units divided into two congruent right angled triangles, from which the trigonometrical ratios of 30° and 60° are obtained.

$$\sin 30^\circ = \frac{1}{2}, \qquad \cos 30^\circ = \frac{\sqrt{3}}{2}, \qquad \tan 30^\circ = \frac{1}{\sqrt{3}}$$

$$\sin 60^\circ = \frac{\sqrt{3}}{2}, \qquad \cos 60^\circ = \frac{1}{2}, \qquad \tan 60^\circ = \sqrt{3}$$

A square of side 1 unit has diagonals of $\sqrt{2}$ units, and in the isosceles triangle of Fig. 21.3 (b)

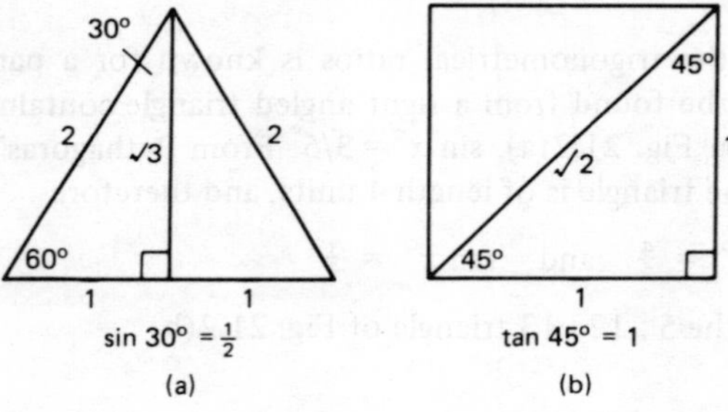

Fig 21.3

$$\sin 45^\circ = \cos 45^\circ = \frac{1}{\sqrt{2}}, \qquad \tan 45^\circ = 1$$

In general, for any acute angle x°

$$\sin x^\circ = \cos (90 - x)^\circ$$

and

$$\tan x^\circ = \frac{1}{\tan (90 - x)^\circ}$$

21.3 FOUR-FIGURE TRIGONOMETRICAL TABLES

While the trigonometrical ratios for a particular angle could be found by constructing a right angled triangle, as in Section 12.7, and measuring the sides, the values have been calculated by other means. Four-figure tables are available, giving the sine, cosine, and tangent of angles from 0° to 90° and many electronic calculators have function keys which give the ratios directly.

The tables are used in the same way as the other four-figure tables described in Chapter 5, the whole number of degrees being found on the left of the table and the decimal fractions 0.0 to 0.9 in the appropriate column.

Example 21.1
Use four-figure tables to evaluate the following functions (a) sine 35.2°, (b) sine 73.75°, (c) cosine 64.5°, (d) cosine 5.45°, (e) tangent 17.3°, (f) tangent 67.65°.

Solution

(a) sine 35.2°: 0.5764. (b) sine 73.75°: 0.9600.
(c) cosine 64.5°: 0.4305. (d) cosine 5.45°: 0.9955.
(e) tangent 17.3°: 0.3115. (f) tangent 67.65°: 2.4322.

Comment The value for 67.65° is half way between 67.6 and 67.7, but results to the nearest 0.1° are sufficiently accurate for most calculations at this level.

To save time when tables are used in calculations, the logarithms of the trigonometrical ratios of acute angles have been tabulated, and these are used in the same way as logarithms of numbers.

Inverse trigonometrical functions
When the value of a trigonometrical ratio is known, the corresponding acute angle is found by using the tables in reverse. Thus if $\sin x = 0.5764$, then $x = 35.2°$ and if $\tan x = 0.3115$, then $x = 17.3°$.

The inverse functions are called arcsin, arccos and arctan and for acute angles they are true functions; if $y = \sin x$, then $x = \arcsin y$. For instance, $30° = \arcsin 1/2$ and $45° = \arctan 1$.

Example 21.2
In a triangle XYZ, ∠Y is a right angle. Calculate the size of the other angles when (a) XY = 3 cm and YZ = 8 cm, (b) XY = 4.2 cm and XZ = 5.7 cm.

Solution The triangles are shown in Fig. 21.4.

(a) $\quad \text{Tan } \angle XZY = 3 \div 8 = 0.375 \Rightarrow \angle XZY = \arctan 0.375$

From the table of natural tangents the nearest values are 0.3739 and 0.3759 for 20.5° and 20.6°. Therefore ∠XZY = 20.55° and ∠ZXY = 69.45° (the complement of 20.55°).
(b) Sin ∠XZY = 4.2 ÷ 5.7 and in this example it is easier to use logarithms

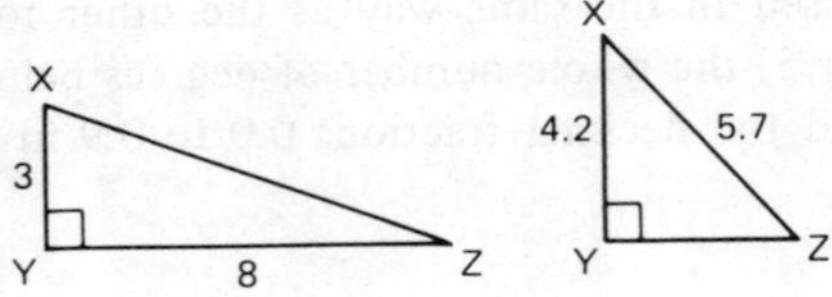

Fig 21.4

Number	Log
4.2	0.6232
5.7	0.7559
	$\overline{1}.8673$

From the table of logarithms of sines, the nearest values are $\overline{1}.8669$ for 47.4° and $\overline{1}.8676$ for 47.5°. Hence, $\angle XZY = 47.45°$ and $\angle ZXY = 42.55°$.

Example 21.3
In Fig 21.5, if $x = 76$ mm, calculate the length of (a) AD, (b) BC.

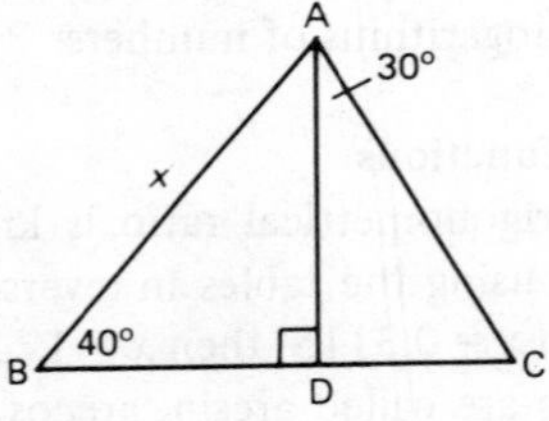

Fig 21.5

Solution (a) In triangle ABD

$$\sin 40° = AD/AB = AD/x$$

Hence

$$AD = x \sin 40° = 76 \sin 40° \text{ mm}$$
$$= \underline{48.9 \text{ mm}} \quad \text{(using a calculator)}$$

(b) $BD = x \cos 40° = 76 \cos 40°$ mm = 58.2 mm
In triangle ADC, $DC/AD = \tan 30° \Rightarrow DC = AD \tan 30°$. Hence DC = 28.2 mm and BC = 58.2 + 28.2 = 86.4 mm.

Exercise 21.1

(1) Write as the ratio of the sides of the triangle in Fig. 21.6(a) (a) $\sin 37^\circ$, (b) $\tan 37^\circ$, (c) $\cos 53^\circ$, (d) $\tan 53^\circ$.

(2) Complete the following statements:
(a) If $\sin x^\circ = 7/25$, (i) $\cos x^\circ =$, (ii) $\tan x^\circ =$,
(b) If $\tan y^\circ = 8/15$, (i) $\sin y^\circ =$, (ii) $\cos y^\circ =$.

(3) Write down the value of (a) $\cos 60^\circ$, (b) $\tan 30^\circ$, (c) $\cos 45^\circ$, (d) $\tan 45^\circ$, (e) $\sin 30^\circ$.

(4) Use four-figure tables or a calculator to find the value of (a) $\cos 35^\circ$, (b) $\sin 62.6^\circ$, (c) $\cos 31.4^\circ$, (d) $\tan 82.8^\circ$, (e) $\tan 28.1^\circ$.

(5) Find an acute angle θ such that (a) $\sin\theta = 0.4446$, (b) $\cos\theta = 0.6794$, (c) $\tan\theta = 2.2148$, (d) $\cos\theta = 0.5835$, (e) $\tan\theta = 1.8266$.

(6) Calculate the value of x in Fig. 21.6(b).

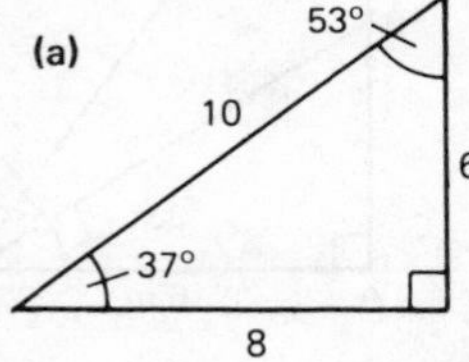

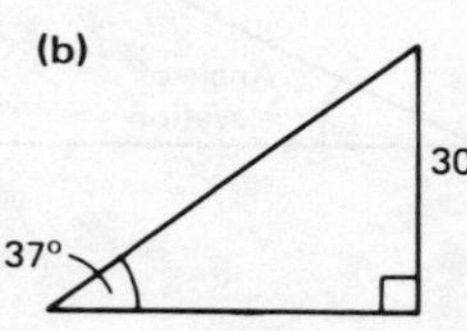

Fig 21.6

(7) Find the lengths and angles denoted by letters in Fig. 21.7.

(8) If two sides of an isosceles triangle are of length 63 mm and the included angle is 78°, calculate the length of the third side.

(9) If the diagonals of a rectangle are 89 mm and the shorter sides are 36 mm long, calculate the angles between the diagonals and the sides of the rectangle.

(10) Find the size of the acute angle between the x-axis and the line (a) $y = 4x - 5$, (b) $2y = 3x - 1$, (c) $3y = 4x + 9$.

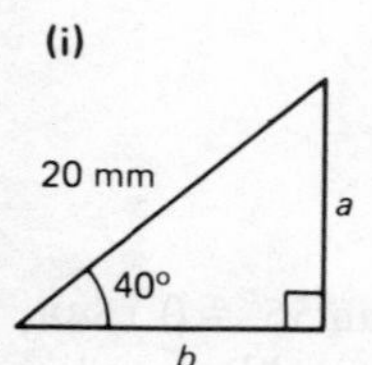

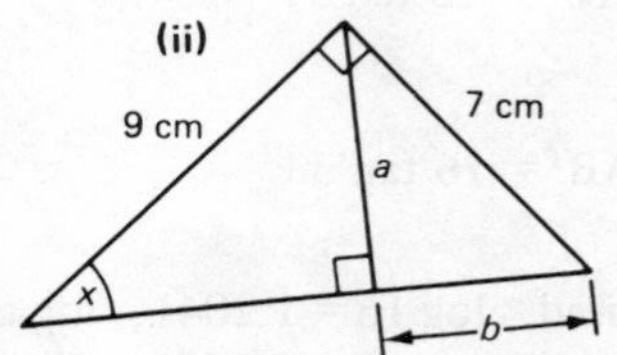

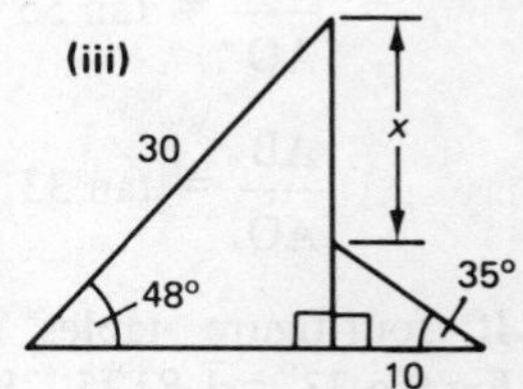

Fig 21.7

21.4 ANGLES OF ELEVATION AND DEPRESSION

In Fig. 21.8(a) AB represents the side of a building, O is a point on the same level as A and O′ is on the same level as B. ∠AOB is called the angle of elevation of B from O and ∠O′BO the angle of depression of O from B. Since BO′ is parallel to AO the angles are equal alternate angles, and for any two points the angle of elevation is always equal to the angle of depression.

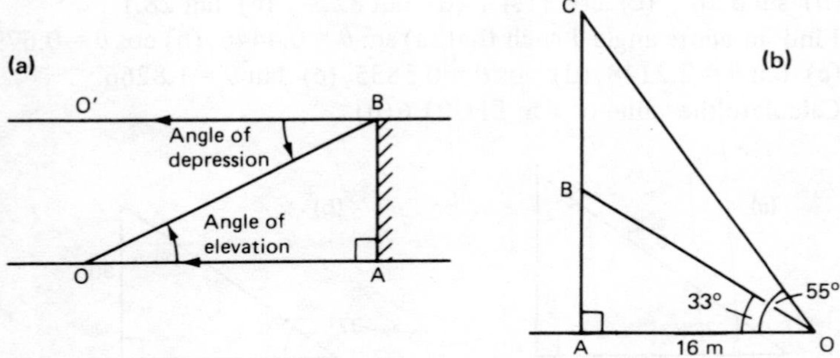

Fig 21.8

Example 21.4
At a point on the ground 16 m from the wall of a church the elevation of the bottom of a window was measured as 33° and the elevation of the top of the same window was 55°. Calculate (a) the height of the window, (b) the height of the bottom of the window above the level of the observation point.

Solution In Fig. 21.8(b), B and C represent the bottom and top of the window, AO = 16 m, ∠AOB = 33°, ∠AOC = 55°.

$$\frac{AC}{AO} = \tan 55° \Rightarrow AC = 16 \tan 55°$$

$$\frac{AB}{AO} = \tan 33° \Rightarrow AB = 16 \tan 33°$$

If four-figure tables are used, $\log 16 = 1.2041$, $\log \tan 55° = 0.1548$, $\log \tan 33° = \bar{1}.8125$. Hence AC = antilog 1.3589 = 22.85; AB = antilog 1.0166 = 10.39. The height of the window was 12.5 m; the height above A was 10.4 m.

Example 21.5

From a point on level ground the elevation of a marker due west on the top of a hill was found to be 40.2°; 50 m further east the elevation of the marker was 25.7°. Calculate the height of the hill.

Solution Suppose the required height is h m and the distance of the first observation point from a point vertically below the marker is d m, as in Fig. 21.9. Neither of these distances could be measured directly. Then

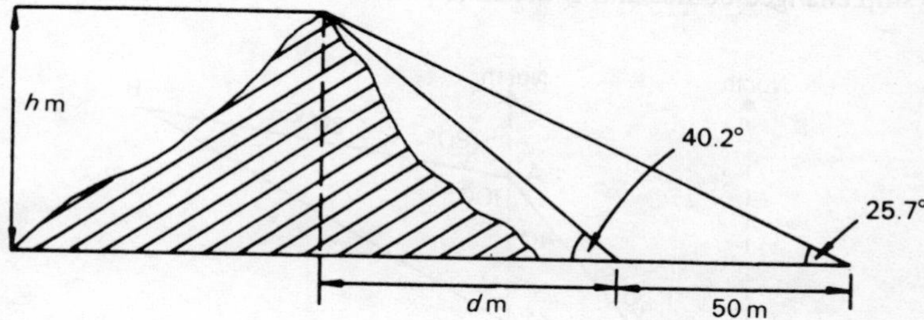

Fig 21.9

$$\frac{h}{d} = \tan 40.2° \quad \text{and} \quad \frac{h}{d + 50} = \tan 25.7°$$

$$\tan 40.2° = 0.8451; \qquad \tan 25.7° = 0.4813$$

$$h = 0.8451d = 0.4813(d + 50) = 0.4813d + 24.06$$

$$(0.8451 - 0.4813)d = 24.06 \Rightarrow d = \frac{24.06}{0.3638}$$

Hence

$$h = 0.8451 \times \frac{24.06}{0.3638} = 55.89$$

The height of the hill was <u>55.9 m</u>.

21.5 BEARINGS

The bearing of a point B from a point A is the angle turned through in a clockwise sense in changing direction from due north of A to the direction AB. Bearings can take values between 0° and 360° and a three-digit notation is used; due north is 000°, due east 090°, due south 180°, due west 270°.

Example 21.6

A ship left harbour and travelled for 20 km on a bearing 040°, then a further 20 km on a bearing 080°. (a) Calculate the final bearing of the ship from the harbour. (b) How far east of the harbour did the ship change course?

Solution In Fig. 21.10 H represents the harbour, A the point at which the ship changed course and B the final position.

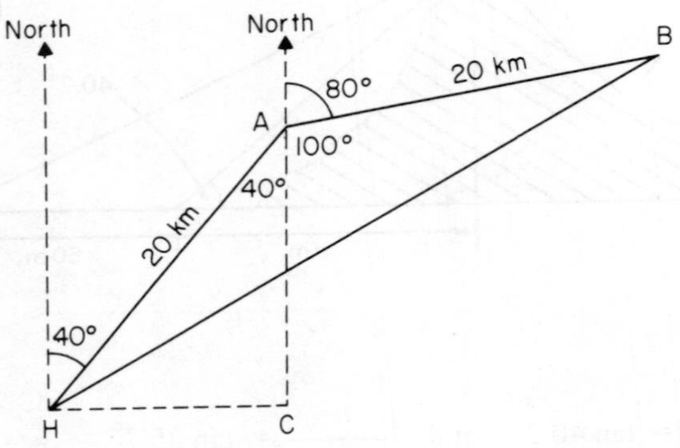

Fig 21.10

(a) HA makes an angle of 40° with the north direction at H, AB makes an angle of 80° with the north direction at A; hence ∠HAB = 140° as shown in the diagram.

Triangle HAB is isosceles, since HA = AB = 20 km. Therefore ∠AHB = ∠ABH = 20°, and HB makes an angle of 40° + 20° with the north direction. The final bearing from the harbour is <u>060°</u>.

(b) In the right angled triangle HAC, ∠AHC = 50°

$$\frac{HC}{HA} = \cos 50^\circ \Rightarrow HC = 20 \cos 50^\circ = 20 \times 0.6428$$

The ship changed course <u>12.9 km</u> east of the harbour.

Example 21.7

From a point P a tower is 200 m away on a bearing 136° and the elevation of the top T of the tower is 22°. From a point Q the tower is on a bearing 046° and the bearing of Q from P is 170°. Calculate (a) the height of the tower, (b) the distance of Q from P.

Solution (a) In Fig. 21.11 M represents the base of the tower, and the height is h m

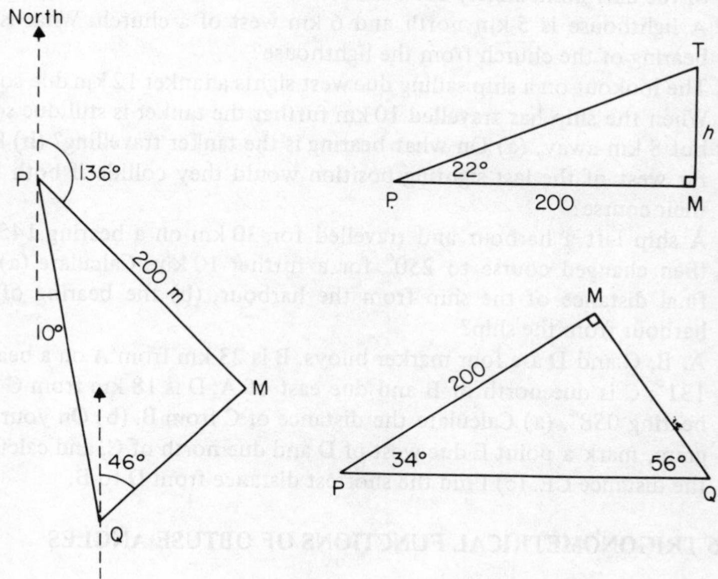

Fig 21.11

$$\frac{h}{200} = \tan 22^\circ \Rightarrow h = 200 \times 0.4040 = 80.8$$

Therefore the height is 80.8 m.

(b) In the triangle PMQ

$$\angle MPQ = 170^\circ - 136^\circ = 34^\circ$$

$$\angle MQP = 46^\circ + 10^\circ = 56^\circ$$

Therefore

the 3rd angle, $\angle PMQ = 90^\circ$ and $\dfrac{PM}{PQ} = \cos 34^\circ$

The distance PQ = 200/cos 34° = 241 m.

Exercise 21.2

(1) From a point 24 m from the outside wall of a block of flats the elevation of the roof was measured as 56°. Calculate the height of the building, assuming the ground was level.

(2) If the angle of depression of a boat from the top of a vertical cliff is 32.6° and the cliff is 80 m high, how far is the boat from the bottom of the cliff immediately below the observer?
(3) A lighthouse is 5 km north and 6 km west of a church. What is the bearing of the church from the lighthouse?
(4) The lookout on a ship sailing due west sights a tanker 12 km due south. When the ship has travelled 10 km further the tanker is still due south but 8 km away. (a) On what bearing is the tanker travelling? (b) How far west of the last sighting position would they collide if both held their course?
(5) A ship left a harbour and travelled for 30 km on a bearing 145°; it then changed course to 250° for a further 10 km. Calculate (a) the final distance of the ship from the harbour, (b) the bearing of the harbour from the ship?
(6) A, B, C and D are four marker buoys. B is 23 km from A on a bearing 131°; C is due north of B and due east of A; D is 18 km from C on a bearing 058°. (a) Calculate the distance of C from B. (b) On your diagram, mark a point E due west of D and due north of C, and calculate the distance CE. (c) Find the shortest distance from D to B.

21.6 TRIGONOMETRICAL FUNCTIONS OF OBTUSE ANGLES

In Fig. 21.12, O is the origin of Cartesian coordinates and P_1 is the point (x, y). The line OP_1 makes an acute angle α with the positive direction of the x-axis and is the length r units. By convention, the value of r is always positive.

From the definitions given in Section 12.1

$$\sin \alpha = \frac{y}{r}, \qquad \cos \alpha = \frac{x}{r}, \qquad \tan \alpha = \frac{y}{x}$$

and they are all positive since x, y and r are positive.

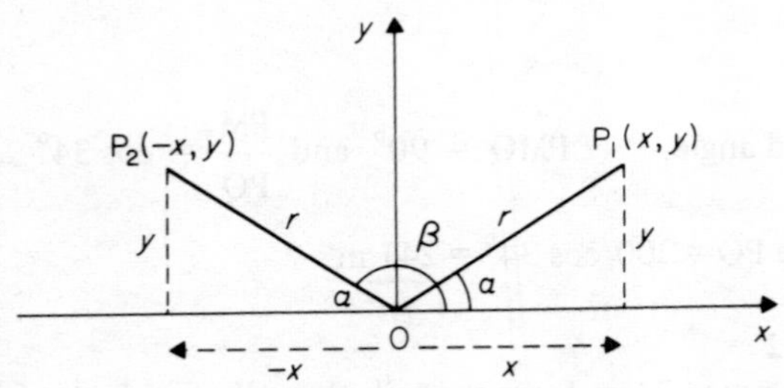

Fig 21.12

P_2 is the point $(-x, y)$, OP_2 is the length r and it makes an obtuse angle β with the positive direction of the x-axis where $\alpha + \beta = 180°$. Then

$$\sin\beta = \frac{y}{r} = \sin\alpha, \qquad \cos\beta = \frac{-x}{r} = -\cos\alpha$$

and

$$\tan\beta = \frac{y}{-x} = -\tan\alpha$$

The values are the same numerically as for the supplementary acute angle, and can be found from tables.

Example 21.8
Use four-figure tables to evaluate the following, (a) sin 130°, (b) cos 152°, (c) cos 97.2°, (d) tan 104.3°, (e) tan 111.6°.

Solution
(a) sin 130°: the supplement is 50°; sin 50° = 0.7660,
(b) cos 152°: the supplement is 28°; −cos 28° = −0.8829,
(c) cos 97.2°: the supplement is 82.8°; −cos 82.8° = −0.1253,
(d) tan 104.3°: the supplement is 75.7°; −tan 75.7° = −3.9232,
(e) tan 111.6°: the supplement is 68.4°; −tan 68.4° = −2.5257.
The negative values may be obtained directly if you have a calculator with trigonometrical function keys.

21.7 THE SOLUTION OF TRIANGLES

To solve a triangle means to calculate the other sides and angles when some are known, and we have already solved some right angled triangles. For other triangles there are special formulae, known as the *sine rule* and the *cosine rule*, which are used to calculate sides and angles; there is also a formula for the area of a triangle.

(a) The sine rule
This can be used when either two angles and one side are known, or two sides and the angle opposite one of them.

In the general triangle ABC, the length of the side opposite the vertex A is called a and A is used to denote the size of the angle at A; similarly for B and C, as in Fig. 21.13.

In triangle ACD, the altitude AD = b sin C; and in triangle ABD, AD = c sin B. Hence

$$b \sin C = c \sin B \text{ or } \frac{b}{\sin B} = \frac{c}{\sin C}$$

By considering the altitude from C, it can be shown in a similar way that

$$\frac{a}{\sin A} = \frac{b}{\sin B}$$

and therefore

$$\frac{a}{\sin A} = \frac{b}{\sin B} = \frac{c}{\sin C}$$

This equation is called the sine rule.

(b) The area rule

In Fig. 21.13, the area of triangle ABC = $\frac{1}{2}$ BC × AD. Since AD = $b \sin C$ or $c \sin B$

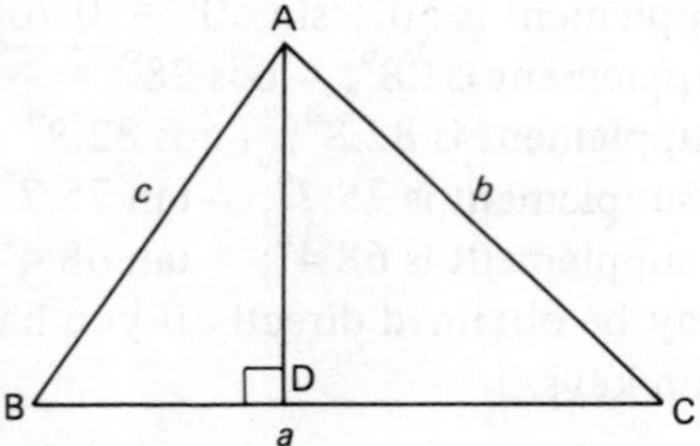

Fig 21.13

$$\text{area ABC} = \tfrac{1}{2}ab \sin C \quad \text{or} \quad \tfrac{1}{2}ac \sin B$$

By considering the altitude from C

$$\text{area ABC} = \tfrac{1}{2}cb \sin A$$

Hence

$$\text{area ABC} = \tfrac{1}{2}bc \sin A \quad \text{or} \quad \tfrac{1}{2}ac \sin B \quad \text{or} \quad \tfrac{1}{2}ab \sin C$$

This is the formula known as the area rule for a triangle; it can be used when two sides and the included angle are given.

Example 21.9

In Fig. 21.14(a) P and Q represent the positions of two coastguard stations and S is a ship. If the distance PQ is 2 km, ∠PQS = 63° and ∠QPS = 46° how far is the ship from each of the stations?

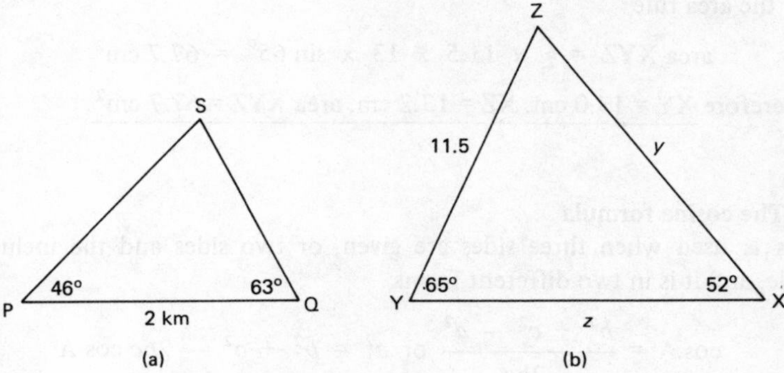

Fig 21.14

Solution

$$\angle PSQ = 180° - 46° - 63° = 71°$$

By the sine rule

$$\frac{2}{\sin 71°} = \frac{PS}{\sin 63°} = \frac{QS}{\sin 46°}$$

from which

$$PS = \frac{2 \sin 63°}{\sin 71°} = 1.885, \quad QS = \frac{2 \sin 46°}{\sin 71°} = 1.522$$

Therefore the distance from P is <u>1.89 km</u> and from Q <u>1.52 km.</u>

Example 21.10

Fig. 21.14(b) shows a triangle XYZ with ∠X = 52°, ∠Y = 65° and YZ = 11.5 cm. Calculate the lengths of the sides XY and XZ and the area of the triangle.

Solution

$$\angle Z = 180° - 65° - 52° = 63°$$

By the sine rule

$$\frac{11.5}{\sin 52°} = \frac{y}{\sin 65°} = \frac{z}{\sin 63°}$$

$$y = \frac{11.5 \sin 65°}{\sin 52°} = 13.2, \quad z = \frac{11.5 \sin 63°}{\sin 52°} = 13.0$$

By the area rule

$$\text{area XYZ} = \tfrac{1}{2} \times 11.5 \times 13 \times \sin 65^\circ = 67.7 \text{ cm}^2$$

Therefore XY = 13.0 cm, XZ = 13.2 cm, area XYZ = 67.7 cm^2.

(c) The cosine formula

This is used when three sides are given, or two sides and the included angle, and it is in two different forms

$$\cos A = \frac{b^2 + c^2 - a^2}{2bc} \quad \text{or} \quad a^2 = b^2 + c^2 - 2bc \cos A$$

$$\cos B = \frac{c^2 + a^2 - b^2}{2ca} \quad \text{or} \quad b^2 = c^2 + a^2 - 2ca \cos B$$

$$\cos C = \frac{a^2 + b^2 - c^2}{2ab} \quad \text{or} \quad c^2 = a^2 + b^2 - 2ab \cos C$$

The formula is given for the examination by most of the Boards, and the proof is not required at this level.

Example 21.11

Calculate the largest angle of a triangle with sides of length 30 mm, 40 mm and 60 mm.

Solution In the standard notation, if $a = 30$, $b = 40$ and $c = 60$, $\angle C$ is the largest angle because it is opposite the longest side.
Using the cosine rule

$$\cos C = \frac{a^2 + b^2 - c^2}{2ab}$$

Substituting the given values

$$\frac{30^2 + 40^2 - 60^2}{2 \times 30 \times 40} = -0.4583$$

Arccos 0.4583 = 62.7°, and since the cosine is negative the angle is obtuse. 180° – 62.7° = 117.3°. The largest angle of the triangle is therefore 117.3°.

Example 21.12

A ship left harbour on a bearing 210° and after 10 km it changed course and travelled for a further 7 km on a bearing 270°. What was the final distance and bearing from the starting position?

Solution In Fig. 21.15, A represents the starting point, B the position when the course was changed and C the final position, and we require the length AC and the angle BAC.

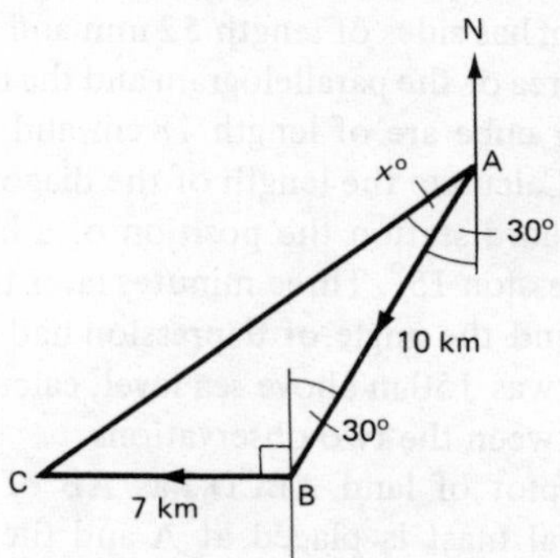

Fig 21.15

AB makes an angle of 30° with the north-south line (210° – 180°), and BC is at right angles to it (bearing 270°); hence ABC = 120°. By the cosine rule

$$AC^2 = 7^2 + 10^2 - 2 \times 7 \times 10 \times \cos 120^\circ = 49 + 100 + 140 \times \tfrac{1}{2}$$

$$AC = \sqrt{219} = 14.8$$

By the sine rule

$$\frac{7}{\sin x^\circ} = \frac{14.8}{\sin 120^\circ} \Rightarrow \sin x^\circ = \frac{7 \sin 60^\circ}{14.8} = 0.4096$$

Hence $x = 24.2$, and the bearing of C from A is 210° + 24.2° = 234.2°. The ship was 14.8 km from harbour on a bearing 234.2°.

Exercise 21.3

(1) Find the value of the function (a) sin 122°, (b) sin 133.5°, (c) cos 105°, (d) cos 170°, (e) tan 100.8°, (f) tan 158°.

(2) With the usual notation for a triangle ABC (i) find b when B = 58°, C = 68° and c = 127 mm, (ii) find a and the area of the triangle when b = 12 cm, c = 14 cm, A = 46°, (iii) find A and B when a = 14 cm, b = 18 cm and c = 25 cm.

(3) If cos P = –0.8236 find the value of (a) sin P, (b) tan P, (c) cos $\tfrac{1}{2}$P.

(4) Calculate the greatest angle of a triangle with sides in the ratio (a) 8 : 9 : 14, (b) 7 : 4 : 8, (c) 9 : 10 : 16.

(5) A triangle XYZ has area 61 cm^2 and two sides of length 14 cm and

10 cm. Calculate the angle between these sides and also the length of the third side.

(6) Two trees were 3 km and 7 km from an observer on bearings 040° and 325° respectively. How far apart were the trees?

(7) A parallelogram has sides of length 52 mm and 120 mm and one angle 70°. Find the area of the parallelogram and the length of the diagonals.

(8) The sides of a cube are of length 18 cm and ABCD and EFGH are parallel faces. Calculate the length of the diagonal AG.

(9) From a coastguard station the position of a boat was noted as bearing 300°, depression 15°. Three minutes later the same boat was on a bearing 030° and the angle of depression had decreased to 9°. If the lookout point was 150 m above sea level, calculate the average speed of the boat between the two observations.

(10) A rectangular plot of land ABCD has AB of length 37 m and AD 22 m. A vertical mast is placed at A and the angle of elevation of the top from D is 28.6°. Calculate (a) the height of the mast, (b) the angle of elevation from C.

(11) The base of a geometrical solid VABC is a triangle ABC with a right angle at C, BC = 54 cm, AB = 90 cm and V is 84 cm vertically above C. Calculate (a) area ABC, (b) $\angle$BAC, (c) $\angle$VAC.

21.8 THE EARTH: LATITUDE AND LONGITUDE

In elementary mathematics the Earth is considered as a sphere with its axis of rotation joining the north and south poles. A great circle has the same radius as the Earth, about 6400 km.

(a) Lines of longitude

These are half circles joining the poles, and called meridians. The meridian through Greenwich in London is the prime meridian and has longitude 0°, other meridians have longitudes up to 180° east or 180° west of the Greenwich meridian.

(b) Lines of latitude

Circles on the surface of the Earth in planes perpendicular to the axis are called circles, or parallels, of latitude. The great circle of latitude is the equator, with latitude 0°, and circles parallel to the equator are called small circles because they have smaller radii than the Earth itself. Latitudes can take values up to 90° north and up to 90° south of the equator.

The latitude of a place on the Earth's surface is equal to the angle subtended at the centre of the Earth by an arc of the meridian joining the equator to the given point. The longitude of a place is equal to the angle subtended at the centre of its parallel of latitude by the arc joining the prime meridian to the given point. In short, latitude is measured on a meridian and longitude on a small circle.

For example, the point marked A in Fig. 21.16(c) has position 30° N, 40° W. B is 30° N, 0°, it is on the prime meridian and C is on the equator, latitude 0° longitude 40° W.

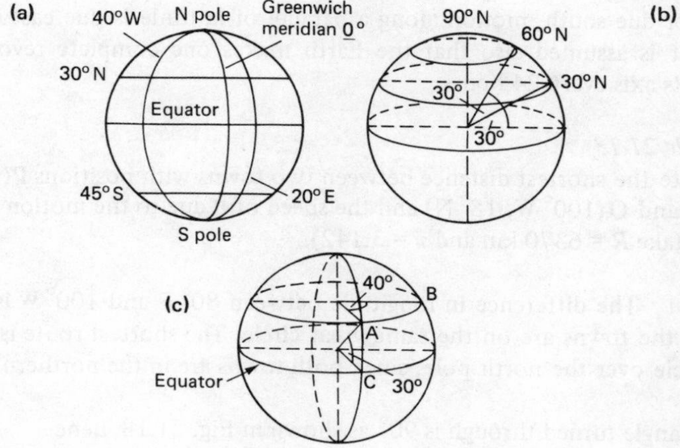

Fig 21.16

(c) The cosine ratio

In Fig. 21.17 R is the radius of the Earth and r is the radius of the small circle of latitude $\alpha°$ N. The values are connected by the equation

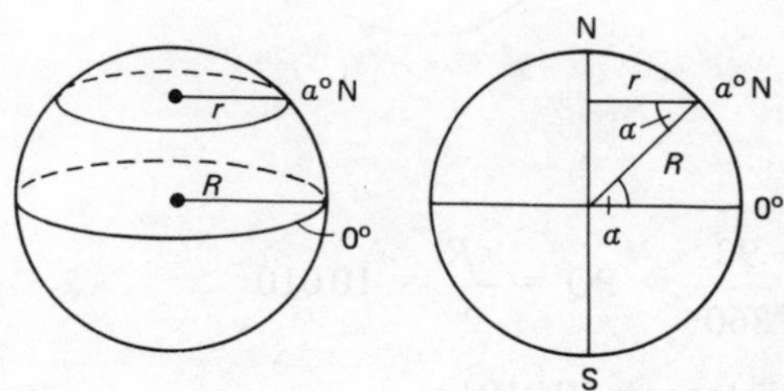

Fig 21.17

$$\frac{r}{R} = \text{cosine } \alpha°$$

On the equator, when $\alpha = 0$, $\cos \alpha° = 1$ and $r = R$. At the poles, when $\alpha = 90°$, $\cos \alpha° = 0$ and $r = 0$.

(d) Solving problems
Examination questions usually require a particular great or small circle to be identified; lengths of arcs and angles subtended by them can then be calculated as in the circle problems of Section 13.1.

The shortest distance on the surface of the Earth between two points is along the great circle joining the points. Motion along a meridian is due north or due south, motion along a parallel of latitude is due east or due west. It is assumed also that the Earth makes one complete revolution about its axis every 24 hours.

Example 21.13
Calculate the shortest distance between two towns with positions P(80° E, 72° N) and Q(100° W, 18° N) and the speed of P due to the motion of the Earth (take $R = 6370$ km and $\pi = 3.142$).

Solution The difference in longitude between 80° E and 100° W is 180° so that the towns are on the same great circle. The shortest route is along this circle over the north pole, since both towns are in the northern hemisphere.

The angle turned through is 90° as shown in Fig. 21.18, hence

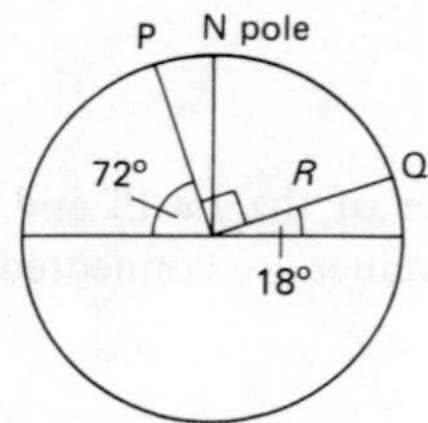

Fig 21.18

$$\frac{PQ}{2\pi R} = \frac{90}{360} \Rightarrow PQ = \frac{\pi R}{2} = 10\,010$$

and the shortest distance is 10 010 km.

The town P moves a distance equal to the circumference of the circle of latitude in 24 hours, and the radius is $R \cos 72°$. The speed is therefore

$$\frac{2\pi R \cos 72°}{24} \text{ km/h} = 515 \text{ km/h}$$

Example 21.14
Two transmitters in the same latitude have longitudes 32° E and 28° W. If

one is 2000 km further east than the other, calculate the latitude assuming the Earth is a sphere of radius 6400 km.

Solution Suppose the latitude is x, then both transmitters lie on a circle of radius $r = 6400 \cos x$ km. The difference in their longitude is $32° + 28° = 60°$, and this is subtended at the centre of the circle of latitude by an arc of length 2000 km

$$\frac{2000}{2\pi r} = \frac{60}{360} \Rightarrow r = \frac{6000}{\pi} \text{ km}$$

$$6400 \cos x = \frac{6000}{\pi} \Rightarrow x = \arccos \frac{15}{16\pi} = 72.6°$$

The latitude of the transmitters is <u>72.6° N</u> or <u>72.6° S</u>.

Exercise 21.4

In this exercise take the radius of the Earth to be 6370 km and π as 22/7.

(1) Calculate the distance from the equator of (a) Newcastle 55° N, (b) Montevideo 35° S.

(2) Calculate the length of the parallel of latitude (a) 50° N, (b) 35° S, (c) 28° N.

(3) Two places in the southern hemisphere have the same latitude but longitudes 30° E and 120° W. If the distance between them along the parallel of latitude is 4300 km, what is the latitude?

(4) A ship travels due east from Rio de Janeiro (22.1° S, 43.2° W) for 3200 km. What is its new position?

(5) Write down the latitude and longitude of the place at the opposite end of the diameter of the great circle from (a) 32° N, 17° W, (b) 28° S, 45° E.

(6) Calculate the speed in km/h, due only to the Earth's rotation, of a place with latitude (a) 49.5° N, (b) 34.6° S.

CHAPTER 22

VECTORS

A north-east wind blowing steadily at a speed of 30 km/h has magnitude (30 km/h) and direction (bearing 225°) and the velocity of the wind is a vector. Displacement, acceleration and force are vector quantities, and we are studying the theory of vectors here because the mathematics of one type of vector can be applied to every other vector.

22.1 VECTOR NOTATION

A vector is represented in a diagram by a straight line in the same sense and direction as the vector, the length of the line being made proportional to the magnitude. In this book, vectors are represented in bold type, either by a single letter or by two capital letters with an arrow to show the direction, as in Fig. 22.1. In written work a bar can be placed above or below the letter(s).

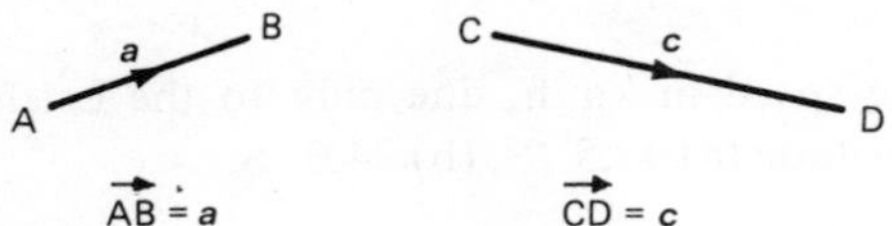

Fig 22.1

Free vectors and position vectors

A position vector is fixed with respect to a point of origin, and we shall consider only two-dimensional space. A free vector is not associated with a position in space and can be made to act at any point.

22.2 VECTOR ADDITION, THE TRIANGLE LAW

In Fig. 22.2 free vectors $\boldsymbol{a}$, $\boldsymbol{b}$, and $\boldsymbol{c}$ are represented by the line segments

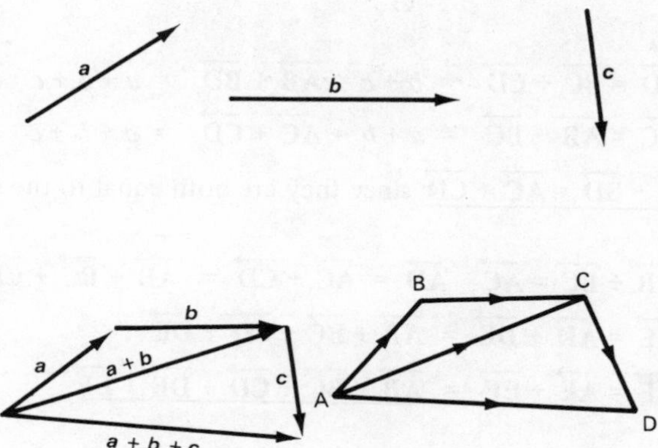

Fig 22.2

shown. To find the sum of $\boldsymbol{a}$ and $\boldsymbol{b}$, $\boldsymbol{b}$ is moved to the end of $\boldsymbol{a}$ as in the diagram so that they form two sides of a triangle with the arrows 'head to tail'. The sum is then represented by the third side of the triangle.

$$\overrightarrow{\mathbf{AB}} + \overrightarrow{\mathbf{BC}} = \overrightarrow{\mathbf{AC}}; \qquad \overrightarrow{\mathbf{AC}} + \overrightarrow{\mathbf{CD}} = \overrightarrow{\mathbf{AD}}$$

Definition Vectors have magnitude and direction and are combined according to the triangle law of addition.

Example 22.1

In Fig. 22.3, (a) show that $\overrightarrow{\mathbf{AB}} + \overrightarrow{\mathbf{BD}} = \overrightarrow{\mathbf{AC}} + \overrightarrow{\mathbf{CD}}$, (b) show that $\overrightarrow{\mathbf{AF}} = \overrightarrow{\mathbf{AB}} + \overrightarrow{\mathbf{BC}} + \overrightarrow{\mathbf{CD}} + \overrightarrow{\mathbf{DE}} + \overrightarrow{\mathbf{EF}}$.

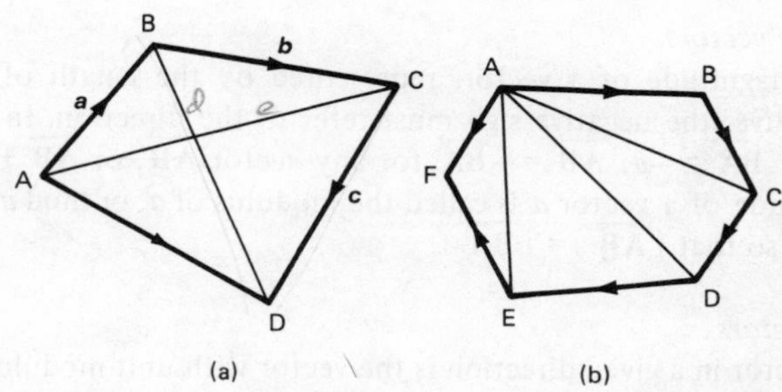

Fig 22.3

Solution

(a)

$$\overrightarrow{\mathbf{BD}} = \overrightarrow{\mathbf{BC}} + \overrightarrow{\mathbf{CD}} = \boldsymbol{b} + \boldsymbol{c} \Rightarrow \overrightarrow{\mathbf{AB}} + \overrightarrow{\mathbf{BD}} = \boldsymbol{a} + \boldsymbol{b} + \boldsymbol{c}$$

$$\overrightarrow{\mathbf{AC}} = \overrightarrow{\mathbf{AB}} + \overrightarrow{\mathbf{BC}} = \boldsymbol{a} + \boldsymbol{b} \Rightarrow \overrightarrow{\mathbf{AC}} + \overrightarrow{\mathbf{CD}} = \boldsymbol{a} + \boldsymbol{b} + \boldsymbol{c}$$

Hence $\underline{\overrightarrow{\mathbf{AB}} + \overrightarrow{\mathbf{BD}} = \overrightarrow{\mathbf{AC}} + \overrightarrow{\mathbf{CD}}}$ since they are both equal to the same vector.

(b)

$$\overrightarrow{\mathbf{AB}} + \overrightarrow{\mathbf{BC}} = \overrightarrow{\mathbf{AC}}, \quad \overrightarrow{\mathbf{AD}} = \overrightarrow{\mathbf{AC}} + \overrightarrow{\mathbf{CD}} = \overrightarrow{\mathbf{AB}} + \overrightarrow{\mathbf{BC}} + \overrightarrow{\mathbf{CD}}$$

$$\overrightarrow{\mathbf{AE}} = \overrightarrow{\mathbf{AD}} + \overrightarrow{\mathbf{DE}} = \overrightarrow{\mathbf{AB}} + \overrightarrow{\mathbf{BC}} + \overrightarrow{\mathbf{CD}} + \overrightarrow{\mathbf{DE}}$$

$$\overrightarrow{\mathbf{AF}} = \overrightarrow{\mathbf{AE}} + \overrightarrow{\mathbf{EF}} = \underline{\overrightarrow{\mathbf{AB}} + \overrightarrow{\mathbf{BC}} + \overrightarrow{\mathbf{CD}} + \overrightarrow{\mathbf{DE}} + \overrightarrow{\mathbf{EF}}}$$

(a) The polygon rule

Example 22.1 illustrates the polygon rule for vector addition.

(i) If a number of vectors are represented in magnitude and direction by the sides of a polygon taken in order, then the sum of the vectors is represented by the remaining side from the starting point which closes the polygon. In Fig. 22.3(a), $\overrightarrow{\mathbf{AD}}$ was the remaining side of the quadrilateral ABCD; in Fig. 22.3(b) $\overrightarrow{\mathbf{AF}}$ was the remaining side of the hexagon ABCDEF

(ii) If the sides of a closed polygon taken in order represent vectors, then the sum of the vectors is the zero vector.

(b) Zero, negative and unit vectors

(i) Zero vector

The zero vector **O** is such that $\boldsymbol{a} + \mathbf{O} = \boldsymbol{a}$ for every vector $\boldsymbol{a}$. It is not the same as the number zero, which is a scalar, as vectors cannot be added to scalars. A scalar quantity has magnitude but not direction, and all real numbers are considered as scalars.

(ii) Negative vectors

Since the magnitude of a vector, represented by the length of a line, is always positive, the negative sign must refer to the direction. In Fig. 22.1 $\overrightarrow{\mathbf{AB}} = \boldsymbol{a}$ and $\overrightarrow{\mathbf{BA}} = -\boldsymbol{a}$. $\overrightarrow{\mathbf{AB}} = -\overrightarrow{\mathbf{BA}}$ for any vector $\overrightarrow{\mathbf{AB}}$, or $\overrightarrow{\mathbf{AB}} + \overrightarrow{\mathbf{BA}} = \mathbf{O}$. The magnitude of a vector $\boldsymbol{a}$ is called the modulus of $\boldsymbol{a}$, or mod $\boldsymbol{a}$, and it is written $|\boldsymbol{a}|$ so that $|\overrightarrow{\mathbf{AB}}| = |\overrightarrow{\mathbf{BA}}|$.

(iii) Unit vectors

The unit vector in a given direction is the vector with unit modulus. If $\mathbf{e}_a$ is the unit vector in the direction of the vector $\boldsymbol{a}$ then $|\mathbf{e}_a| = 1$.

(c) Vector addition is commutative and associative

(i) $\boldsymbol{a} + \boldsymbol{b} = \boldsymbol{b} + \boldsymbol{a}$

(ii) $\boldsymbol{a} + (\boldsymbol{b} + \boldsymbol{c}) = (\boldsymbol{a} + \boldsymbol{b}) + \boldsymbol{c}$.

This is true for all vectors $\boldsymbol{a}$, $\boldsymbol{b}$ and $\boldsymbol{c}$.

Example 22.2

Find in terms of the vectors $\boldsymbol{a}$, $\boldsymbol{b}$ and $\boldsymbol{c}$ the vectors (a) $\overrightarrow{\mathbf{PQ}}$, (b) $\overrightarrow{\mathbf{AC}}$ and $\overrightarrow{\mathbf{BD}}$ shown in Fig. 22.4.

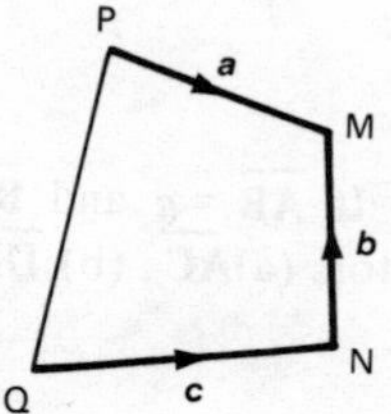

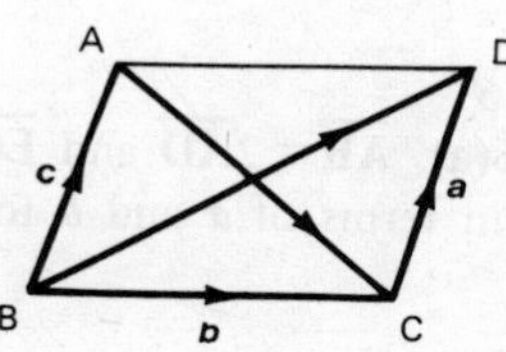

Fig 22.4

Solution

(a) $\overrightarrow{\mathbf{PQ}} = \overrightarrow{\mathbf{PM}} + \overrightarrow{\mathbf{MN}} + \overrightarrow{\mathbf{NQ}}$

$= \overrightarrow{\mathbf{PM}} - \overrightarrow{\mathbf{NM}} - \overrightarrow{\mathbf{QN}}$

$= \underline{\boldsymbol{a} - \boldsymbol{b} - \boldsymbol{c}}$

(b) $\overrightarrow{\mathbf{AC}} = \overrightarrow{\mathbf{AB}} + \overrightarrow{\mathbf{BC}} = -\overrightarrow{\mathbf{BA}} + \overrightarrow{\mathbf{BC}}$

$= -\boldsymbol{c} + \boldsymbol{b}$ or $\underline{\boldsymbol{b} - \boldsymbol{c}}$

$\overrightarrow{\mathbf{BD}} = \overrightarrow{\mathbf{BC}} + \overrightarrow{\mathbf{CD}} = \underline{\boldsymbol{b} + \boldsymbol{a}}$

22.3 MULTIPLICATION OF A VECTOR BY A SCALAR

In Fig. 22.5, $\mathbf{e}_a$ represents the unit vector in the direction of $\boldsymbol{a}$, and $\mathbf{e}_b$ represents the unit vector in the direction of $\boldsymbol{b}$. If $\mathbf{e}_a$ is of length 5 mm and

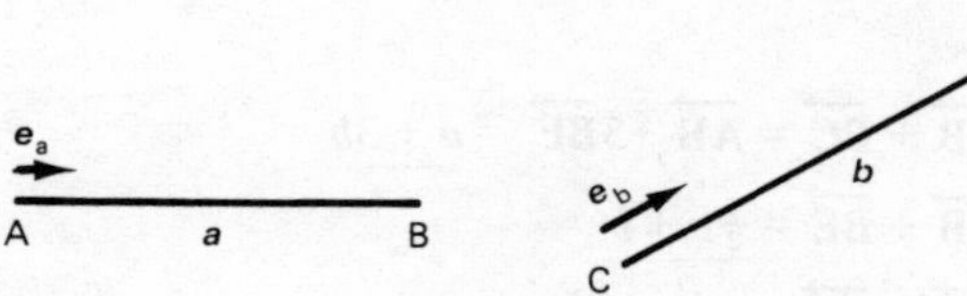

Fig 22.5

AB is of length 4 cm, then $\boldsymbol{a} = 8$ N if the unit of force is the newton, or $\boldsymbol{a} = 8$ m/s if the unit of velocity is 1 m/s. Similarly, if $\mathbf{e}_b$ is of length 10 mm and CD 50 mm then $|\overrightarrow{\mathbf{CD}}| = 5$.

(a) Parallel vectors

Every vector can be regarded as the product of a unit vector and a scalar magnitude; if one vector is a scalar multiple of another vector the two have the same unit vector, and hence they are parallel. For example, if the velocity of one car is represented by $\boldsymbol{a}$ and the velocity of a second car by $-2\,\boldsymbol{a}$, the second car has twice the speed of the first but in the opposite direction.

(b) Multiplication by a scalar is distributive over vector addition

$$k\,(\boldsymbol{b} + \boldsymbol{c}) = k\boldsymbol{b} + k\boldsymbol{c} \text{ for any number } k$$

Example 22.3

In Fig. 22.6(a), $\overrightarrow{\mathbf{AB}} = 2\overrightarrow{\mathbf{AD}}$ and $\overrightarrow{\mathbf{EC}} = 2\overrightarrow{\mathbf{BE}}$. If $\overrightarrow{\mathbf{AB}} = \boldsymbol{a}$ and $\overrightarrow{\mathbf{BE}} = \boldsymbol{b}$ find expressions in terms of $\boldsymbol{a}$ and $\boldsymbol{b}$ for the vectors (a)$\overrightarrow{\mathbf{AC}}$, (b) $\overrightarrow{\mathbf{DE}}$, (c) $\overrightarrow{\mathbf{CD}}$, (d) $\overrightarrow{\mathbf{EA}}$.

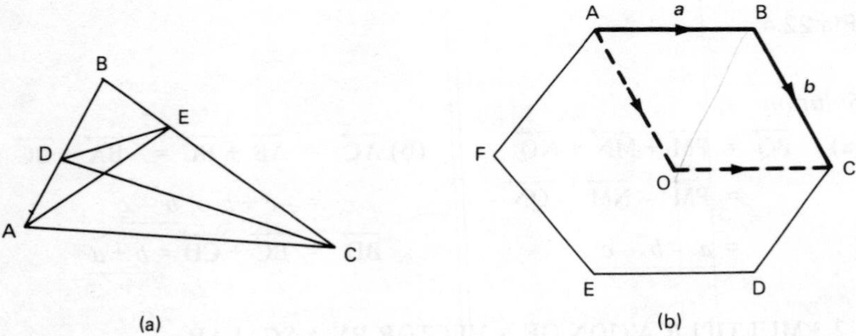

Fig 22.6

Solution

(a) $\overrightarrow{\mathbf{AC}} = \overrightarrow{\mathbf{AB}} + \overrightarrow{\mathbf{BC}} = \overrightarrow{\mathbf{AB}} + 3\overrightarrow{\mathbf{BE}} = \underline{\boldsymbol{a} + 3\boldsymbol{b}}$

(b) $\overrightarrow{\mathbf{DE}} = \overrightarrow{\mathbf{DB}} + \overrightarrow{\mathbf{BE}} = \underline{\tfrac{1}{2}\boldsymbol{a} + \boldsymbol{b}}$

(c) $\overrightarrow{\mathbf{CD}} = \overrightarrow{\mathbf{CB}} + \overrightarrow{\mathbf{BD}} = \underline{-3\boldsymbol{b} - \tfrac{1}{2}\boldsymbol{a}}$

(d) $\overrightarrow{\mathbf{EA}} = \overrightarrow{\mathbf{EB}} + \overrightarrow{\mathbf{BA}} = \underline{-\boldsymbol{b} - \boldsymbol{a}}$

Example 22.4

In Fig. 22.6(b) ABCDEF is a regular hexagon, $\overrightarrow{\mathbf{AB}} = \boldsymbol{a}$ and $\overrightarrow{\mathbf{BC}} = \boldsymbol{b}$. Express in terms of $\boldsymbol{a}$ and $\boldsymbol{b}$ the vectors (a) $\overrightarrow{\mathbf{FD}}$, (b) $\overrightarrow{\mathbf{FE}}$, (c) $\overrightarrow{\mathbf{BE}}$.

Solution From the symmetry of a regular hexagon $\overrightarrow{\mathbf{FD}} = \overrightarrow{\mathbf{AC}}$, $\overrightarrow{\mathbf{FE}} = \overrightarrow{\mathbf{BC}}$, $\overrightarrow{\mathbf{OC}} = \overrightarrow{\mathbf{AB}}$. O is the centre of symmetry.

(a) $\overrightarrow{\mathbf{FD}} = \overrightarrow{\mathbf{AC}} = \overrightarrow{\mathbf{AB}} + \overrightarrow{\mathbf{BC}} = \underline{\boldsymbol{a} + \boldsymbol{b}}$

(b) $\overrightarrow{\mathbf{FE}} = \overrightarrow{\mathbf{BC}} = \underline{\boldsymbol{b}}$

(c) $\overrightarrow{\mathbf{BE}} = 2\overrightarrow{\mathbf{BO}} = 2(\overrightarrow{\mathbf{BC}} + \overrightarrow{\mathbf{CO}}) = \underline{2\boldsymbol{b} - 2\boldsymbol{a}}$

Exercise 22.1

(1) Given that $\overrightarrow{\mathbf{AB}} = \boldsymbol{b}$, $\overrightarrow{\mathbf{BC}} = \boldsymbol{c}$ and $\overrightarrow{\mathbf{CA}} = \boldsymbol{a}$, express as a single vector (a) $\boldsymbol{a} + \boldsymbol{b}$, (b) $\boldsymbol{c} + \boldsymbol{a}$, (c) $-\boldsymbol{a} - \boldsymbol{b}$, (d) $\boldsymbol{a} + \boldsymbol{b} + \boldsymbol{c}$.

(2) Find an expression for $\overrightarrow{\mathbf{AC}}$ given that $\overrightarrow{\mathbf{AB}} = \boldsymbol{a}$ and $\overrightarrow{\mathbf{CB}} = \boldsymbol{b}$. What relation would show that A, B, and C are in a straight line?

(3) ABCD is a trapezium with $\overrightarrow{\mathbf{BC}} = 2\boldsymbol{a}$, $\overrightarrow{\mathbf{AD}} = 3\boldsymbol{a}$ and $\overrightarrow{\mathbf{CD}} = \boldsymbol{b}$. Write in terms of the vectors $\boldsymbol{a}$ and $\boldsymbol{b}$ (a) $\overrightarrow{\mathbf{BD}}$, (b) $\overrightarrow{\mathbf{AC}}$, (c) $\overrightarrow{\mathbf{AB}}$. K is a point on AD such that AK/KD = 3/5; find an expression for $\overrightarrow{\mathbf{BK}}$.

(4) ABCDE is a pentagon in which $\overrightarrow{\mathbf{AB}} = \boldsymbol{v}$, $\overrightarrow{\mathbf{CB}} = \boldsymbol{w}$, $\overrightarrow{\mathbf{CD}} = \boldsymbol{x}$ and $\overrightarrow{\mathbf{DE}} = \boldsymbol{y}$. Write in terms of $\boldsymbol{v}, \boldsymbol{w}, \boldsymbol{x}, \boldsymbol{y}$ (a) $\overrightarrow{\mathbf{AE}}$, (b) $\overrightarrow{\mathbf{DA}}$, (c) $\overrightarrow{\mathbf{CE}}$, (d) $\overrightarrow{\mathbf{CA}}$.

(5) The position vectors of A and B referred to O are $\boldsymbol{a}$ and $\boldsymbol{b}$ and P and Q are the mid-points of OA and OB respectively. Write in terms of $\boldsymbol{a}$ and $\boldsymbol{b}$ (a) $\overrightarrow{\mathbf{AB}}$, (b) $\overrightarrow{\mathbf{PQ}}$, and deduce a geometrical connection between PQ and AB.

(6) Lines OA and OB represent the vectors $\boldsymbol{p}$ and $\boldsymbol{q}$ respectively. F is a point on OA such that 3AF = FO and a line through F parallel to OB meets AB at G. Express in terms of $\boldsymbol{p}$ and $\boldsymbol{q}$ (a) $\overrightarrow{\mathbf{AB}}$, (b) $\overrightarrow{\mathbf{FG}}$, (c) $\overrightarrow{\mathbf{FB}}$.

(7) The position vectors of points P and Q referred to O are $\boldsymbol{p}$ and $\boldsymbol{q}$. Which line segment represents $\boldsymbol{p} - \boldsymbol{q}$? Draw the vector $\overrightarrow{\mathbf{OT}} = \boldsymbol{p} + \boldsymbol{q}$ and show that OQ = PT.

(8) A vector $\boldsymbol{r}$ can be expressed in terms of two non-parallel vectors $\boldsymbol{s}$ and $\boldsymbol{t}$ by $\boldsymbol{r} = a\boldsymbol{s} + b\boldsymbol{t}$ or $\boldsymbol{r} = (1 - 3a)\boldsymbol{s} + [1 - (b/4)]\boldsymbol{t}$. Calculate the value of the scalars a and b.

22.4 VECTORS IN THE CARTESIAN SYSTEM

Unit vectors in the positive direction of the x and y axes are denoted by **i** and **j** respectively.

If O is the origin of coordinates and A is the point (2, 4) in Fig. 22.7, the position vector $\overrightarrow{\mathbf{OA}} = \boldsymbol{a} = 2\mathbf{i} + 4\mathbf{j}$. Similarly, if B has coordinates (7, 3) then $\overrightarrow{\mathbf{OB}} = \boldsymbol{b} = 7\mathbf{i} + 3\mathbf{j}$.

The displacement vector $\overrightarrow{\mathbf{CD}}$ can be regarded as a translation of 4 units in the x direction and 5 units in the y direction, and hence $\overrightarrow{\mathbf{CD}} = \boldsymbol{c} = 4\mathbf{i} + 5\mathbf{j}$. By the triangle law of addition

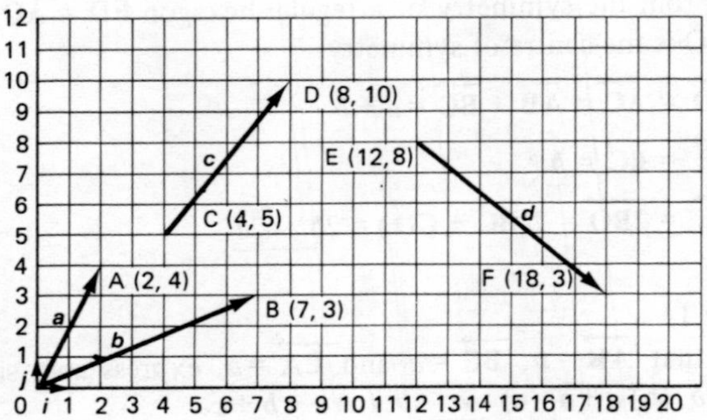

Fig 22.7

$$\overrightarrow{\mathbf{OC}} + \overrightarrow{\mathbf{CD}} = \overrightarrow{\mathbf{OD}}$$

therefore

$$\overrightarrow{\mathbf{CD}} = \overrightarrow{\mathbf{OD}} - \overrightarrow{\mathbf{OC}}$$
$$= (8\mathbf{i} + 10\mathbf{j}) - (4\mathbf{i} + 5\mathbf{j})$$
$$= 4\mathbf{i} + 5\mathbf{j}$$

A displacement vector is therefore the difference between two position vectors. Using a similar argument

$$\overrightarrow{\mathbf{OE}} = 12\mathbf{i} + 8\mathbf{j}, \quad \overrightarrow{\mathbf{OF}} = 18\mathbf{i} + 3\mathbf{j}$$

therefore

$$\overrightarrow{\mathbf{EF}} = \overrightarrow{\mathbf{OF}} - \overrightarrow{\mathbf{OE}} = 6\mathbf{i} - 5\mathbf{j}$$

6**i** is called the component of $\overrightarrow{\mathbf{EF}}$ in the direction of x, and $-5\mathbf{j}$ is the component in the y direction. The modulus of $\overrightarrow{\mathbf{EF}}$, the length of the line EF, is given by Pythagoras' theorem as $|\overrightarrow{\mathbf{EF}}| = \sqrt{(6^2 + 5^2)}$. The acute angle between the direction $\overrightarrow{\mathbf{EF}}$ and the x-axis is arctan 5/6.

In general, if $\boldsymbol{a} = x\mathbf{i} + y\mathbf{j}$, $|\boldsymbol{a}| = \sqrt{(x^2 + y^2)}$ and arctan y/x is the angle between the direction of the vector $\boldsymbol{a}$ and the positive direction of the x-axis.

The unit vector in the direction of $\boldsymbol{a}$ can be expressed as

$$\frac{\boldsymbol{a}}{|\boldsymbol{a}|} = \frac{x\mathbf{i} + y\mathbf{j}}{\sqrt{(x^2 + y^2)}}$$

Example 22.5

Write down an expression for the magnitude of $\boldsymbol{V}$ and the unit vector in the direction of $\boldsymbol{V}$ when (a) $\boldsymbol{V} = -8\mathbf{i} + 15\mathbf{j}$, (b) $\boldsymbol{V} = 24\mathbf{i} - 7\mathbf{j}$.

Solution

(a) $V = -8\mathbf{i} + 15\mathbf{j} \Rightarrow |V| = \sqrt{(8^2 + 15^2)} = \sqrt{(64 + 225)} = \underline{17}$

The unit vector $\mathbf{e}_V = \underline{\dfrac{-8}{17}\mathbf{i} + \dfrac{15}{17}\mathbf{j}.}$

(b) $V = 24\mathbf{i} - 7\mathbf{j} \Rightarrow |V| = \sqrt{(24^2 + 7^2)} = \sqrt{(576 + 49)} = 25$

The unit vector $\mathbf{e}_V = \underline{\dfrac{24}{25}\mathbf{i} - \dfrac{7}{25}\mathbf{j}.}$

Example 22.6

Given that $\boldsymbol{p} = 2\mathbf{i} - 3\mathbf{j}$ and $\boldsymbol{q} = \mathbf{i} + 5\mathbf{j}$ find (a) the magnitude of the vector $\boldsymbol{p} + 2\boldsymbol{q}$, (b) the unit vector in the direction of $2\boldsymbol{p} - \boldsymbol{q}$. (Leave the answer in surd form.)

Solution

(a) $\boldsymbol{p} + 2\boldsymbol{q} = 2\mathbf{i} - 3\mathbf{j} + 2\mathbf{i} + 10\mathbf{j} = 4\mathbf{i} + 7\mathbf{j}$

The magnitude is $\sqrt{(4^2 + 7^2)} = \underline{\sqrt{65}.}$

(b) $2\boldsymbol{p} - \boldsymbol{q} = 4\mathbf{i} - 6\mathbf{j} - \mathbf{i} - 5\mathbf{j} = 3\mathbf{i} - 11\mathbf{j}$

The magnitude is $\sqrt{(3^2 + 11^2)} = \sqrt{130}$. The unit vector is

$$\underline{\frac{3}{\sqrt{130}}\mathbf{i} - \frac{11}{\sqrt{130}}\mathbf{j}.}$$

22.5 VECTORS IN MATRIX FORM

A vector in the Cartesian form $x\mathbf{i} + y\mathbf{j}$ can be written as a 2 by 1 column matrix $\binom{x}{y}$. For example, in Fig. 22.7, A is the point (2, 4) and the position vector $\overrightarrow{\mathbf{OA}}$ is $\binom{2}{4}$. Similarly B is (7, 3) and $\overrightarrow{\mathbf{OB}} = \binom{7}{3}$.

Example 22.7

If $\boldsymbol{p} = \binom{4}{7}$ and $\boldsymbol{q} = \binom{-3}{5}$, find the column vector which represents (a) $3\boldsymbol{p} + 2\boldsymbol{q}$, (b) $2\boldsymbol{p} - 3\boldsymbol{q}$.

Solution

(a) $3\boldsymbol{p} = 3\begin{pmatrix}4\\7\end{pmatrix} = \begin{pmatrix}12\\21\end{pmatrix}, \qquad 2\boldsymbol{q} = 2\begin{pmatrix}-3\\5\end{pmatrix} = \begin{pmatrix}-6\\10\end{pmatrix}$

$$3\boldsymbol{p} + 2\boldsymbol{q} = \begin{pmatrix}12\\21\end{pmatrix} + \begin{pmatrix}-6\\10\end{pmatrix} = \underline{\begin{pmatrix}6\\31\end{pmatrix}}$$

(b) $$2\boldsymbol{p} - 3\boldsymbol{q} = 2\begin{pmatrix}4\\7\end{pmatrix} - 3\begin{pmatrix}-3\\5\end{pmatrix} = \begin{pmatrix}8\\14\end{pmatrix} - \begin{pmatrix}-9\\15\end{pmatrix} = \underline{\begin{pmatrix}17\\-1\end{pmatrix}}$$

Displacement vectors

By the triangle rule for addition, $\overrightarrow{\mathbf{OA}} + \overrightarrow{\mathbf{AB}} = \overrightarrow{\mathbf{OB}}$ and hence $\overrightarrow{\mathbf{AB}} = \overrightarrow{\mathbf{OB}} - \overrightarrow{\mathbf{OA}}$. Referring to Fig. 22.7

$$\overrightarrow{\mathbf{AB}} = \begin{pmatrix}7\\3\end{pmatrix} - \begin{pmatrix}2\\4\end{pmatrix} = \begin{pmatrix}5\\-1\end{pmatrix}$$

Displacement by a vector $\binom{a}{b}$ is equivalent to addition of the vector $\binom{a}{b}$ to a given position vector.

Example 22.8

If $\boldsymbol{p} = \binom{3}{1}$ and $\boldsymbol{q} = \binom{5}{9}$, find the coordinates of the image A′ of A(7, –3) after displacement by the vector $2\boldsymbol{p} - \boldsymbol{q}$.

Solution

$$2\boldsymbol{p} - \boldsymbol{q} = \begin{pmatrix}6\\2\end{pmatrix} - \begin{pmatrix}5\\9\end{pmatrix} = \begin{pmatrix}1\\-7\end{pmatrix}$$

Adding the displacement vector $\binom{1}{-7}$ to the position vector of A

$$\begin{pmatrix}7\\-3\end{pmatrix} + \begin{pmatrix}1\\-7\end{pmatrix} = \begin{pmatrix}8\\-10\end{pmatrix}$$

The image of A is $\underline{\text{A}'\ (8, -10)}$.

Example 22.9

The coordinates of the points O, P, Q, R are (0, 0), (2, –1), (3, 2) and (13, 4) respectively, and $\overrightarrow{\mathbf{OR}} = a\,\overrightarrow{\mathbf{OP}} + b\,\overrightarrow{\mathbf{OQ}}$. Find the value of the scalars a and b.

$$\overrightarrow{\mathbf{OR}} = \begin{pmatrix}13\\4\end{pmatrix}, \quad \overrightarrow{\mathbf{OP}} = \begin{pmatrix}2\\-1\end{pmatrix}, \quad \overrightarrow{\mathbf{OQ}} = \begin{pmatrix}3\\2\end{pmatrix}$$

$$a\begin{pmatrix}2\\-1\end{pmatrix} + b\begin{pmatrix}3\\2\end{pmatrix} = \begin{pmatrix}13\\4\end{pmatrix} \Rightarrow \begin{cases}2a + 3b = 13\\-a + 2b = 4\end{cases}$$

The solution of these simultaneous equations is $\underline{a = 2, b = 3.}$

Exercise 22.2

(1) If $\boldsymbol{x} = \binom{2}{-5}$, $\boldsymbol{y} = \binom{-4}{6}$, $\boldsymbol{z} = \binom{2}{3}$, what column vector is represented by (a) $\boldsymbol{p} = 3\boldsymbol{x} - \boldsymbol{y}$, (b) $\boldsymbol{q} = 3\boldsymbol{z} + \boldsymbol{x}$, (c) $\boldsymbol{r} = \boldsymbol{x} - \boldsymbol{y} + \boldsymbol{z}$?

(2) For the vectors $\boldsymbol{x}$, $\boldsymbol{y}$, $\boldsymbol{z}$ of question 1 find the image of the point (2, −4) after displacement by (a) $2\boldsymbol{x} - \boldsymbol{y}$, (b) $\boldsymbol{y} + \boldsymbol{z}$, (c) $3\boldsymbol{x} - \frac{1}{2}\boldsymbol{y}$.

(3) Find the magnitude of the vectors represented by (a) $\begin{pmatrix} 5 \\ 12 \end{pmatrix}$, (b) $\begin{pmatrix} 3 \\ 4 \end{pmatrix}$, (c) $\begin{pmatrix} 5 \\ 1 \end{pmatrix} + \begin{pmatrix} -1 \\ 3 \end{pmatrix}$.

(4) A (3, 2) and B(−2, 4) are positions of points referred to the origin O. Find the coordinates of (a) the point C if $\overrightarrow{\mathbf{OC}} = \overrightarrow{\mathbf{OA}} + \overrightarrow{\mathbf{AB}}$, (b) the point D when $\overrightarrow{\mathbf{OD}} = \overrightarrow{\mathbf{OA}} - \overrightarrow{\mathbf{OB}}$.

(5) Find the value, in surd form, of the magnitude of the vector $\boldsymbol{v}$. (a) $\boldsymbol{v} = -3\mathbf{i} + 10\mathbf{j}$, (b) $\boldsymbol{v} = 4\mathbf{i} - 6\mathbf{j}$, (c) $\boldsymbol{v} = -2\mathbf{i} - 3\mathbf{j}$.

(6) Calculate the angle made with the positive direction of the x-axis by the vector (a) $\boldsymbol{v} = 3\mathbf{i} + 4\mathbf{j}$, (b) $\boldsymbol{r} = 8.9\mathbf{i} + 4.8\mathbf{j}$.

(7) Which of the following vectors are parallel to the vector $\begin{pmatrix} 4 \\ 3 \end{pmatrix}$? (a) $\begin{pmatrix} 3 \\ 4 \end{pmatrix}$, (b) $\begin{pmatrix} 8 \\ 4 \end{pmatrix}$, (c) $\begin{pmatrix} 8 \\ 6 \end{pmatrix}$, (d) $\begin{pmatrix} -6 \\ -4 \end{pmatrix}$, (e) $\begin{pmatrix} -12 \\ -9 \end{pmatrix}$.

(8) A is the point (−3, 2) and the vectors $\overrightarrow{\mathbf{AB}}$, $\overrightarrow{\mathbf{DA}}$, $\overrightarrow{\mathbf{AC}}$ are represented by the column vectors $\begin{pmatrix} 2 \\ 1 \end{pmatrix}$, $\begin{pmatrix} -1 \\ 2 \end{pmatrix}$, $\begin{pmatrix} 3 \\ 4 \end{pmatrix}$, respectively. Find the coordinates of B, D, and C and the length of the line BD.

(9) Show that the vectors $\boldsymbol{a} = \begin{pmatrix} 4 \\ 2 \end{pmatrix}$ and $\boldsymbol{b} = \begin{pmatrix} -2 \\ 4 \end{pmatrix}$ have the same magnitude and are perpendicular to each other. (Hint: find the angle each makes with the x-axis.)

(10) The point A has coordinates (4, 1), $\overrightarrow{\mathbf{AB}} = \begin{pmatrix} -1 \\ 9 \end{pmatrix}$, $\overrightarrow{\mathbf{BC}} = \begin{pmatrix} 5 \\ -4 \end{pmatrix}$. Find the coordinates of B and the column vector representing $\overrightarrow{\mathbf{CA}}$.

TRANSFORMATIONS OF THE CARTESIAN PLANE

Transformation problems can be solved by graphical methods and by algebraic methods using matrices, but you may not be given a choice in a particular question.

23.1 TRANSLATIONS

A *translation* of the Cartesian plane occurs when the same displacement vector is added to the position vector of every point; the plane is *transformed* by the vector. If the displacement is represented by $a\mathbf{i} + b\mathbf{j}$, the column vector $\binom{a}{b}$ is the matrix of the transformation. If a point (x, y) is translated to (x', y') then (x', y') is called the *image* of (x, y) under the translation, and $\binom{x}{y} + \binom{a}{b} = \binom{x'}{y'}$.

Example 23.1
Find the image of the point $(3, -2)$ and the point which is transformed to $(3, -2)$ under translation by the vector $2\mathbf{i} + 5\mathbf{j}$.

Solution The matrix of the translation is $\binom{2}{5}$

$$\begin{pmatrix}3\\-2\end{pmatrix} + \begin{pmatrix}2\\5\end{pmatrix} = \begin{pmatrix}5\\3\end{pmatrix}$$

and the image of $(3, -2)$ is $\underline{(5, 3)}$.

Suppose $(3, -2)$ is the image of (x, y). Then

$$\begin{pmatrix}x\\y\end{pmatrix} + \begin{pmatrix}2\\5\end{pmatrix} = \begin{pmatrix}3\\-2\end{pmatrix}$$

and

$$\begin{pmatrix}x\\y\end{pmatrix} = \begin{pmatrix}3\\-2\end{pmatrix} - \begin{pmatrix}2\\5\end{pmatrix} = \begin{pmatrix}1\\-7\end{pmatrix}$$

The image of (1, −7) is (3, −2).

Any figure is transformed to a congruent figure by a translation. For example, if the triangle formed by A(1, 1), B(3, 2), C(4, 7) is translated by the vector $\binom{5}{4}$, the image triangle is A′(6, 5), B′(8, 6), C′(9, 11) and if the two triangles are drawn on graph paper it is evident that they are congruent (Section 11.5).

Combination of translations

The image of a point (x, y) under successive translations T_1, T_2 is the same as the image under a single translation $T = T_1 + T_2$.

$$T\begin{pmatrix} x \\ y \end{pmatrix} = T_2 T_1 \begin{pmatrix} x \\ y \end{pmatrix}$$

is taken to mean transformation first by T_1 and then by T_2. Addition of matrices is commutative (Section 10.2) and so for translations the order is not important.

Example 23.2

The unit square O(0, 0), A(1, 0), B(1, 1), C(0, 1) is transformed successively by vectors $3\mathbf{i} + 2\mathbf{j}$, $a\mathbf{i} + b\mathbf{j}$, and $-4\mathbf{i} - 5\mathbf{j}$. Calculate the value of a and b if the image square is O′(1, 1), A′(2, 1), B′(2, 2), C′(1, 2).

Since the origin is mapped to (1, 1) the matrix of the combined translation is $\binom{1}{1}$. Hence

$$\begin{pmatrix} 3 \\ 2 \end{pmatrix} + \begin{pmatrix} a \\ b \end{pmatrix} + \begin{pmatrix} -4 \\ -5 \end{pmatrix} = \begin{pmatrix} 1 \\ 1 \end{pmatrix} \Rightarrow \begin{pmatrix} a \\ b \end{pmatrix} = \begin{pmatrix} 2 \\ 4 \end{pmatrix}$$

Therefore $a = 2$, $b = 4$.

23.2 TRANSFORMATIONS REPRESENTED BY MATRIX OPERATORS

Transformations of the plane which leave the origin unchanged can be represented by a 2 × 2 matrix, since $\begin{pmatrix} a & b \\ c & d \end{pmatrix}\begin{pmatrix} 0 \\ 0 \end{pmatrix} = \begin{pmatrix} 0 \\ 0 \end{pmatrix}$ for all values of a, b, c and d.

Such transformations include rotations about the origin, reflections in lines through the origin and enlargements centred at the origin. The matrix representing a particular transformation can be found by locating the images of the points (1, 0) and (0, 1), since

$$\begin{pmatrix} a & b \\ c & d \end{pmatrix}\begin{pmatrix} 1 \\ 0 \end{pmatrix} = \begin{pmatrix} a \\ c \end{pmatrix} \quad \text{and} \quad \begin{pmatrix} a & b \\ c & d \end{pmatrix}\begin{pmatrix} 0 \\ 1 \end{pmatrix} = \begin{pmatrix} b \\ d \end{pmatrix}$$

For every 2 × 2 matrix M, $M\begin{pmatrix} 1 & 0 \\ 0 & 1 \end{pmatrix} = M$ (Section 10.3). The inverse transformation is represented by the inverse matrix. If $B = MA$ then $A = M^{-1}\,B$ (Section 10.4).

Example 23.3
Find the image P′ of the point P(3, 5) and the coordinates of the point Q which has image Q′(7, 6) under transformation by the matrix operator $M = \begin{pmatrix} 2 & 3 \\ 1 & 4 \end{pmatrix}$.

Solution

$$\begin{pmatrix} 2 & 3 \\ 1 & 4 \end{pmatrix} \begin{pmatrix} 3 \\ 5 \end{pmatrix} = \begin{pmatrix} 6 + 15 \\ 3 + 20 \end{pmatrix} = \begin{pmatrix} 21 \\ 23 \end{pmatrix}$$

Therefore P′ = (21, 23). Suppose Q = (x, y), then

$$\begin{pmatrix} 2 & 3 \\ 1 & 4 \end{pmatrix} \begin{pmatrix} x \\ y \end{pmatrix} = \begin{pmatrix} 7 \\ 6 \end{pmatrix} \Rightarrow \begin{pmatrix} x \\ y \end{pmatrix} = \frac{1}{5} \begin{pmatrix} 4 & -3 \\ -1 & 2 \end{pmatrix} \begin{pmatrix} 7 \\ 6 \end{pmatrix}$$

$$= \frac{1}{5} \begin{pmatrix} 28 - 18 \\ -7 + 12 \end{pmatrix} = \begin{pmatrix} 2 \\ 1 \end{pmatrix}$$

Therefore Q = (2, 1).

Combination of transformations
Matrix multiplication is not commutative, and it is important that the order in which the operations are performed is stated.

Exercise 23.1

(1) Under a translation T the image of the point (2, −5) is (−1, 3). Find the image of the point (5, −4).

(2) A translation of the Cartesian plane maps A(3, −4) to A′(6, −1) and B(1, 5) to B′. What are the coordinates of B′?

(3) $T = \begin{pmatrix} 1 \\ 3 \end{pmatrix}$, $T_1 = \begin{pmatrix} -5 \\ 2 \end{pmatrix}$, $T_2 = \begin{pmatrix} -1 \\ -3 \end{pmatrix}$ represent translations of the Cartesian plane. Find the image of the point P(2, 3) under translations (a) T_1, (b) T_1T_2, (c) TT_1, (d) TT.

(4) D(1, 5) is mapped by a translation to D′(3, 4). If the line $y = x + 5$ is mapped on to the line $y = ax + b$ by the same translation, find the value of the constants a and b. (Hint: two points define a straight line.)

(5) A triangle has vertices at A(1, −2), B(3, −4), C(2, 3). Find the image triangle A′B′C′ under a translation T when $T(1, 2) = (-2, 3)$.

(6) Find the image of the point G(2, −1) and the point H which has image H′(3, 5) under transformation by the matrix operator $M = \begin{pmatrix} 3 & -1 \\ 2 & 3 \end{pmatrix}$.

(7) Find the matrix operator M which transforms A(3, 5) to A′(3, −4)

and B(1, 1) to B′(1, −2). Find also the image of the point C(−1, −2) when transformed by the same matrix M.

(8) $M = \begin{pmatrix} 2 & 1 \\ -3 & 4 \end{pmatrix}$ and $N = \begin{pmatrix} 5 & -2 \\ -1 & 4 \end{pmatrix}$ represent transformations of the Cartesian plane. Find the coordinates of the image of the point (1, −2) under transformations represented by (a) MN, (b) NM, (c) M^2.

23.3 ENLARGEMENTS

(a) Enlargements centred at the origin

An enlargement of K units centred at the origin increases the coordinates of any point in the ratio $K : 1$. K is called the scale factor (SF) of the transformation, and it can be positive, negative or fractional.

In Fig. 23.1 (a) the position vector $\binom{1}{0}$ is enlarged to $\binom{K}{0}$ and the position vector $\binom{0}{1}$ is enlarged to $\binom{0}{K}$. The matrix of the enlargement is

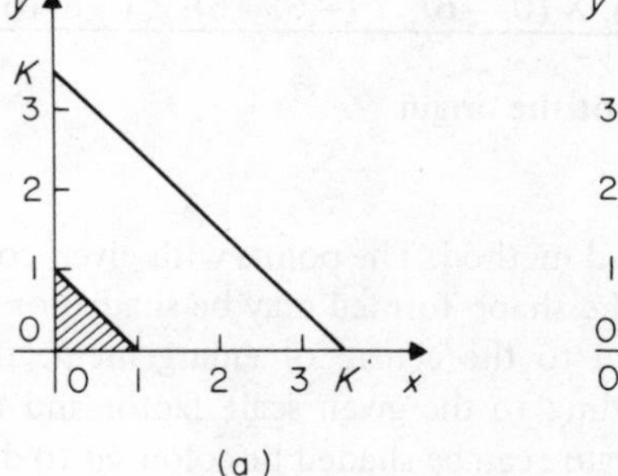

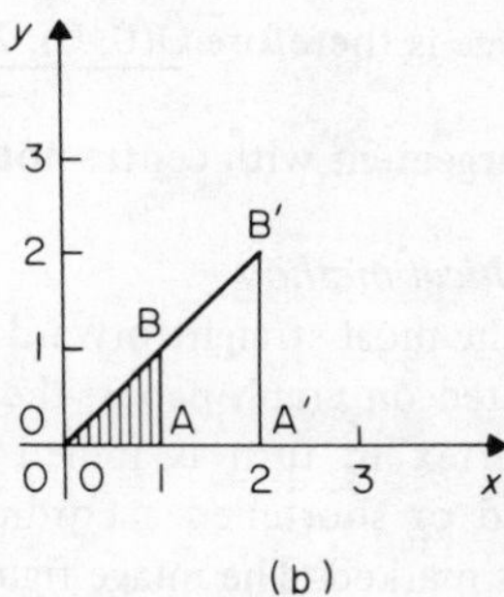

Fig 23.1

$$E = E\begin{pmatrix} 1 & 0 \\ 0 & 1 \end{pmatrix} = \begin{pmatrix} K & 0 \\ 0 & K \end{pmatrix}$$

and for any point (x, y) in the plane

$$E\begin{pmatrix} x \\ y \end{pmatrix} = \begin{pmatrix} K & 0 \\ 0 & K \end{pmatrix}\begin{pmatrix} x \\ y \end{pmatrix} = \begin{pmatrix} Kx \\ Ky \end{pmatrix}$$

The image of any figure is similar to the original figure, and the areas are in the ratio $K^2 : 1$ (see Section 13.3.).

Example 23.4

Draw Cartesian axes and mark the points A(1, 0), B(1, 1) and the images A′ and B′ under an enlargement with scale factor 2. Show that area OAB is one-quarter the area OA′B′.

Solution The points are shown in Fig. 23.1(b), A′ has coordinates (2, 0) and B′(2, 2).

$$\text{Area OAB} = \tfrac{1}{2} \times 1 \times 1 = \tfrac{1}{2} \text{ sq. unit;} \qquad \text{area OA'B'} = \tfrac{1}{2} \times 2 \times 2$$

Therefore area OAB = $\frac{1}{4}$ area OA′B′.

Example 23.5
Find the image of the square with vertices at O(0, 0), X(0, 2), Y(2, 2), Z(2, 0) under an enlargement with scale factor −3 centred at the origin.

Solution The matrix of the transformation is $\begin{pmatrix} -3 & 0 \\ 0 & -3 \end{pmatrix}$ and the coordinates of the vertices can be written as a 2 × 4 matrix.

$$\begin{pmatrix} -3 & 0 \\ 0 & -3 \end{pmatrix} \begin{pmatrix} 0 & 0 & 2 & 2 \\ 0 & 2 & 2 & 0 \end{pmatrix} = \begin{pmatrix} 0 & 0 & -6 & -6 \\ 0 & -6 & -6 & 0 \end{pmatrix}$$

The image is therefore O(0, 0), X′(0, −6), Y′(−6, −6), Z′(−6, 0).

(b) Enlargement with centre not the origin

(i) Graphical method
This is the most straightforward method. The points with given coordinates are plotted on graph paper; the shape formed may be shaded or coloured. Each vertex in turn is joined to the centre of enlargement, the line is extended or shortened according to the given scale factor and the image vertex is marked. The image figure can be shaded or coloured to distinguish it from the original shape.

Example 23.6
Draw the square OXYZ of Example 23.5 and the image O′X′Y′Z′ under an enlargement of 3 units centred at (a) (1, 0), (b) (3, 1).

Solution A sketch of the transformations is shown in Fig. 23.2. The image of each vertex is 3 times the original distance from the centre of enlargement.
(a) The image square has vertices O′(−2, 0), X′(−2, 6), Y′(4, 6) and Z′(4, 0).
(b) The image square is O′(−6, −2), X′(−6, 4), Y′(0, 4), Z′(0, −2). Notice that the areas are in the ratio $36 : 4 = 3^2 : 1$.

When the image figure is given, the centre of enlargement can be found by joining each vertex to its image and extending the lines to cut each other; the point of intersection is the centre of the enlargement.

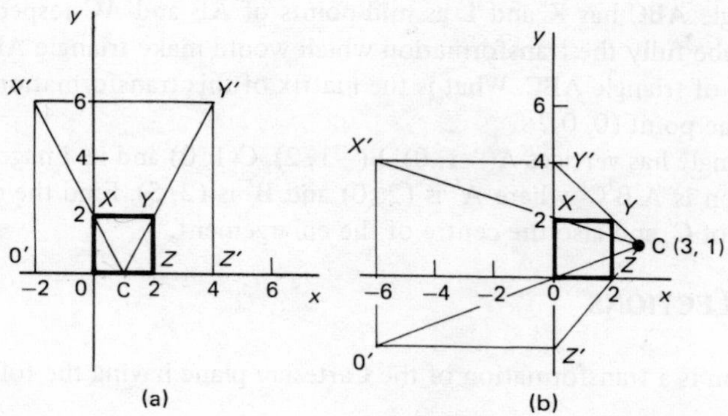

Fig 23.2

(ii) Algebraic method

When the centre of enlargement is not the origin the graphical method is recommended, but the matrix solution of Example 23.6(b) is given here to show the algebraic method.

First the origin is transferred to the centre C(3, 1) by subtracting the vector $\binom{3}{1}$ from the position vector of each vertex. The square

$$\begin{pmatrix} 0 & 0 & 2 & 2 \\ 0 & 2 & 2 & 0 \end{pmatrix} \text{ becomes } \begin{pmatrix} -3 & -3 & -1 & -1 \\ -1 & 1 & 1 & -1 \end{pmatrix}$$

Transforming by the matrix of the enlargement $\begin{pmatrix} 3 & 0 \\ 0 & 3 \end{pmatrix}$

$$\begin{pmatrix} 3 & 0 \\ 0 & 3 \end{pmatrix} \begin{pmatrix} -3 & -3 & -1 & -1 \\ -1 & 1 & 1 & -1 \end{pmatrix} \Rightarrow \begin{pmatrix} -9 & -9 & -3 & -3 \\ -3 & 3 & 3 & -3 \end{pmatrix}$$

Finally the vector $\binom{3}{1}$ is added to return the origin to (0, 0). The image square is obtained as $\begin{pmatrix} -6 & -6 & 0 & 0 \\ -2 & 4 & 4 & -2 \end{pmatrix}$ which was shown in Fig. 23.2(b).

Exercise 23.2

(1) Find the image of the point (1, −3) under an enlargement with scale factor (a) 2.5, (b) −1.7, centred at the origin.

(2) An enlargement centred at (0, 2) maps the point (3, 5) to (5, t). What is the value of t?

(3) The line segment AB is mapped to AC by an enlargement centred at A. If AC is of length 21 mm find the length of AB when the scale factor is (a) 1.5, (b) −3.5.

(4) Find the image of the rectangle A(−2, 2), B(2, 2), C(2, −1), D(−2, −1) under an enlargement of 4 units centred at (a) (0, 0), (b) (−1, −2).

(5) Triangle ABC has K and L as mid-points of AB and AC respectively. Describe fully the transformation which would make triangle AKL the image of triangle ABC. What is the matrix of this transformation when A is the point (0, 0)?

(6) A triangle has vertices A(−1, 0), B(−1, 2), C(1, 0) and its image under dilation is A′B′C′ where A′ is (2, 0) and B′ is (2, 5). Find the coordinates of C′ and also the centre of the enlargement.

23.4 REFLECTIONS

A reflection is a transformation of the Cartesian plane having the following properties.

(1) The line joining a point (x, y) to the reflected image (x', y') is perpendicular to the mirror line and bisected by the mirror line.

(2) Any point on the mirror line is the same distance from a point A as from the image A′.

(3) Points on the mirror line are unchanged; the mirror line is called the invariant line of the transformation. In figure 23.3, BB′ ⊥ ML, BL = LB′, AN = NA′, AM = MA′, ∠ AMN = ∠A′MN.

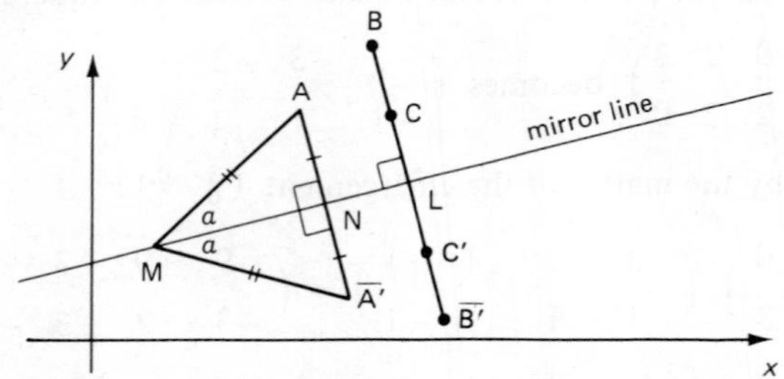

Fig 23.3

(4) The image of a figure formed by reflection is oppositely congruent to the original figure. In figure 23.4 ΔABC ≡ ΔA′B′C′. The mirror line is the perpendicular bisector of the line joining a point to its image under reflection.

Problems involving reflection in axes through the origin (0, 0) may be solved by drawing or by using the matrix of the reflection. When the mirror line is not through the origin the graphical method is better.

(a) Reflection in the coordinate axes

(*i*) Reflection in the x-axis, $y = 0$. The x-coordinate remains unchanged and the y-coordinate changes sign, for example $(3, 4) \rightarrow (3, -4)$ and $(1, -6) \rightarrow (1, 6)$. In general $(x, y) \rightarrow (x, -y)$. The matrix of the

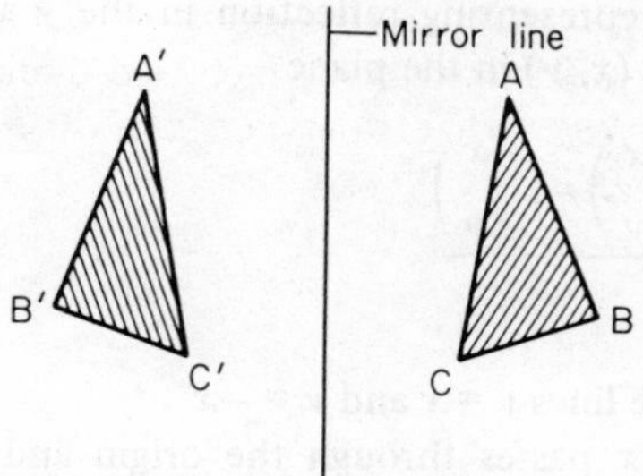

Fig 23.4

transformation is found as follows. In figure 23.5(a) the point (1, 0) is on the mirror line and is unchanged, the point B(0, 1) is reflected to B′ (0, −1). If the matrix is M, then

$$M = M\begin{pmatrix}1 & 0\\ 0 & 1\end{pmatrix} = \begin{pmatrix}1 & 0\\ 0 & -1\end{pmatrix}$$

For any point (x, y) in the plane

$$M\begin{pmatrix}x\\ y\end{pmatrix} = \begin{pmatrix}1 & 0\\ 0 & -1\end{pmatrix}\begin{pmatrix}x\\ y\end{pmatrix} = \begin{pmatrix}x\\ -y\end{pmatrix}$$

(*ii*) Reflection in the y-axis, $x = 0$. In this case the y-coordinate remains unchanged and the x-coordinate changes sign and in general $(x, y) \to (-x, y)$. The matrix of the transformation is found as follows. In Fig. 23.5 (b), the point A(1, 0) is reflected to A′(−1, 0) and the point B(0, 1) is on the mirror line and is unchanged. Hence

the matrix $M = M\begin{pmatrix}1 & 0\\ 0 & 1\end{pmatrix} = \begin{pmatrix}-1 & 0\\ 0 & 1\end{pmatrix}$

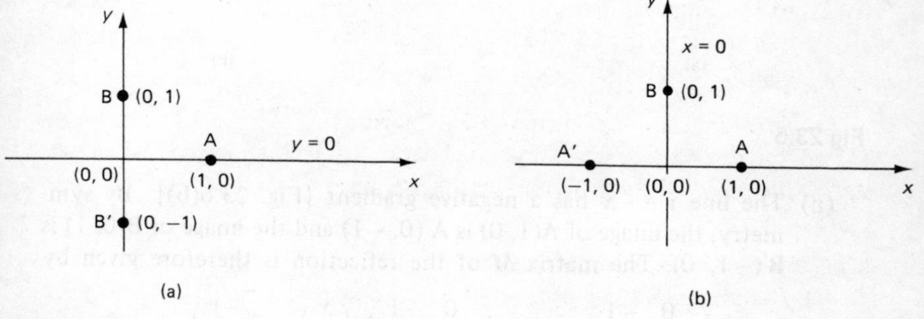

Fig 23.5

The matrix representing reflection in the y-axis is $\begin{pmatrix}-1 & 0\\ 0 & 1\end{pmatrix}$ and for any point (x, y) in the plane

$$\begin{pmatrix}-1 & 0\\ 0 & 1\end{pmatrix}\begin{pmatrix}x\\ y\end{pmatrix}=\begin{pmatrix}-x\\ y\end{pmatrix}$$

(b) Reflections in the lines $y = x$ and $y = -x$

(*i*) The line $y = x$ passes through the origin and makes equal angles with the coordinate axes [Fig. 23.6(a)]. By symmetry, A(1, 0) is reflected to B(0, 1) and also A is the image of B. Hence the matrix of the reflection is

$$M = M\begin{pmatrix}1 & 0\\ 0 & 1\end{pmatrix} = \begin{pmatrix}0 & 1\\ 1 & 0\end{pmatrix}$$

and for every point in the plane

$$\begin{pmatrix}0 & 1\\ 1 & 0\end{pmatrix}\begin{pmatrix}x\\ y\end{pmatrix}=\begin{pmatrix}y\\ x\end{pmatrix}$$

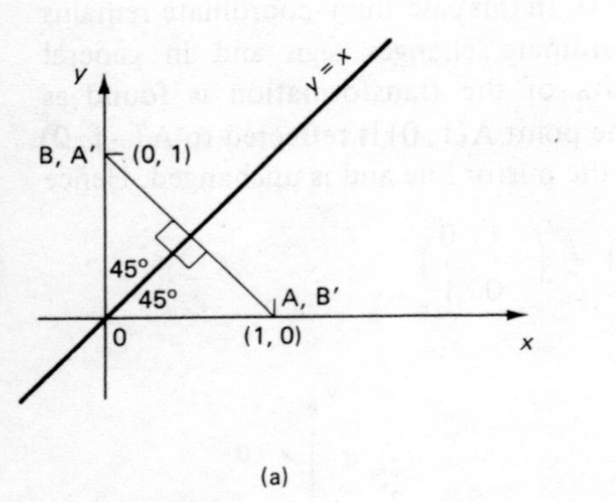

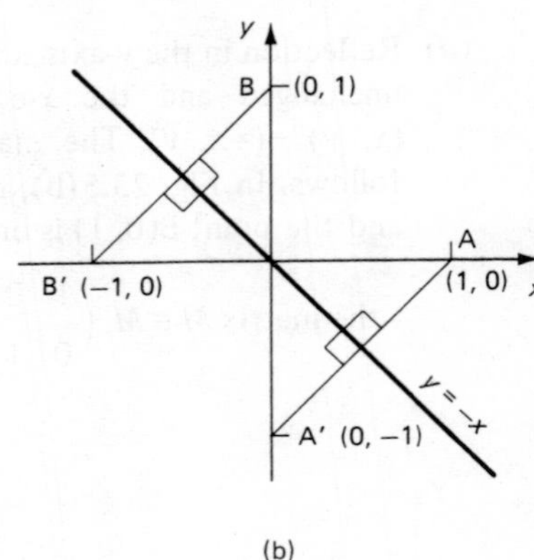

Fig 23.6

(ii) The line $y = -x$ has a negative gradient [Fig. 23.6(b)]. By symmetry, the image of A(1, 0) is A′(0, −1) and the image of B(0, 1) is B′(−1, 0). The matrix M of the reflection is therefore given by

$$M = \begin{pmatrix}0 & -1\\ -1 & 0\end{pmatrix} \quad \text{and} \quad \begin{pmatrix}0 & -1\\ -1 & 0\end{pmatrix}\begin{pmatrix}x\\ y\end{pmatrix}=\begin{pmatrix}-y\\ -x\end{pmatrix}$$

Example 23.7

ABC is a triangle with vertices at the points A(2, 0), B(2, 2), C(0, 2). If *Mx* represents reflection in the *x*-axis and *My* reflection in *y*-axis find the image of ABC under the transformation (a) *Mx*, (b) *My*, (c) *MxMy*.

Solution Graphical method: The triangle is drawn on graph paper and shaded as shown in the sketch [Fig. 23.7(a)].

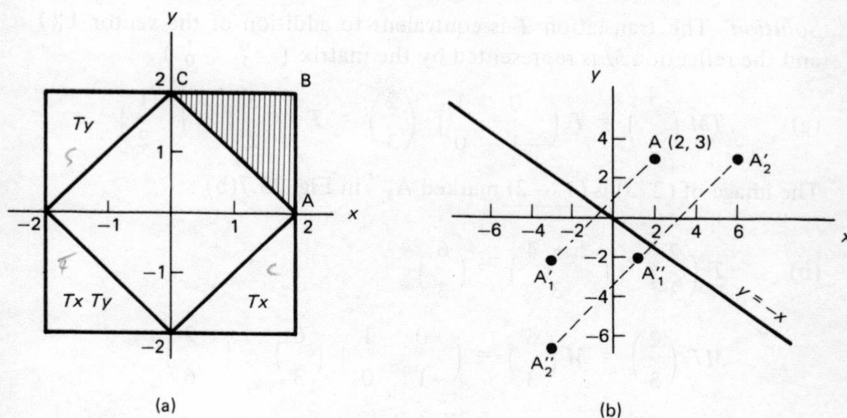

Fig 23.7

The image in the *x*-axis is marked *Tx*, the image in the *y*-axis is marked *Ty* and the image of *Ty* in the *x*-axis is marked *TxTy*.

(a) The image triangle has coordinates (2, 0), (2, −2), (0, −2)

(b) The image triangle has coordinates (−2, 0), (−2, 2), (0, 2)

(c) The image triangle has coordinates (−2, 0), (−2, −2), (0, −2).

Algebraic method:

(a) The matrix of the transformation *Mx* is $\begin{pmatrix} 1 & 0 \\ 0 & -1 \end{pmatrix}$

$$\begin{pmatrix} 1 & 0 \\ 0 & -1 \end{pmatrix}\begin{pmatrix} 2 & 2 & 0 \\ 0 & 2 & 2 \end{pmatrix} = \begin{pmatrix} 2 & 2 & 0 \\ 0 & -2 & -2 \end{pmatrix}$$

(b) The matrix of the reflection *My* is $\begin{pmatrix} -1 & 0 \\ 0 & 1 \end{pmatrix}$

$$\begin{pmatrix} -1 & 0 \\ 0 & 1 \end{pmatrix}\begin{pmatrix} 2 & 2 & 0 \\ 0 & 2 & 2 \end{pmatrix} = \begin{pmatrix} -2 & -2 & 0 \\ 0 & 2 & 2 \end{pmatrix}$$

(c) $$\begin{pmatrix} 1 & 0 \\ 0 & -1 \end{pmatrix}\begin{pmatrix} -1 & 0 \\ 0 & 1 \end{pmatrix}\begin{pmatrix} 2 & 2 & 0 \\ 0 & 2 & 2 \end{pmatrix} = \begin{pmatrix} -2 & -2 & 0 \\ 0 & -2 & -2 \end{pmatrix}$$

The image triangles are evidently the same as those obtained by the graphical method.

Example 23.8
T represents a translation of +4 units parallel to the x-axis and M a reflection in the line $y = -x$. Find the image of the point A(2, 3) under the transformation represented by (a) TM, (b) MT.

Solution The translation T is equivalent to addition of the vector $\binom{4}{0}$ and the reflection M is represented by the matrix $\begin{pmatrix} 0 & -1 \\ -1 & 0 \end{pmatrix}$.

(a) $$TM\begin{pmatrix} 2 \\ 3 \end{pmatrix} = T\begin{pmatrix} 0 & -1 \\ -1 & 0 \end{pmatrix}\begin{pmatrix} 2 \\ 3 \end{pmatrix} = T\begin{pmatrix} -3 \\ -2 \end{pmatrix} = \begin{pmatrix} 1 \\ -2 \end{pmatrix}$$

The image of (2, 3) is (1, −2) marked A_1'' in Fig. 23.7(b).

(b) $$T\begin{pmatrix} 2 \\ 3 \end{pmatrix} = \begin{pmatrix} 2+4 \\ 3 \end{pmatrix} = \begin{pmatrix} 6 \\ 3 \end{pmatrix}$$

$$MT\begin{pmatrix} 2 \\ 3 \end{pmatrix} = M\begin{pmatrix} 6 \\ 3 \end{pmatrix} = \begin{pmatrix} 0 & -1 \\ -1 & 0 \end{pmatrix}\begin{pmatrix} 6 \\ 3 \end{pmatrix} = \begin{pmatrix} -3 \\ -6 \end{pmatrix}$$

The image is (−3, −6) marked A_2'' in the figure.

Summary of Reflection Transformations

Mirror line	*Matrix*	*Image of* (x, y)
$x = 0$	$\begin{pmatrix} -1 & 0 \\ 0 & 1 \end{pmatrix}$	$(-x, y)$
$y = 0$	$\begin{pmatrix} 1 & 0 \\ 0 & -1 \end{pmatrix}$	$(x, -y)$
$y = x$	$\begin{pmatrix} 0 & 1 \\ 1 & 0 \end{pmatrix}$	(y, x)
$y = -x$	$\begin{pmatrix} 0 & -1 \\ -1 & 0 \end{pmatrix}$	$(-y, -x)$

Exercise 23.3

(1) Draw the trapezium with coordinates A(2, 1), B(3, 1), C(3, 5), D(2, 4), and find its image under reflection in the line (a) $y = 0$, (b) $x = 0$, (c) $y = x$, (d) $y = 0$ and then $y = -x$.

(2) For the trapezium ABCD of question 1 find the image under translation by the vector $\binom{2}{2}$ followed by reflection in the line $y = -x$.

(3) Find the coordinates of the image when the given point is reflected in the given line. (a) (2, 4), $x = 0$, (b) (3, −1), $y = 0$, (c) (2, −3), $y = x$, (d) (−1, 2), $y = -x$, (e) (−2, 4), $x = 5$, (f) (3, 2), $x = -2$, (g) (−4, 3), $y = 4$, (h) (1, 2), $y = x + 2$.

(4) Find by a graphical method the equation of the mirror line for the transformation (a) (1, 2) → (5, 2), (b) (−2, 1) → (−2, 5), (c) (2, 0) → (−2, 4).

23.5 ROTATIONS

When the line joining any point in the plane to a fixed point C is rotated about C, the transformation of the plane is called a *rotation* about the centre C. In other words, the position vectors relative to the centre of rotation are all turned through the same angle.

(a) Rotations about the origin

In the Cartesian system anticlockwise rotation is positive and clockwise rotation is negative.

Rotation through an angle θ

In Fig. 23.8(a) the unit vector $\overrightarrow{\mathbf{OA}}$ has been rotated through an angle θ anticlockwise to $\overrightarrow{\mathbf{OA}}'$, so that OA = OA′ = 1. The coordinates of A′ are ON = $\cos\theta$ and A′N = $\sin\theta$. Hence, the vector $\binom{1}{0}$ becomes $\binom{\cos\theta}{\sin\theta}$.

In a similar way the unit vector $\overrightarrow{\mathbf{OB}}$ is rotated through an angle θ to $\overrightarrow{\mathbf{OB}}'$, and the coordinates of B′ are B′M = $-\sin\theta$ and OM = $\cos\theta$. Hence the vector $\binom{0}{1}$ becomes $\binom{-\sin\theta}{\cos\theta}$.

The matrix of the rotation R is

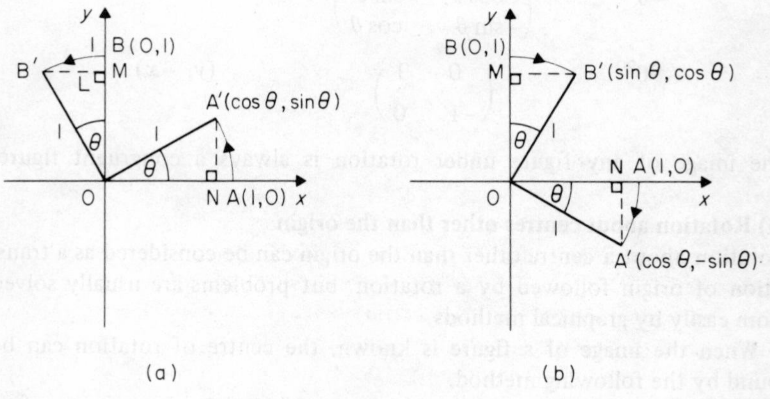

Fig 23.8

$$R\begin{pmatrix}1 & 0\\ 0 & 1\end{pmatrix} = \begin{pmatrix}\cos\theta & -\sin\theta\\ \sin\theta & \cos\theta\end{pmatrix}$$

In Fig. 23.8(b) the rotation is clockwise through an angle θ. The coordinates of A′ are ON = $\cos\theta$, A′N = $-\sin\theta$. Similarly the coordinates of B′ are B′M = $\sin\theta$, OM = $\cos\theta$. Hence

$$\begin{pmatrix}1\\ 0\end{pmatrix} \rightarrow \begin{pmatrix}\cos\theta\\ -\sin\theta\end{pmatrix} \quad \text{and} \quad \begin{pmatrix}0\\ 1\end{pmatrix} \rightarrow \begin{pmatrix}\sin\theta\\ \cos\theta\end{pmatrix}$$

and the matrix for rotation through a negative angle is

$$R = R\begin{pmatrix}1 & 0\\ 0 & 1\end{pmatrix} = \begin{pmatrix}\cos\theta & \sin\theta\\ -\sin\theta & \cos\theta\end{pmatrix}$$

The matrices representing rotation through 90° and 180° may be obtained by substituting for θ.

Summary of rotation transformations

Rotation	*Matrix*	*Image of* (x, y)
θ	$\begin{pmatrix}\cos\theta & -\sin\theta\\ \sin\theta & \cos\theta\end{pmatrix}$	
90°	$\begin{pmatrix}0 & -1\\ 1 & 0\end{pmatrix}$	$(-y, x)$
180°	$\begin{pmatrix}-1 & 0\\ 0 & -1\end{pmatrix}$	$(-x, -y)$
$-\theta$	$\begin{pmatrix}\cos\theta & \sin\theta\\ -\sin\theta & \cos\theta\end{pmatrix}$	
−90°	$\begin{pmatrix}0 & 1\\ -1 & 0\end{pmatrix}$	$(y, -x)$

The image of any figure under rotation is always a congruent figure.

(b) Rotation about centres other than the origin

Rotation about a centre other than the origin can be considered as a translation of origin followed by a rotation, but problems are usually solved more easily by graphical methods.

When the image of a figure is known, the centre of rotation can be found by the following method.

(i) Each vertex is joined to its image by a straight line

(ii) The perpendicular bisector is drawn of each line

(iii) These perpendicular bisectors meet at the centre of rotation.

Example 23.9

Describe the transformation associated with $M = \begin{pmatrix} 0 & -1 \\ 1 & 0 \end{pmatrix}$ and $N = \begin{pmatrix} 8 & 0 \\ 0 & 8 \end{pmatrix}$. If $P = \begin{pmatrix} 2 & -2 \\ 2 & 2 \end{pmatrix}$, show that $P^2 = MN$ and describe the transformation associated with P.

Solution M represents a rotation of 90° about the origin, and N represents an enlargement centred at the origin with scale factor 8.

$$MN = \begin{pmatrix} 0 & -1 \\ 1 & 0 \end{pmatrix} \begin{pmatrix} 8 & 0 \\ 0 & 8 \end{pmatrix} = \begin{pmatrix} 0 & -8 \\ 8 & 0 \end{pmatrix}$$

$$P^2 = \begin{pmatrix} 2 & -2 \\ 2 & 2 \end{pmatrix} \begin{pmatrix} 2 & -2 \\ 2 & 2 \end{pmatrix} = \begin{pmatrix} 0 & -8 \\ 8 & 0 \end{pmatrix} = MN$$

Hence P^2 represents an enlargement of 8 units followed by rotation through 90°, and P represents an enlargement of $\sqrt{8}$ units followed by rotation through 45°. *Check*

$$P = \begin{pmatrix} \cos 45° & -\sin 45° \\ \sin 45° & \cos 45° \end{pmatrix} \begin{pmatrix} \sqrt{8} & 0 \\ 0 & \sqrt{8} \end{pmatrix} = \begin{pmatrix} 2 & -2 \\ 2 & 2 \end{pmatrix}$$

since

$$\sin 45° = \cos 45° = \frac{1}{\sqrt{2}} \text{ and } \frac{\sqrt{8}}{\sqrt{2}} = 2$$

Example 23.10

If $P = \begin{pmatrix} 0 & -1 \\ 1 & 0 \end{pmatrix}$, describe the transformations of the plane represented by $P\begin{pmatrix} x \\ y \end{pmatrix}$ and $P^2 \begin{pmatrix} x \\ y \end{pmatrix}$ and hence write P^8 as a single matrix.

Solution

$$P\begin{pmatrix} x \\ y \end{pmatrix} = \begin{pmatrix} 0 & -1 \\ 1 & 0 \end{pmatrix} \begin{pmatrix} x \\ y \end{pmatrix} = \begin{pmatrix} -y \\ x \end{pmatrix};$$

$$P^2\begin{pmatrix} x \\ y \end{pmatrix} = \begin{pmatrix} 0 & -1 \\ 1 & 0 \end{pmatrix} \begin{pmatrix} -y \\ x \end{pmatrix} = \begin{pmatrix} -x \\ -y \end{pmatrix}$$

P represents rotation of 90° and P^2 rotation of 180° about the origin. P^8 therefore represents the identity transformation, since rotating through 180° an even number of times brings every point back to its starting position. Hence

$$P^8 = \begin{pmatrix} 1 & 0 \\ 0 & 1 \end{pmatrix}$$

Example 23.11
The triangle with vertices at A(−3, 2), B(−2, 4), C(−4, 4) is mapped onto the triangle A″(5, 2), B″(3, 1), C″(3, 3) by a rotation followed by reflection in the line $x = 2$.

Draw a graph showing the position of ABC, A″B″C″ and also the intermediate triangle A′B′C′ and determine the centre of the rotation.

Solution The reflection of triangle A″B″C″ in the line $x = 2$ is marked A′B′C′ in Fig. 23.9. B′C′ is at right angles to BC and it is evident from the

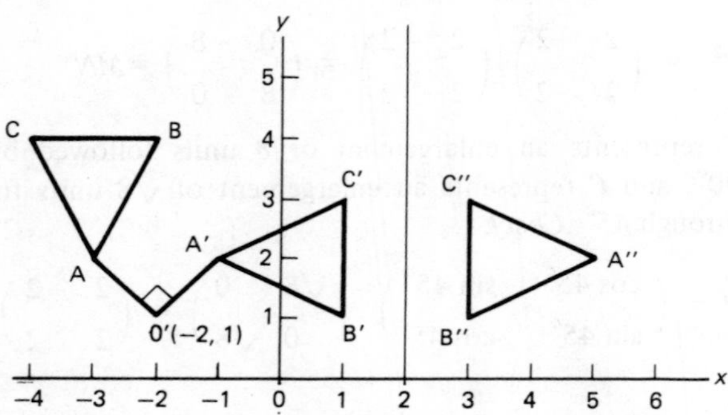

Fig 23.9

diagram that triangle ABC has been rotated through 90° clockwise to the position A′(−1, 2), B′(1, 1), C′(1, 3). The centre of the rotation O′ is such that O′A = O′A′ and also O′A ⊥ O′A′ since the rotation was 90°, and the coordinates are seen from the graph to be (−2, 1). The centre of rotation is therefore <u>(−2, 1).</u>

23.6 **SHEARS AND STRETCHES**

(a) Shears parallel to the coordinate axes

A shear transforms a rectangle into a parallelogram of the same height and area.

(i) A shear of K units parallel to the y-axis transforms the point (1, 0) to (1, K) and leaves the point (0, 1) unchanged since the y-axis is the invariant line. The matrix of the shear is $H = \begin{pmatrix} 1 & 0 \\ K & 1 \end{pmatrix}$

(ii) A shear of K units parallel to the x-axis leaves the point (1, 0) unchanged and transforms the point (0, 1) to (K, 1). The matrix representing the shear is therefore given by

$$H = H\begin{pmatrix} 1 & 0 \\ 0 & 1 \end{pmatrix} = \begin{pmatrix} 1 & K \\ 0 & 1 \end{pmatrix}$$

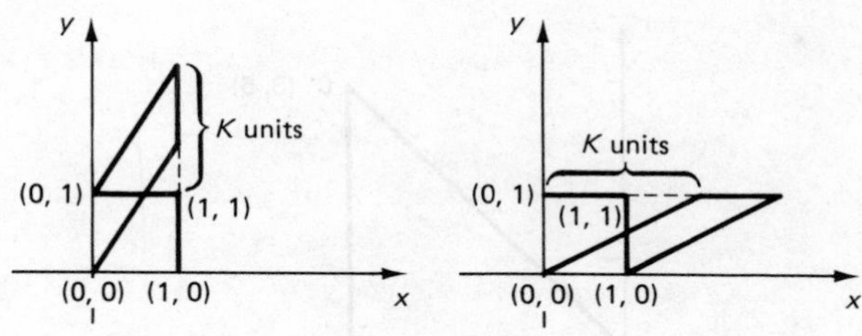

Fig 23.10

and for any point (x, y) in the plane

$$\begin{pmatrix} 1 & K \\ 0 & 1 \end{pmatrix} \begin{pmatrix} x \\ y \end{pmatrix} = \begin{pmatrix} x + Ky \\ y \end{pmatrix}; \begin{pmatrix} 1 & 0 \\ K & 1 \end{pmatrix} \begin{pmatrix} x \\ y \end{pmatrix} = \begin{pmatrix} x \\ Kx + y \end{pmatrix}$$

Example 23.12

The matrix $P\ \left(\begin{smallmatrix} 3 & 0 \\ 3 & 3 \end{smallmatrix}\right)$ represents a transformation of the Cartesian plane. (a) Find the image of the unit square A(0, 0) B(1, 0) C(1, 1) D(0, 1) under the transformation $\left(\begin{smallmatrix} x' \\ y' \end{smallmatrix}\right) = P\ \left(\begin{smallmatrix} x \\ y \end{smallmatrix}\right)$. (b) Show that it is a parallelogram and calculate its area. (c) Describe the geometrical transformation represented by the matrix P.

Solution

(a) $$P(\text{ABCD}) = \begin{pmatrix} 3 & 0 \\ 3 & 3 \end{pmatrix} \begin{pmatrix} 0 & 1 & 1 & 0 \\ 0 & 0 & 1 & 1 \end{pmatrix} = \begin{pmatrix} 0 & 3 & 3 & 0 \\ 0 & 3 & 6 & 3 \end{pmatrix}$$

The image A′B′C′D′ has vertices as (0, 0), (3, 3), (3, 6), (0, 3). The square and its image are shown in Fig. 23.11.

(b) A′D′ and B′C′ are parallel to the y-axis and of length 3 units, and so the image figure is a parallelogram (one pair of sides equal and parallel) of area 9 square units.

(c) It can be seen from the diagram that the square of side 3 units was transformed by a shear of 1 unit parallel to the y-axis. The matrix P therefore represents an enlargement of 3 units centred at the origin followed by a shear of 1 unit parallel to the y-axis

$$\begin{pmatrix} 1 & 0 \\ 1 & 1 \end{pmatrix} \begin{pmatrix} 3 & 0 \\ 0 & 3 \end{pmatrix} = \begin{pmatrix} 3 & 0 \\ 3 & 3 \end{pmatrix}$$

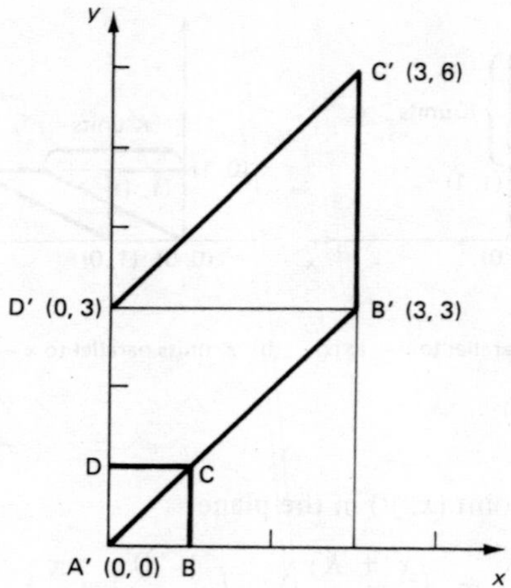

Fig 23.11

(b) Shears not parallel to the axes

Example 23.13
Find the image of the points (0, 0), (1, 0), (0, 1) (−1, 1), (1, −1), (1, 1) under transformation by the matrix $H = \begin{pmatrix} 2 & 1 \\ -1 & 0 \end{pmatrix}$. Describe geometrically the line which is left invariant by the transformation.

Solution

$$\begin{pmatrix} 2 & 1 \\ -1 & 0 \end{pmatrix} \begin{pmatrix} 0 & 1 & 0 & -1 & 1 & 1 \\ 0 & 0 & 1 & 1 & -1 & 1 \end{pmatrix} = \begin{pmatrix} 0 & 2 & 1 & -1 & 1 & 3 \\ 0 & -1 & 0 & 1 & -1 & -1 \end{pmatrix}$$

The points unchanged are (0, 0), (−1, 1) and (1, −1) which are on the line $y = -x$ and this is the invariant line of the transformation.

Fig. 23.12 shows the original right angled triangle transformed by a shear parallel to the line $y = -x$ to a triangle of equal area.

(c) Stretches parallel to the axes
A stretch is an enlargement in one direction only. A stretch of K units parallel to the x-axis increases the x-coordinate of a point to K times the original value but leaves the y-coordinate unchanged. The matrix of the transformation is

$$S_x = \begin{pmatrix} K & 0 \\ 0 & 1 \end{pmatrix}$$

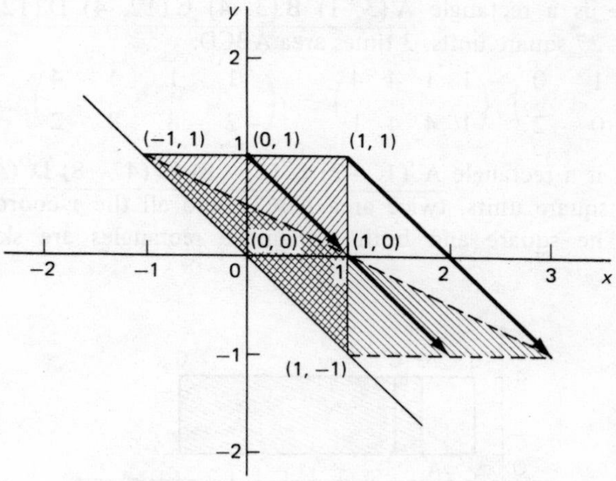

Fig 23.12

and

$$S_x \begin{pmatrix} x \\ y \end{pmatrix} = \begin{pmatrix} K & 0 \\ 0 & 1 \end{pmatrix} \begin{pmatrix} x \\ y \end{pmatrix} = \begin{pmatrix} Kx \\ y \end{pmatrix}$$

A stretch parallel to the y-axis increases the y-coordinate of any point and leaves the x-coordinate unchanged. The matrix representing it is

$$S_y = \begin{pmatrix} 1 & 0 \\ 0 & K \end{pmatrix}$$

and

$$S_y \begin{pmatrix} x \\ y \end{pmatrix} = \begin{pmatrix} x \\ Ky \end{pmatrix} \quad \text{for any point } (x, y) \text{ in the plane}$$

Example 23.14

Find the image of the square with coordinates at A(1, 1) B(1,4) C(4, 4) D(4, 1) under a stretch (a) 3 units parallel to the x-axis, (b) -2 units parallel to the y-axis. Find the area of ABCD and the image in each case.

Solution

$$\begin{pmatrix} 3 & 0 \\ 0 & 1 \end{pmatrix} \begin{pmatrix} 1 & 1 & 4 & 4 \\ 1 & 4 & 4 & 1 \end{pmatrix} = \begin{pmatrix} 3 & 3 & 12 & 12 \\ 1 & 4 & 4 & 1 \end{pmatrix}$$

The image is a rectangle A′(3, 1) B′(3, 4) C′(12, 4) D′(12, 1). Area A′B′C′D′ = 27 square units, 3 times area ABCD.

(b)

$$\begin{pmatrix} 1 & 0 \\ 0 & -2 \end{pmatrix} \begin{pmatrix} 1 & 1 & 4 & 4 \\ 1 & 4 & 4 & 1 \end{pmatrix} = \begin{pmatrix} 1 & 1 & 4 & 4 \\ -2 & -8 & -8 & -2 \end{pmatrix}$$

The image is a rectangle A″(1, −2) B″(1, −8) C″(4, −8) D″(4, −2). Its area is 18 square units, twice area ABCD, and all the y-coordinates are negative. The square and both the image rectangles are sketched in Fig. 23.13.

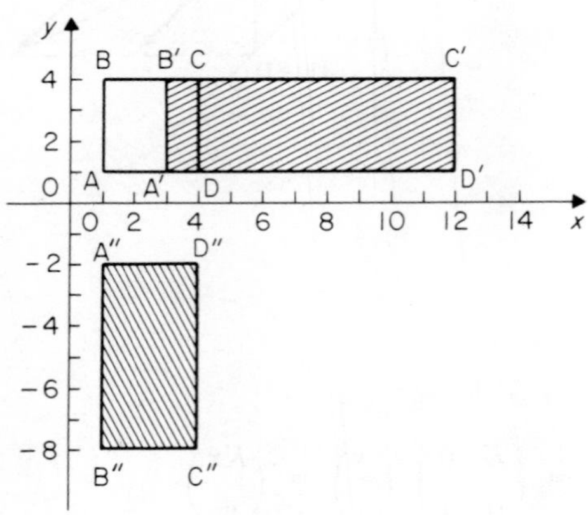

Fig 23.13

Exercise 23.4

(1) Find the coordinates of the image of the point (3, 4) under a clockwise rotation of 90° and an anticlockwise rotation of 90° centred at (a) (0, 0), (b) (1, 3), (c) (−1, 2).

(2) What is the centre of rotation when (a) the line $y = 2x + 2$ is rotated to $y = x - 1$, (b) the line segment joining P(2, 1) and Q(3, −1) is transformed to P′(3, 4) and Q′(5, 5)?

(3) The unit square ABCD has vertices at A(0, 0), B(1, 0), C(1, 1), D(0, 1). Find the coordinates of A′B′C′D′, the image of ABCD under transformation represented by the matrices (a) $\begin{pmatrix} 2 & 0 \\ 0 & -2 \end{pmatrix}$, (b) $\begin{pmatrix} 1 & 0 \\ 0 & -1 \end{pmatrix}$, (c) $\begin{pmatrix} 1 & 0 \\ 0 & 1 \end{pmatrix}$, (d) $\begin{pmatrix} 1 & 0 \\ 2 & 1 \end{pmatrix}$, (e) $\begin{pmatrix} 0 & -1 \\ 1 & 0 \end{pmatrix}$.

(4) Find the images of A(1, 1) and B(3, 3) under the transformation (a) $\begin{pmatrix} -1 & 0 \\ 0 & 1 \end{pmatrix}$, (b) $\begin{pmatrix} 0 & -1 \\ 1 & 0 \end{pmatrix}$, (c) $\begin{pmatrix} 4 & 0 \\ 0 & 4 \end{pmatrix}$, (d) $\begin{pmatrix} 1/\sqrt{2} & -1/\sqrt{2} \\ 1/\sqrt{2} & 1/\sqrt{2} \end{pmatrix}$, (e) $\begin{pmatrix} 4 & -4 \\ 4 & 4 \end{pmatrix}$, (f) $\begin{pmatrix} -1 & 0 \\ 0 & -1 \end{pmatrix}$.

(5) Determine which of the matrices in Question 4 represent the following geometric transformations:

(i) anticlockwise rotation of 90° about the origin

(ii) reflection in the y-axis
(iii) anticlockwise rotation of 45° about the origin
(iv) reflection in the line $y = -x$
(v) enlargement of 4 units centred at the origin
(vi) rotation through 45° and enlargement of $4\sqrt{2}$ units centred at the origin.

(6) Transformation A is an enlargement of scale factor 3 and centre C$(-1, 2)$. B is a clockwise rotation of 90° about the origin. If P is the point $(3, -1)$ find the position vector of its image when transformed by (a) B^2, (b) AB, (c) BA.

(7) Find, by drawing, the image of the triangle with vertices at A(4, 2), B(2, 5), C(6, 6) under successive transformations (i) anticlockwise rotation of 90° about (0, 8), (ii) enlargement with centre A and scale factor 2, (iii) translation $(x, y) \to (x + 2, y - 3)$. State the coordinates of the image triangle.

CHAPTER 24

FURTHER GEOMETRY

Traditional or Euclidean geometry (named after Euclid) comprises a few basic *axioms* which are so obviously true that no proof is necessary, and a sequence of *theorems*, the proof of each theorem following from those earlier in the sequence. Formal proofs of theorems are not given here, since they are no longer required in the syllabus of the major Examining Boards, but a number of the important theorems are stated, with examples to show how they are applied to problems in geometry.

In solving geometrical problems any knowledge may be used; solutions may be obtained by traditional means such as congruent triangles, by symmetry, by vectors and transformations, by trigonometry, or by a combination of these methods.

Some Euclidean axioms and theorems have been met already in Chapters 11 and 13, and they are listed here for easy reference, labelled 1 to 19.

24.1 LINES AND ANGLES

(1) Angles at a point total 360° — Section 11.2(a)
(2) Angles on a straight line total 180° — 11.2(b)
(3) Vertically opposite angles are equal — 11.3(a)
(4) Parallel lines make equal angles with a fixed line
(a) Corresponding angles are equal
(b) Alternate angles are equal. — 11.3(b)

24.2 POLYGONS

(5) The sum of the exterior angles of a convex polygon is 360°
(6) The sum of the interior angles is $(n - 2)$ straight angles where n is the number of sides. — 11.4(a)

24.3 TRIANGLES

(7) The angle sum of a triangle is 180° Example 11.6(i)
(8) Any two sides of a triangle are together greater than the third
(9) The largest angle is opposite the longest side Section 11.5
(10) An exterior angle of a triangle is equal to the sum of the interior opposite angles Example 11.6(ii)
(11) The base angles of an isosceles triangle are equal Section 11.5
(12) Triangles on the same base and between the same parallels are equal in area 13.2(c)
(13) Pythagoras' theorem: in any right angled triangle, the square on the hypotenuse is equal to the sum of the squares on the other two sides Figure 11.10
(14) Similar triangles have angles the same size and corresponding sides in the same ratio Section 11.6(a)
(15) Congruent triangles are equal in all respects. 11.6(b)

24.4 PARALLELOGRAMS

(16) Opposite sides of a parallelogram are equal 11.7
(17) Opposite angles of a parallelogram are equal 11.7
(18) The diagonals of a parallelogram bisect each other 11.7
(19) Parallelograms on the same base and between the same parallels are equal in area. 13.2(b)

24.5 RATIO THEOREMS OF TRIANGLES

(20) A line drawn parallel to one side of a triangle divides the other two sides in the same ratio
(21) The line joining the mid-points of two sides of a triangle is parallel to the third side and equal to half its length.

In Fig. 24.1(a)

$$\frac{AX}{XB} = \frac{AY}{YC} \quad \text{(triangles AXY and ABC are similar)}$$

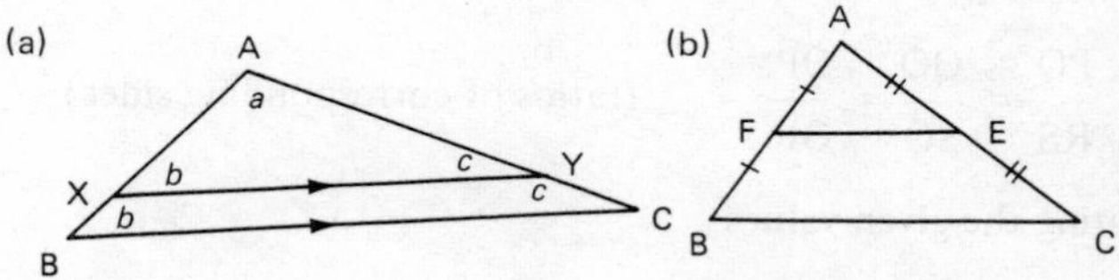

Fig 24.1

In Fig. 24.1(b)

$$\frac{AF}{AB} = \frac{AE}{AC} = \frac{FE}{BC} = \frac{1}{2} \qquad \text{(triangles AFE and ABC are similar)}$$

For a proof of these theorems see Example 11.9.

Example 24.1
In Fig. 24.2 P, Q, R, S are points on the sides of a quadrilateral ABCD. PR and QS intersect at O and ∠PQS = ∠QSR. If PQ = 3.6 cm, SR = 4 cm, SO = 2.5 cm, OP = 2.7 cm, calculate the length of QO and OR.

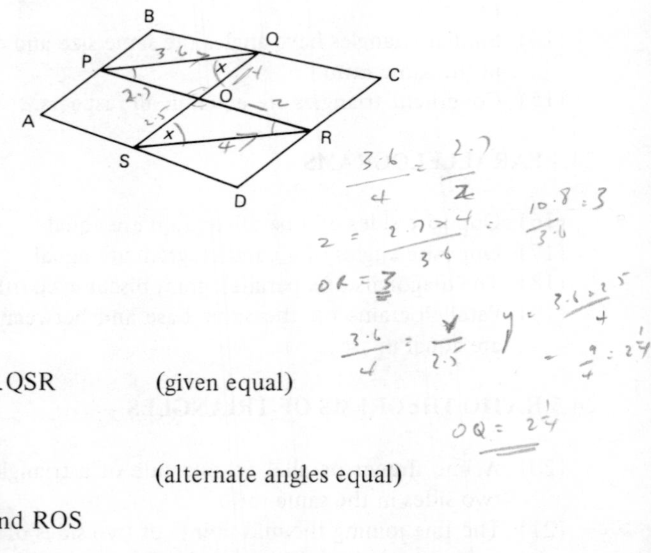

Fig 24.2

Solution

∠PQS = ∠QSR (given equal)

therefore

PQ ∥ SR (alternate angles equal)

In triangles POQ and ROS

∠PQO = ∠OSR (QOS is a straight line)

∠QPO = ∠SRO (alternate angles)

∠POQ = ∠ROS (vertically opposite angles)

therefore triangles POQ and ROS are similar and hence

$$\frac{PQ}{RS} = \frac{QO}{SO} = \frac{OP}{OR} \qquad \text{(ratios of corresponding sides)}$$

Substituting the given values

$$\frac{3.6}{4} = \frac{QO}{2.5} = \frac{2.7}{OR} \Rightarrow QO = \frac{2.5 \times 3.6}{4} = \underline{2.25 \text{ cm}}$$

and

$$OR = \frac{2.7 \times 4}{3.6} = \underline{3 \text{ cm}}$$

(22) Angle bisector theorems

(a) The bisector of an interior angle of a triangle divides the opposite side in the ratio of the sides containing the angle.

(b) The bisector of an exterior angle of a triangle divides the opposite side externally in the ratio of the sides containing the angle. In both triangles in Fig. 24.3 AD is the bisector of angle A of the triangle ABC and DB/DC = AB/AC.

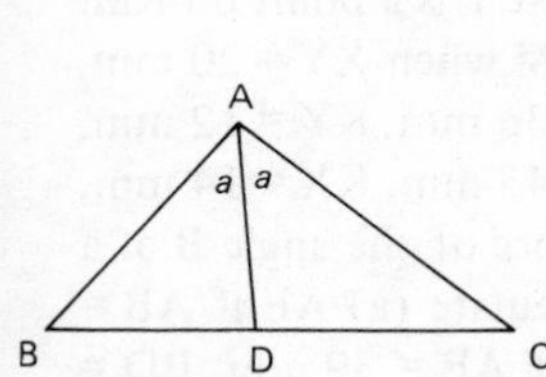

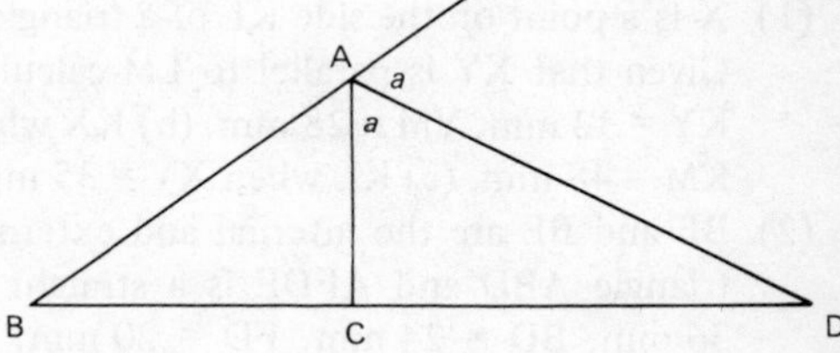

Fig 24.3

Example 24.2

In Fig. 24.4, AD is the bisector of ∠BAC. Calculate the value of x in each triangle, when all the lengths are given in centimetres.

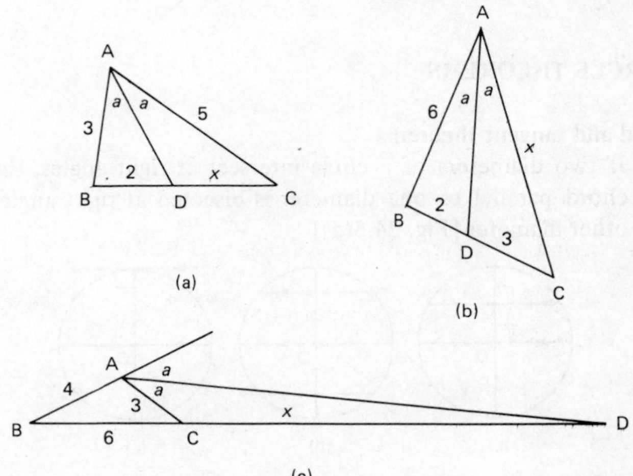

Fig 24.4

Solution From the angle bisector theorem DB/DC = AB/AC.

(a) $\frac{2}{x} = \frac{3}{5} \Rightarrow \underline{x = 3\frac{1}{3}}$

(b) $\frac{2}{3} = \frac{6}{x} \Rightarrow \underline{x = 9}$

(c) $\frac{6+x}{x} = \frac{4}{3} \Rightarrow 18 + 3x = 4x \Rightarrow \underline{x = 18}$

Exercise 24.1

(1) X is a point on the side KL of a triangle KLM and Y is a point on KM. Given that XY is parallel to LM calculate (a) LM when XY = 20 mm, KY = 32 mm, YM = 28 mm, (b) KX when KL = 36 mm, KY = 12 mm, KM = 48 mm, (c) KL when XY = 35 mm, LM = 45 mm, KX = 14 mm.

(2) BF and BE are the internal and external bisectors of the angle B of a triangle ABD and AFDE is a straight line. Calculate (a) AF if AB = 36 mm, BD = 24 mm, FD = 20 mm, (b) AE if AB = 39 mm, BD = 26 mm, DE = 36 mm, (c) BD if AF = 12 mm, FD = 36 mm, AB = 45 mm, (d) AD if DE = 90 mm, AB = 80 mm, BD = 50 mm.

(3) ABCD is a quadrilateral with $\angle$CDA = $\angle$CAB = 90°, CD = DA = 6 cm and CA = AB. Calculate (a) AB, (b) BC.

(4) The side QR of a triangle PQR is extended to S and PS is the bisector of the exterior angle QPR. Given that QS = 12 cm, QR = 4 cm and PQ = 9 cm, calculate (a) the length of PR, (b) $\angle$RPS.

24.6 CIRCLE THEOREMS

(a) Chord and tangent theorems

(23) If two diameters of a circle intersect at right angles, then every chord parallel to one diameter is bisected at right angles by the other diameter [Fig. 24.5(a)]

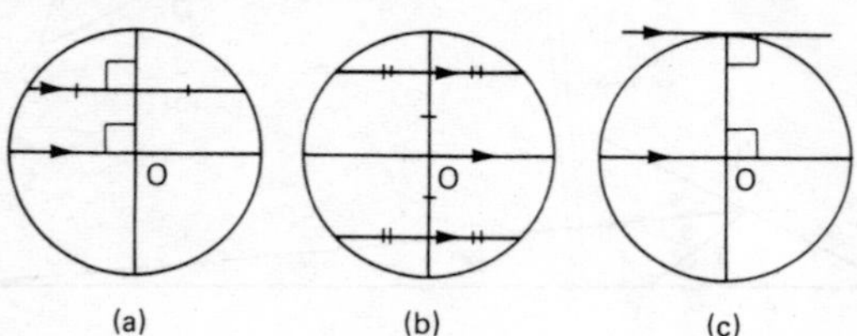

Fig 24.5

(24) Chords which are equidistant from the centre of a circle are equal in length. Conversely, equal chords are equidistant from the centre [Fig. 24.5(b)]

(25) A tangent to a circle is perpendicular to the radius drawn at the point of contact [Figure 24.5(c)]

(26) The two tangents which can be drawn to a circle from an external point are equal in length [Fig. 24.6(a)]. This is proved by showing that triangles APO and BPO are congruent [see Example 24.3(a)].

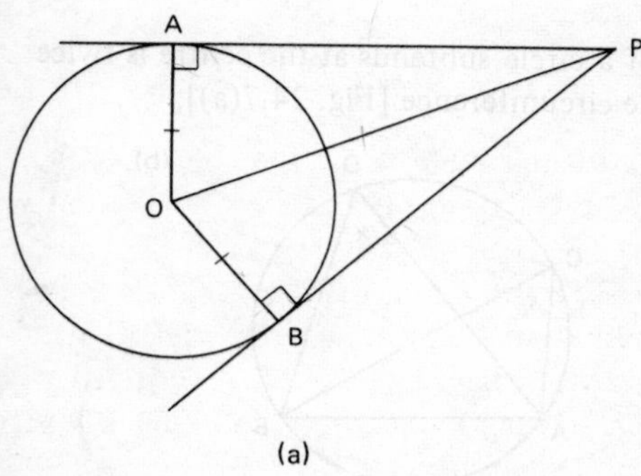

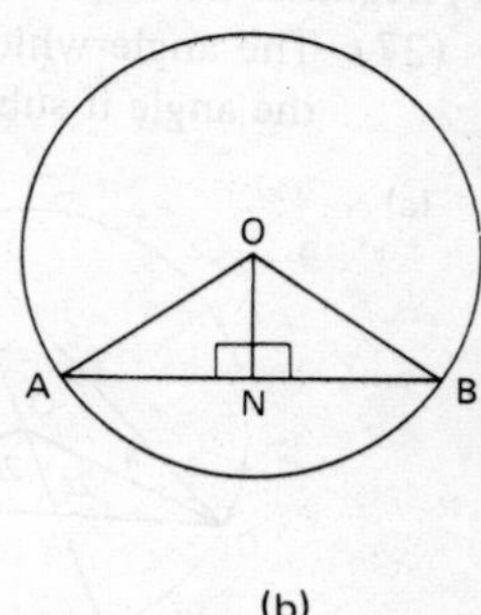

Fig 24.6

Comment To construct the tangents from an external point P to a circle with centre at O, a circle is drawn on OP as diameter. This cuts the first circle at the points of contact of the tangents.

Example 24.3

(a) In Fig. 24.6(a) PA and PB are tangents to the circle with centre O. Show that PA = PB.

Solution In triangles OAP and OBP

OA = OB (radii of circle)

$\angle OAP = \angle OBP = 90°$ (tangents ⊥ radii)

OP is common

Hence

$\Delta OAP \equiv \Delta OBP$ (R-H-S)

In particular PA = PB.

(b) In Fig. 24.6(b) the chord AB is of length 10 cm and its distance from the centre of the circle is $\sqrt{11}$ cm. Calculate the radius of the circle and the size of $\angle AOB$.

Solution

$AN = NB = 5$ cm (radius ⊥ chord bisects the chord)

Applying Pythagoras' theorem to ΔAON

$$OA^2 = ON^2 + AN^2 = 11 + 25 = 36$$

Therefore the radius OA = 6 cm.

$$\sin \angle AON = AN/OA = 5/6 \Rightarrow \angle AON = 56.45^\circ, \angle AOB = 112.9^\circ.$$

(b) Segment theorems

(27) The angle which an arc of a circle subtends at the centre is twice the angle it subtends at the circumference [Fig. 24.7(a)],

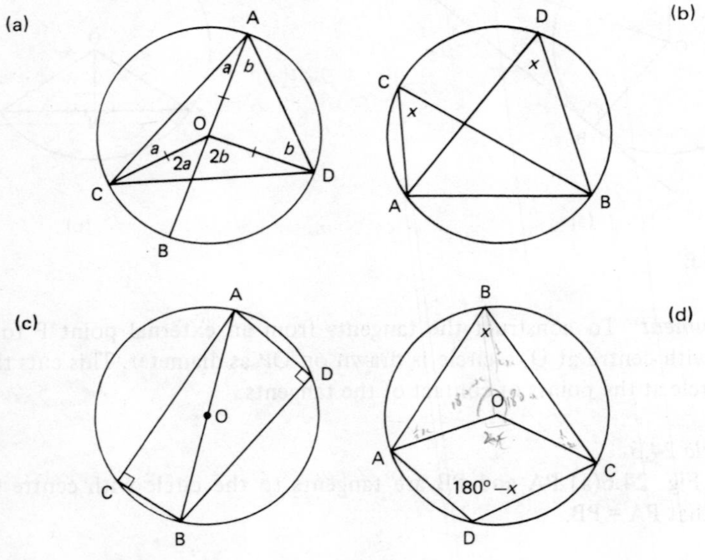

Fig 24.7

(28) Angles in the same segment are equal to each other [Fig. 24.7(b)],
(29) Angles in a semicircle are right angles [Fig. 24.7(c)],
(30) The opposite angles of a cyclic quadrilateral are supplementary. [Figure 24.7(d).]

Comment Points which lie on the same circle are called *concyclic* and the vertices of a cyclic quadrilateral are concyclic.

Example 24.4
(a) In Fig. 24.8(a) BE and CF intersect at G. Show that A, E, G, F are concyclic points and express $\angle BCF$ in terms of x.

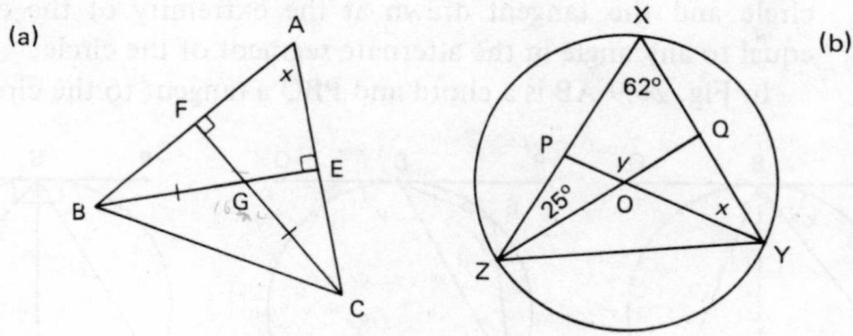

Fig 24.8

Solution

$\angle AEG = \angle AFG = 90°$ (Given)

But these are opposite angles of the quadrilateral AEGF, which is therefore a cyclic quadrilateral [converse of (30)]. Hence the points A, E, G, F are concylic.

$\angle EAF = x$ (Given)

$\angle EGF = 180° - x$ (Opposite angles of AEGF)

Therefore

$\angle BGC = 180° - x$ (vertically opposite angles)

$\angle GBC + \angle GCB = x$ (angle sum of triangle 180°)

But triangle BGC is isosceles (BG = CG given)
Therefore $\angle CBG = \angle BCG = \frac{1}{2}x$.

(b) In Fig. 24.8(b) O is the centre of the circle and YOP, ZOQ are straight lines. Calculate the angles marked x and y.

Solution

$\angle YQZ = 62° + 25° = 87°$ (exterior angle of triangle XQZ)

$\angle YOZ = 2\,\angle YXZ = 124°$ (angle at centre on same arc)

Therefore $\angle POQ = y = 124°$. (vertically opposite angles equal)

In triangle QYO

$\angle YOQ = 56°$ (supplement of $\angle POQ$)

$\angle OQY = \angle YQZ = 87°$

Therefore $\angle QYO = x = 37°$ $(180° - 56° - 87°)$.

(31) The alternate segment theorem The angle between a chord of a

circle and the tangent drawn at the extremity of the chord is equal to any angle in the alternate segment of the circle.

In Fig. 24.9 AB is a chord and PBQ a tangent to the circle. The

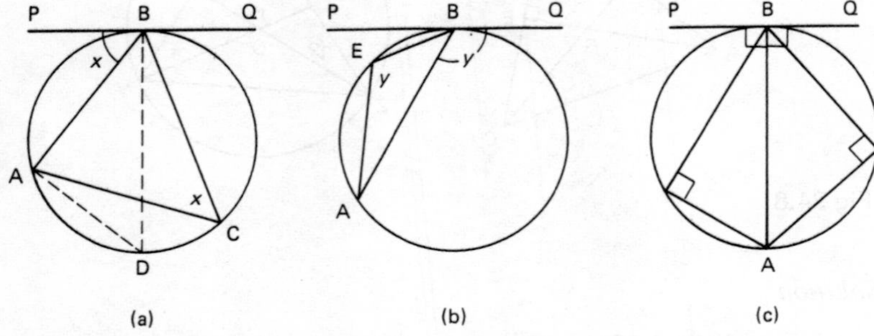

Fig 24.9

acute angle ABP is equal to the angle ACB in the major segment, on the other side of the chord. The obtuse angle ABQ is equal to the angle AEB in the minor segment.

In the special case when the chord is a diameter, both angles are 90° and the angle in a semicircle is a right angle.

Example 24.5

(a) In Fig. 24.10(a), PTQ is a tangent at T, $\angle PTA = 3x^\circ$, $\angle ATB = 50^\circ$ and $\angle TAB = 2x^\circ$. Calculate the value of x and $\angle QTB$.

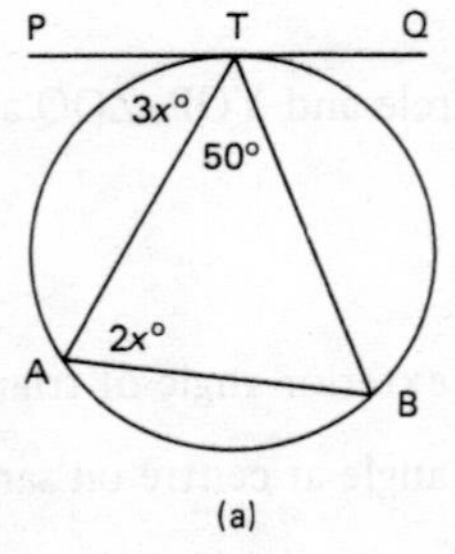

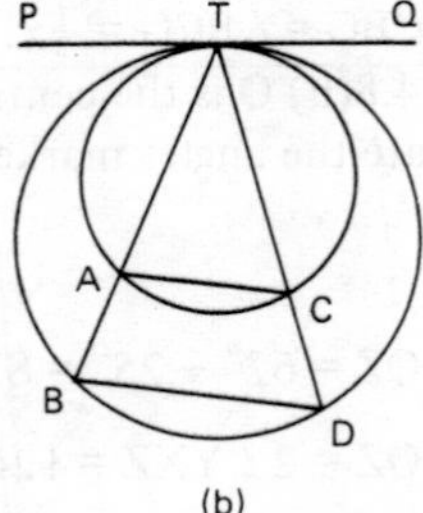

Fig 24.10

Solution

$\angle TBA = \angle PTA = 3x^\circ$ (alternate segment)

$50^\circ + 2x^\circ + 3x^\circ = 180^\circ$ (angle sum of ΔTAB)

Therefore $\underline{x = 26}$

$\angle QTB = \angle TAB = 2x°$ (alternate segment)

Therefore $\underline{\angle QTB = 52°}$.

(b) In Fig. 24.10(b), PTQ is a tangent at T to both circles, and TAB and TCD are straight lines. If $\angle PTB = 65°$ calculate $\angle BDT$ and show that AC is parallel to BD.

Solution

$\angle BDT = \angle PTB = \underline{65°}$ (alternate segment)

But

$\angle ACT = \angle ATP = 65°$ (alternate segment of small circle)

Therefore $\angle BDT = \angle ACT$. Since these are corresponding angles AC is parallel to BD.

Exercise 24.2

(1) In Fig. 24.11 O is the centre of the circle and PTR is a tangent at T. Find the size of the angles x, y and z.

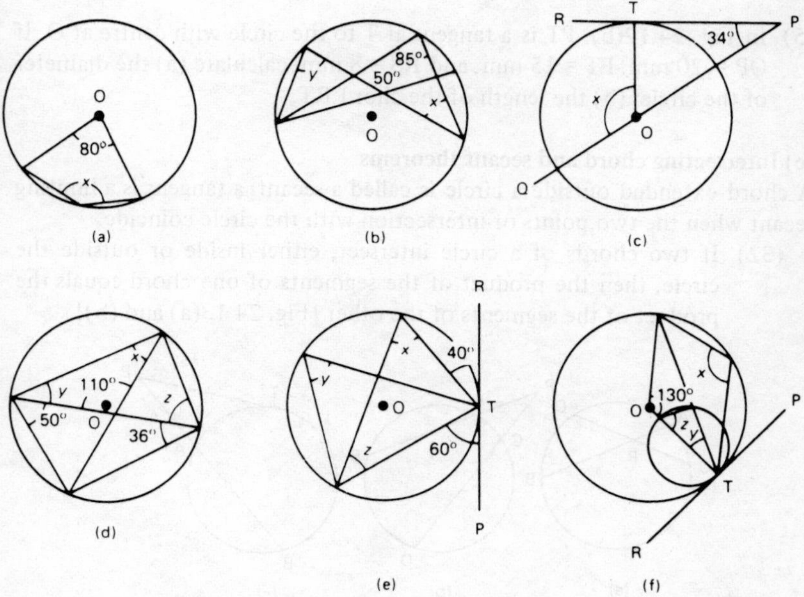

Fig 24.11

(2) ABCD is a cyclic quadrilateral such that CB is a diameter and AD is not. CA is extended to E and EB = BC. Prove that AE = AC, and if ∠EBA = 50° find the size of angle CDA.

(3) PT and PR are tangents at T and R to a circle with centre O and radius 8 cm. Given that ∠TOR = 120°, calculate (a) ∠TPR, (b) the length of TR. If PO is extended to cut the circle at X and TR at Y, calculate the length of XY.

(4) In Fig. 24.12(a) find the size of (a) ∠PRS, (b) ∠SPR.

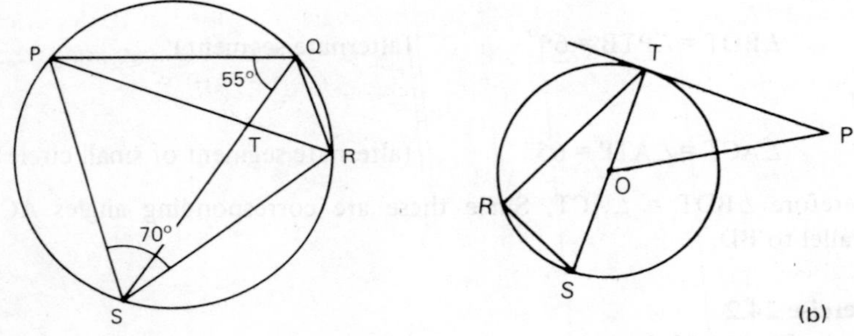

Fig 24.12

(5) In Fig. 24.12(b), PT is a tangent at T to the circle with centre at O. If OP = 20 mm, PT = 15 mm, and RS = 8 mm, calculate (a) the diameter of the circle, (b) the length of the chord RT.

(c) Intersecting chord and secant theorems

A chord extended outside a circle is called a secant; a tangent is a limiting secant when the two points of intersection with the circle coincide.

(32) If two chords of a circle intersect, either inside or outside the circle, then the product of the segments of one chord equals the product of the segments of the other [Fig. 24.13(a) and (b)],

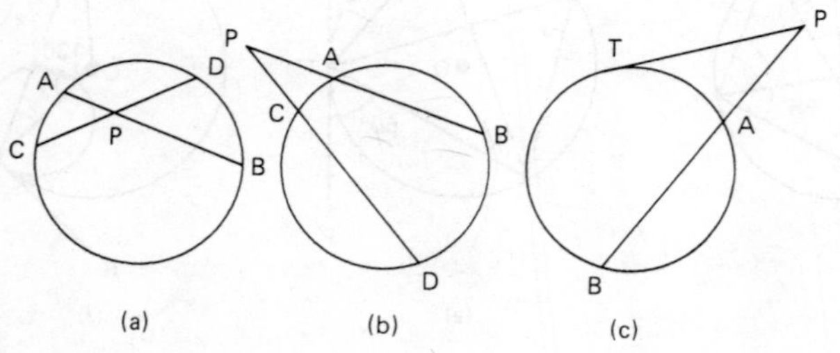

Fig 24.13

(33) If, from a point outside a circle, a secant and a tangent are drawn, then the product of the segments of the secant equals the square of the tangent [Fig. 24.13(c)]. In Fig. 24.13(a) and (b)

$$PA \times PB = PC \times PD$$

In Fig. 24.13(c)

$$PA \times PB = PT \times PT = PT^2$$

Example 24.6

AB and CD are chords of a circle intersecting at P inside the circle. If CD = 11 cm, AB = 10 cm and CP = 3 cm, calculate the two possible lengths of BP and the ratio of the areas of the triangles CAP and BDP.

Solution

$$PD = CD - CP = 8 \text{ cm}$$

If BP = x cm, then AP = $10 - x$ cm.
By the intersecting chord theorem

$$AP \times PB = CP \times PD$$

$$(10 - x)(x) = 3 \times 8 = 24 \Rightarrow 10x - x^2 = 24$$

$$x^2 - 10x + 24 = 0 \Rightarrow x = 4 \text{ or } x = 6$$

Therefore BP = 4 cm or 6 cm.
The two positions of P are shown in Fig. 24.14(a) and (b).

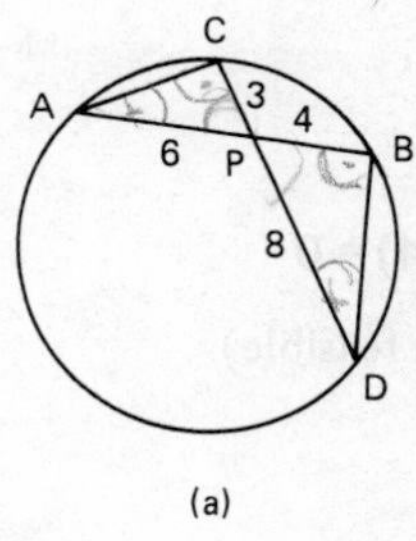

(a)

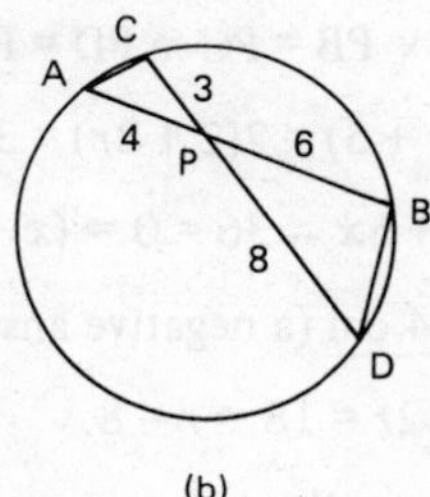

(b)

Fig 24.14

In triangles ACP and BDP

$\angle ACD = \angle ABD$ (angles in same segment)

$\angle CAB = \angle CDB$ (angles in same segment)

$\angle CPA = \angle BPD$ (vertically opposite angles)

Therefore ΔCAP is similar to ΔBDP (three angles equal) and

$$\text{area CAP} : \text{area BDP} = CP^2 : BP^2$$

$$CP = 3 \text{ cm}$$

(a) When BP = 4 cm the areas are in the ratio 9 : 16.
(b) When BP = 6 cm the areas are in the ratio 9 : 36 or 1 : 4.

Example 24.7
In Fig. 24.15 O is the centre of the circle and PT is a tangent at T. If PC = 2 cm, AB = 5 cm, PT = 6 cm, calculate (a) PA, (b) the radius of the circle.

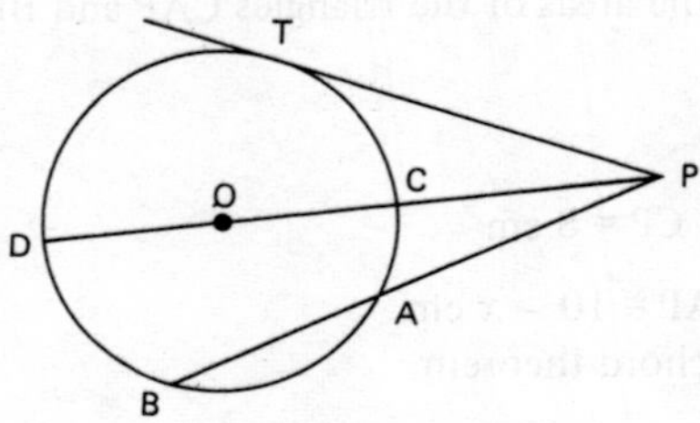

Fig 24.15

Solution Let PA = x cm and the radius r cm. Then

$$PB = PA + AB = x + 5 \text{ cm}, \; PD = PC + CD = 2 + 2r \text{ cm}$$

By the intersecting secant theorem

$$PA \times PB = PC \times PD = PT^2$$

$$x(x + 5) = 2(2 + 2r) = 36$$

(a) $x^2 + 5x - 36 = 0 \Rightarrow (x + 9)(x - 4) = 0$

Hence PA = 4 cm (a negative answer is not feasible).

(b) $2 + 2r = 18 \Rightarrow r = 8.$

Therefore the radius is 8 cm.

Exercise 24.3
(1) In Fig. 24.16(a) EG and HJ intersect at F, HF = 20 mm, FJ = 18 mm, FG = 12 mm. Calculate EF.
(2) In Fig. 24.16(b) MN = 3 cm, NP = 12 cm, MQ = 4 cm. Calculate MR.
(3) In Fig. 24.16(c) SX = 4 cm, SY = 13 cm, VW = 9 cm. Calculate (a) SV, (b) ST.

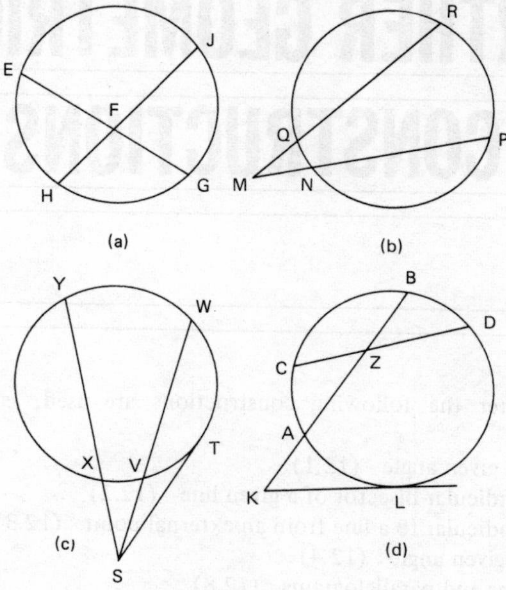

Fig 24.16

(4) In Fig. 24.16(d) KA = 3 cm, CZ = 2 cm, ZD = 4 cm, BZ = 4 cm. Find the length of the tangent KL.

(5) AOB is the diameter of a circle with centre at O, and a chord PQ of length 40 mm cuts AOB at X, the mid-point of PQ. If XB is 5 mm calculate the radius of the circle.

(6) O is the centre of a circle, PAOB is a secant and PQR is a tangent at Q. Show that $PQ^2 = PO^2 - OB^2$ and if it is given that $\angle QBO = 30°$, find the value of (a) $\angle QPO$, (b) $\angle RQB$.

(7) AB is a tangent at B to a circle with centre O and ACOD is a secant. If AB is 50 mm and the radius is 20 mm, calculate (a) AC, (b) the area of the triangle ABD.

(8) Two chords of a circle, AB and CD, intersect at X, AX = BX, CX = 4 cm and DX = 9 cm. Calculate (a) AB, (b) the radius of the circle, given that it has an integer value. Y is on the chord DC and DY = 5 cm. If AY is extended to cut the circle at Z and AZ = 14 cm, calculate YZ.

(9) Two circles intersect at M and N. The straight line ABXCD cuts the circle at A, B, C, D and the line MN at X. If AB = 40 mm, BX = 10 mm and CD = 50 mm, calculate the length XC.

CHAPTER 25

FURTHER GEOMETRICAL CONSTRUCTIONS

In this chapter the following constructions are used, as described in Chapter 12.

Bisecting a given angle (12.1)
The perpendicular bisector of a given line (12.2)
The perpendicular to a line from an external point (12.3)
Copying a given angle (12.4)
Parallel lines and parallelograms (12.8).

25.1 TO DIVIDE A LINE IN A GIVEN RATIO

The method used is based on the theorem that lines drawn parallel to one side of a triangle divide the other two sides in the same ratio.

Example 25.1
Divide the given straight line AB at a point P such that AP : PB = a : b.

The given line AB is shown in Fig. 25.1.

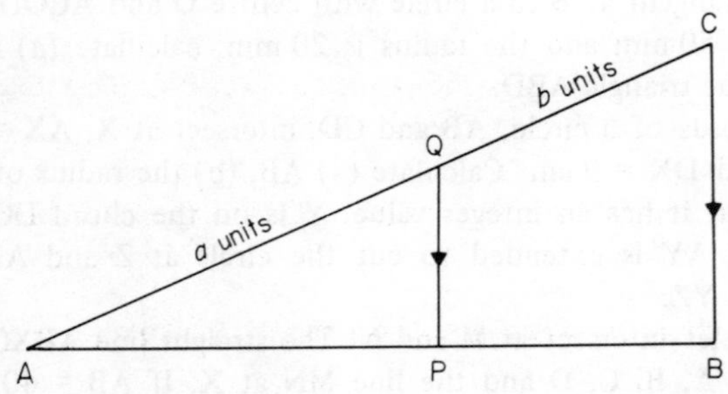

Fig 25.1

Method of construction

(i) Draw a second line AC making an acute angle with AB, the length of AC being $a + b$ measured in convenient units

(ii) Join C and B

(iii) Mark a point Q on AC such that AQ = a units and QC = b

(iv) Using a ruler and set square, draw a line through Q parallel to CB to meet AB at P.

Then AP : PB = $a : b$.

Comment The lines AB and AC should be of comparable length. For example, if a line of length 11 cm is to be divided in the ratio 5 : 8, 13 cm is a suitable length for the second line. To divide a line of 5 cm in the ratio 3 : 4 : 5, 6 cm is a more suitable length than 12 cm.

A similar method can be used to divide a line externally.

Example 25.2

Divide the given straight line AB externally at a point P such that AP : PB = $a : b$.

The given line AB is shown in Fig. 25.2.

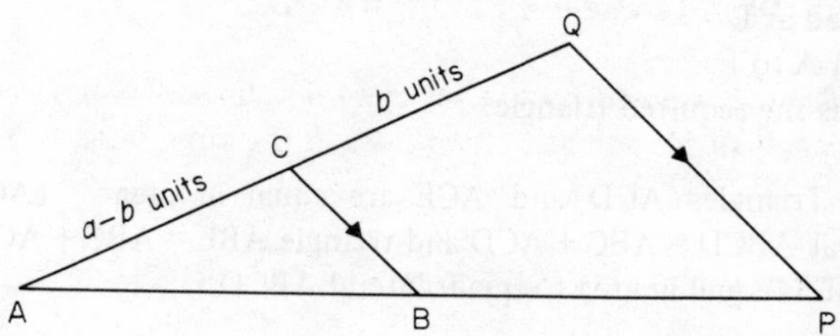

Fig 25.2

Method of construction

(i) Draw a line AQ, at an acute angle to AB, of length a units,

(ii) Mark a point C on AQ, b units from Q,

(iii) Join C to B,

(iv) Draw a line through Q parallel to CB, to meet AB produced at P.

Then AP : PB = $a : b$.

25.2 TO CONSTRUCT A TRIANGLE EQUAL IN AREA TO A GIVEN QUADRILATERAL

The construction is based on the theorem that triangles on the same base and between the same parallels are equal in area.

Example 25.3
Construct a triangle ABE equal in area to the given quadrilateral ABCD.

The quadrilateral ABCD is shown in Fig. 25.3.

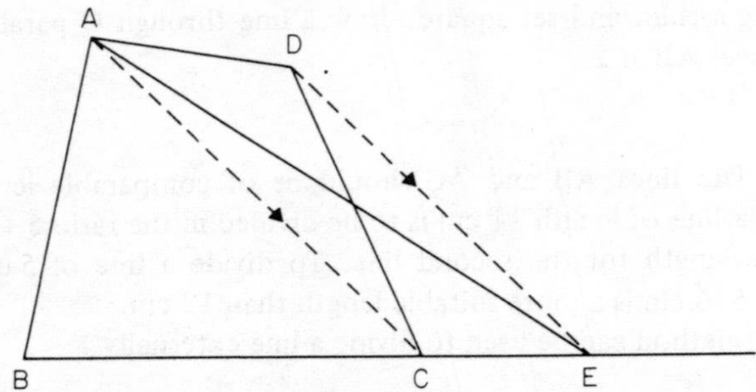

Fig 25.3

Method of construction

(i) Draw the diagonal AC of the quadrilateral ABCD
(ii) Through the vertex D draw a line parallel to AC to meet BC produced at E
(iii) Join A to E.

Then ABE is the required triangle.

Comment Triangles ACD and ACE are equal in area (AC || DE). Quadrilateral ABCD = ABC + ACD and triangle ABE = ABC + ACE. Hence triangle ABE is equal in area to quadrilateral ABCD.

25.3 TO CONSTRUCT A SQUARE EQUAL IN AREA TO A GIVEN RECTANGLE

This construction is based on the theorem that a diameter perpendicular to a chord bisects the chord, and on the intersecting chord theorem.

Example 25.4
Construct a square BEFG equal in area to the given rectangle ABCD.

The given rectangle ABCD is shown in Fig. 25.4.

Method of construction

(i) With centre at B and radius BC, draw an arc to cut AB produced at H, so that BH = BC

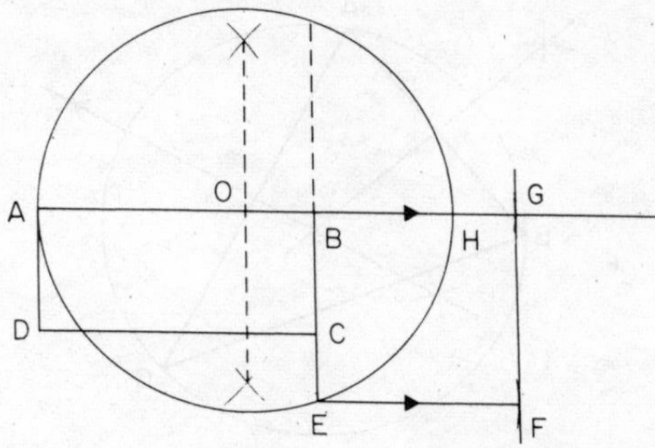

Fig 25.4

(ii) Bisect AH and draw the circle with the mid-point O as centre and AH as diameter

(iii) Extend BC to meet this circle at E.

Then BE is one side of the required square.

(iv) Through E draw a line parallel to AH

(v) With centre at B and radius BE, draw an arc to cut AH produced at G

(vi) With centre at E and the same radius draw an arc to cut the parallel line through E at F.

Then BEFG is the required square.

Comment B is the mid-point of the chord perpendicular to the diameter AH. By the intersecting chord theorem, AB × BH = BE × BE. Hence AB × BC = BE^2 and area ABCD = area BEFG.

25.4 TO CONSTRUCT THE CIRCUMSCRIBED CIRCLE OF A TRIANGLE

There is only one circle that can be drawn through three given points; it is called the circumscribed circle of the triangle with vertices at the three points.

Example 25.5

Construct the circumscribed circle of the given triangle ABC.

The given triangle ABC is shown in Fig. 25.5.

Method of construction

(i) Draw the perpendicular bisector of the side AB of the given triangle ABC

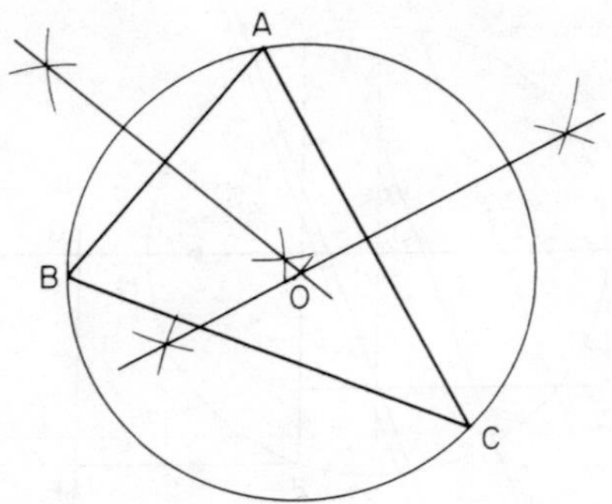

Fig 25.5

(ii) Draw the perpendicular bisector of the side AC, to meet the bisector of AB at O

(iii) With centre at O and radius OA draw a circle.

This is the required circle, and it passes through B and C. O is the circumcentre of the triangle ABC.

Comment Since the point O lies on the perpendicular bisector of AB it is equidistant from A and B; similarly it is equidistant from A and C. Therefore OA = OB = OC = R, the radius of the circumscribed circle.

There is a connection between the value of R and the sine formula for the triangle ABC

$$\frac{a}{\sin A} = \frac{b}{\sin B} = \frac{c}{\sin C} = 2R$$

25.5 TO CONSTRUCT THE INSCRIBED CIRCLE OF A TRIANGLE

The inscribed circle of a triangle touches all three sides; its centre I is equidistant from the sides of the triangle and is called the incentre.

Example 25.6

Construct the inscribed circle of the given triangle ABC.

Fig. 25.6 shows the triangle ABC.

Method of construction

(i) Draw the bisector of ∠ABC of the given triangle

(ii) Draw the bisector of ∠ACB to meet the first bisector at I

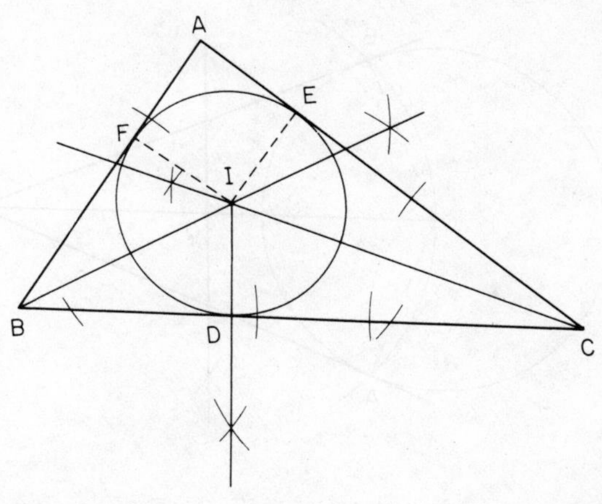

Fig 25.6

(iii) From I erect a perpendicular to BC, to meet BC at D
(iv) With centre at I and radius ID, draw a circle.

This circle is the required inscribed circle; it touches AC at E and AB at F.

Comment The bisector of ∠ABC is the locus of points equidistant from AB and CB, therefore IF = ID. The bisector of ∠ACB is equidistant from AC and BC, therefore IE = ID. Hence ID = IE = IF = r, the radius of the inscribed circle of the triangle ABC.

The value of r is connected to the area of the triangle by the equation area ABC = rs, where s is the semi-perimeter $\frac{1}{2}(a + b + c)$.

25.6 TO CONSTRUCT THE TANGENTS TO A CIRCLE FROM A GIVEN EXTERNAL POINT

This construction is based on the theorems that the angle in a semicircle is a right angle, and that a tangent is perpendicular to the radius drawn at the point of contact.

Example 25.7
Construct the tangents from the given point P to the given circle with centre at O.

Method of construction

(i) Join P to O
(ii) Bisect OP, and mark the mid-point Q

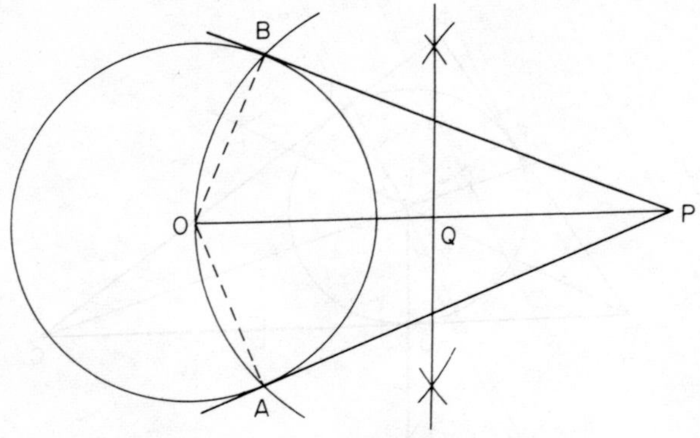

Fig 25.7

(iii) With centre at Q and radius OQ, draw an arc to cut the given circle at A and B
(iv) Join P to A and P to B.

Then PA and PB are the required tangents to the given circle.

Comment Angles OAP and OBP are right angles since they are on the circle with OP as diameter, and so OA ⊥ AP and OB ⊥ BP. But OA and OB are radii of the given circle, and therefore PA and PB are tangents.

25.7 TO CONSTRUCT THE SEGMENT OF A CIRCLE CONTAINING A GIVEN ANGLE

The construction is based on the alternate segment theorem.

Example 25.8
On the given line AB, construct a segment of a circle containing an angle equal to the given angle x.

Fig. 25.8 shows the given line AB and the given angle x.

Method of construction

(i) Draw a line AC such that BAC is equal to the given angle x
(ii) Erect a line AD perpendicular to AC on the other side of AB from C
(iii) Draw the perpendicular bisector of AB to meet AD at O
(iv) With centre at O and radius OA, draw an arc APB on the same side of AB as O.

Then $\angle APB = x$ and APB is the required segment.

Comment Since the radius OA is perpendicular to AC, AC is a tangent at A. The angle x between the chord AB and the tangent AC is equal to the angle APB in the alternate segment.

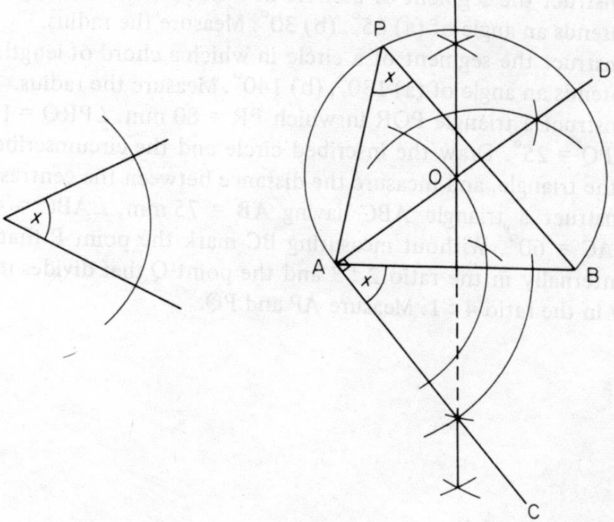

Fig 25.8

Exercise 25.1

(1) Draw a line AB of length 120 mm and by construction obtain a point P dividing AB internally in the ratio (a) 3 : 5, (b) 4 : 7. Measure AP.

(2) Draw a line AB of length 80 mm and divide it externally at P in the ratio (a) 5 : 3, (b) 7 : 4. Measure AP.

(3) Construct a quadrilateral ABCD having AB = 3 cm, BC = 5 cm, CD = 6 cm, AD = 4 cm, BD = 6 cm and measure angle BCD. Construct a triangle AXD equal in area to ABCD and calculate its area.

(4) Construct a square equal in area to a given rectangle PQRS and measure the diagonal of the square when (a) PQ = 95 mm and QR = 62 mm, (b) PQ = 70 mm and QR = 30 mm.

(5) Draw an equilateral triangle with sides of length 80 mm. Construct the circumscribed circle and measure its radius.

(6) Construct a triangle PQR with PR = PQ = 74 mm and RQ = 52 mm. Construct (a) the inscribed circle, (b) the circumscribed circle, and measure the radius of both circles.

(7) Calculate the radius of the circumscribed circle of the triangle ABC with sides AB = 51 mm, BC = 28 mm, AC = 63 mm. (Hint: use the cosine formula to find one angle.)

(8) Draw a circle with centre O and radius 38 mm, and construct the tangents from a point P such that the distance PO is (a) 71 mm, (b) 93 mm. Measure the length of the tangents.

(9) Construct the segment of a circle in which a chord of length 60 mm subtends an angle of (a) 65°, (b) 30°. Measure the radius.

(10) Construct the segment of a circle in which a chord of length 75 mm subtends an angle of (a) 130°, (b) 140°. Measure the radius.

(11) Construct a triangle PQR in which PR = 80 mm, ∠PRQ = 120° and ∠RPQ = 25°. Draw the inscribed circle and the circumscribed circle of the triangle, and measure the distance between the centres.

(12) Construct a triangle ABC having AB = 75 mm, ∠ABC = 45° and ∠BAC = 60°. Without measuring BC mark the point P that divides it internally in the ratio 2 : 3 and the point Q that divides it externally in the ratio 4 : 1. Measure AP and PQ.

PROGRESS TEST 5

A

(1) In a triangle ABC, BE is the altitude from B, AB = 12 cm, $\angle$BAC = 54° and $\angle$BCA = 48°. Calculate (a) BE, (b) AE, (c) AC. **(2)** If sin x° = 9/13 and $0 < x < 180$, calculate without using tables (a) cos x°, (b) tan x°. **(3)** Evaluate using tables (a) sin 145°, (b) cos 138°, (c) tan 99°, (d) tan 165°. **(4)** From a point Q on level ground, the bearing of a mast M is 040°. AT P, 450 m due east of Q, the bearing is 290°. If the angle of elevation from Q of the top T of the mast is 42° calculate (a) MP, (b) MQ, (c) MT. **(5)** Solve the triangle ABC and find its area, when (a) $\angle$A = 120°, b = 7 cm, c = 9 cm, (b) $\angle$B = 76°, a = 10 cm, b = 14 cm. **(6)** The sides PQ, QR, and SR of a quadrilateral PQRS represent the vectors, $\boldsymbol{p}$, $\boldsymbol{q}$ and $\boldsymbol{r}$ respectively. X and Y are points on PQ and QR such that PX = 2XQ and QY = 3YR. Express in terms of $\boldsymbol{p}$, $\boldsymbol{q}$ and $\boldsymbol{r}$ the vectors represented by (a) PS, (b) PY, (c) XY, (d) XS. **(7)** Find an expression in surd form for (i) the magnitude, (ii) the direction of the vector (a) $\boldsymbol{u} - 2\boldsymbol{v}$, (b) $2\boldsymbol{u} + \boldsymbol{v}$ when $\boldsymbol{u} = 3\mathbf{i} + 2\mathbf{j}$ and $\boldsymbol{v} = \mathbf{i} - 3\mathbf{j}$. **(8)** T represents a translation and M reflection in the line $y = x$. If the transformation TM maps the point A (3, 1) to (2, 5), find the column vector T and the image of A under the transformation MT. **(9)** POQ is a diameter of a circle with centre O, and the radius OR is parallel to the tangent PT. The line TO cuts the circle at S and PTO = 34°. Calculate (a) $\angle$RQS, (b) $\angle$SQO, (c) $\angle$SPT.

B

(1) If $81 = 144 + 25 - 120 \cos B$ and $B < 180°$, evaluate (a) cos B, (b) tan $2B$, (c) sin $\frac{1}{2}B$. **(2)** A ship travelled 10 km from a pier on a bearing 035° and a further 8 km on a bearing 205°. Calculate the final distance and bearing of the pier from the ship. **(3)** Calculate (a) the area, (b) the length of the diagonals, of a parallelogram PQRS when PQ = 14 cm, QR = 11 cm and $\angle$PQR = 62°. **(4)** Taking the Earth to be a sphere of radius 6370 km and π as 22/7, calculate (a) the shortest distance between positions 40° S 18° E and 10° N 18° E, (b) the radius of the circle of latitude (i) 45.4° N, (ii) 70.6° S. **(5)** $\overrightarrow{OP}$ and $\overrightarrow{OQ}$ represent the position vectors of points P(2, 6) and Q(8, 4); M and N are the mid-points of OQ and PQ respectively. Find (i) the column vector, (ii) the magnitude of the vector, represented by (a) $\overrightarrow{ON}$, (b) $\overrightarrow{PQ}$, (c) $\overrightarrow{MN}$. **(6)** Describe the single transformation P which, when applied twice in succession, maps the point (1, –2) to (4, –8). Find the matrix representing P and a column vector X such that $PX = \begin{pmatrix} -2 \\ 10 \end{pmatrix}$. **(7)** Find (a) the centre, (b) the angle, of the single rotation which transforms the triangle A (6, 2) B (5, 4) C (7, 4) to the image A′ (1, 3) B′ (–1, 2) C′ (–1, 4). **(8)** The side ZX of a triangle XYZ is extended to S and the bisector of $\angle$SXY meets ZY produced at R. If ZY = 60 mm, ZX = 70 mm and YR = 110 mm, calculate (a) XY, (b) $\angle$SXR. **(9)** TD is a diameter, AT a tangent and ABC a secant of a circle, and the chord BC cuts TD at X. If TD = 8 cm, AB = 9 cm, BX = 3 cm and DX = 2 cm, calculate the length of the tangent AT.

ANSWERS TO EXERCISES

EXERCISE 1.1

(1) (a) 4^3, (b) 10^4, (c) 5^2. (2) (a) Two hundred and seventy-four, (b) Three thousand and sixty-three, (c) Ten thousand one hundred. (3) (a) 2306, (b) 305 409. (4) (a) 4800, (b) 30 700, (c) 268 200. (5) (a) 69, (b) 30. (6) 250.

EXERCISE 1.2

(1) (a) 582, (b) 6529, (c) 1995. (2) (a) 228, (b) 111, (c) 1257, (d) 928. (3) (a) 3399. (4) (a) 3097, (b) 3485, (c) 7586. (5) (a) 648, 893, 1093, (b) 2634 litres.

EXERCISE 1.3

(1) (a) 980, (b) 132 672, (c) 29, (d) 28. (2) (a) 294, (b) 405. (3) (a) 8, (b) 3, (c) 9. (4) (a) 1750, (b) 35, 10. (5) (a) 5095, (b) 339.

EXERCISE 1.4

(1) 19. (2) 14. (3) 13. (4) 5. (5) 1. (6) 29. (7) 16. (8) 12.

EXERCISE 1.5

(1) 2×19. (2) $2 \times 2 \times 3 \times 3$. (3) $3 \times 5 \times 5$. (4) $2 \times 2 \times 2 \times 11$. (5) 18. (6) 24. (7) 35. (8) 84. (9) 726. (10) 9. (11) 4. (12) 14. (13) 420.

EXERCISE 1.6

(1) $-4<2<3$. (2) $-8<-1<0$. (3) $-2<-1<15$. (4) True. (5) False. (6) True. (7) True. (8) False.

EXERCISE 1.7

(1) 19. (2) 8. (3) 3. (4) 0. (5) -12. (6) -5. (7) -8. (8) 20. (9) 4. (10) 5. (11) 28. (12) 1. (13) -5. (14) 41. (15) -5.

EXERCISE 1.8

(1) -6. (2) -24. (3) 2. (4) 0. (5) 80. (6) -5. (7) -6. (8) 3. (9) -16. (10) 6. (11) 24.

EXERCISE 2.1

(1) 3/8. (2) 2/3. (3) 3/8. (4) 1/4. (5) 3/5. (6) 6/7. (7) 4/11. (8) 6/7. (9) 21/35, 20/35; $4/7<3/5$. (10) 9/24, 10/24; $3/8<5/12$. (11) 15/45, 10/45, 36/45; $2/9<1/3<4/5$. (12) 45/120, 40/120, 48/120; $1/3<3/8<2/5$. (13) 14/20, 15/20, 16/20; $7/10<3/4<4/5$. (14) 10/18, 6/18, 15/18; $1/3<5/9<5/6$.

EXERCISE 2.2

(1) $2\frac{1}{2}$. (2) $3\frac{1}{3}$. (3) $2\frac{1}{4}$. (4) $3\frac{1}{7}$. (5) $5\frac{3}{8}$. (6) $1\frac{3}{8}$. (7) 9/4. (8) 14/3. (9) 29/9. (10) 47/7. (11) 77/13. (12) 247/24.

EXERCISE 2.3

(1) 11/15. (2) $1\frac{5}{18}$. (3) $2\frac{3}{10}$. (4) $5\frac{71}{84}$. (5) $4\frac{29}{42}$. (6) $3\frac{53}{56}$. (7) 5/8. (8) $2\frac{2}{9}$. (9) $3\frac{11}{56}$. (10) $2\frac{9}{10}$. (11) 7/8. (12) 1/6. (13) 1/6. (14) $1\frac{5}{12}$. (15) $5\frac{41}{60}$. (16) 1/6. (17) $3\frac{5}{6}$.

EXERCISE 2.4

(1) 1/2. (2) $1\frac{3}{4}$. (3) 1/12. (4) $3\frac{1}{16}$. (5) $52\frac{1}{2}$. (6) 7/50. (7) $1\frac{1}{3}$. (8) £36. (9) £176 500. (10) (a) £140, (b) £70.

EXERCISE 2.5

(1) $1\frac{1}{2}$. (2) 2. (3) $1\frac{1}{3}$. (4) 4/9. (5) 7/22. (6) $13\frac{1}{4}$. (7) $1\frac{1}{2}$. (8) 1/2. (9) 6/7. (10) $1\frac{1}{2}$. (11) 5/12. (12) $11\frac{2}{3}$. (13) 1. (14) $2\frac{9}{143}$.

EXERCISE 3.1

(1) 3/5. (2) 3/4. (3) $2\frac{1}{8}$. (4) $4\frac{3}{8}$. (5) 0.75. (6) 0.3125. (7) $1.\dot{3}$. (8) 0.24. (9) (a) 0.51, (b) 0.507. (10) (a) 0.01, (b) 0.008 59. (11) (a) 41.12, (b) 41.1. (12) (a) 34.83, (b) 34.8. (13) (a) 1985.03, (b) 1990. (14) (a) 3895.72, (b) 3900. (15) 47 900. (16) 3000. (17) 5.01. (18) 130. (19) 0.175. (20) 3.143. (21) 1.190. (22) $8.\dot{7}$. (23) 0.929.

EXERCISE 3.2

(1) 5×10^{-2}. (2) 4.9×10^{-3}. (3) 6.275×10^{1}. (4) 5.972×10^{3}. (5) $2.730\,51 \times 10^{5}$. (6) 11.00. (7) 2.555. (8) 20.02. (9) 7.28. (10) 2.34. (11) 1×10^{3}. (12) 5.6×10^{-3}. (13) 6.13×10^{3}. (14) 3.19×10^{-2}. (15) 2.49×10^{4}. (16) 341.1 m, 8.9 m.

EXERCISE 3.3

(1) 1.1. (2) 0.0042. (3) 0.3. (4) 0.9. (5) 6.386. (6) 0.484. (7) 0.057 36. (8) 6.2. (9) 3.6. (10) 32. (11) 0.54. (12) 0.04. (13) 1.41. (14) £38.38. (15) £587.11. (16) 55. (17) (a) £32.93, (b) £10.66. (18) £1.21.

EXERCISE 4.1

(1) 10. (2) 18. (3) 20. (4) 2. (5) 2. (6) 3. (7) 2. (8) 5. (9) 9. (10) 1/4. (11) $2\sqrt{3}$. (12) $5\sqrt{2}$. (13) $4\sqrt{2}$. (14) $2\sqrt{2}$. (15) 3^{1}. (16) 2^{3}. (17) $5^{-3/4}$. (18) $3^{7/2}$. (19) 7^{2}. (20) 2^{-5}. (21) $5^{1/2}$. (22) $3^{-1/2}$. (23) $6^{2/3}$. (24) $7^{1/2}$.

EXERCISE 4.2

(1) 10.1. (2) 7.84×10^{-4}. (3) 1.84×10^{5}. (4) 2.25×10^{-4} (5) 0.208. (6) 0.0159. (7) 1.92. (8) 41.7. (9) 42.6. (10) 0.112. (11) 0.254. (12) 5.76.

EXERCISE 4.3

(1) 2.19. (2) 8.54. (3) 0.762. (4) 0.078. (5) 0.0826. (6) 3.84.
(7) 0.787. (8) (a) 0.2, (b) 0.4, (c) 0.08. (9) (a) 23.664, (b) 0.23664,
(c) 0.023 664.

EXERCISE 4.4

(1) 0.6866. (2) 1.5816. (3) 4.8627. (4) $\overline{1}$.7490. (5) $\overline{3}$.9161.
(6) $\overline{4}$.4554. (7) $\overline{19}$.2304. (8) 3.796. (9) 703.1. (10) 0.1747.
(11) 0.081 68. (12) 1 000 000 or 10^6.

EXERCISE 4.5

(1) 5.0141. (2) $\overline{3}$.5324. (3) 4.3603. (4) $\overline{1}$.3422. (5) 1.7721.
(6) 85.96. (7) 0.380. (8) 0.0777. (9) 726. (10) 0.0864.
(11) 0.0522 mm. (12) 5.45. (13) £617.30. (14) 14.8 km/l.

EXERCISE 4.6

(1) $\overline{6}$.9030. (2) $\overline{16}$.8626. (3) $\overline{10}$.0970. (4) 1.6201. (5) 0.1590.
(6) $\overline{1}$.2205. (7) $\overline{1}$.7186. (8) $\overline{1}$.3206. (9) $\overline{1}$.4851. (10) 83.3.
(11) 0.173. (12) 1.02. (13) 8.85. (14) 0.374. (15) 0.229.
(16) 0.541. (17) 166. (18) 7.30×10^5. (19) £1220.
(20) 2.67×10^3 joules. (21) 3280.

EXERCISE 5.1

(1) 25%. (2) $37\frac{1}{2}$%. (3) 80%. (4) $233\frac{1}{3}$%. (5) $412\frac{1}{2}$%.
(6) 28%. (7) 1270%. (8) 1/2. (9) 7/25. (10) 2/3.
(11) 157/200. (12) £14.73. (13) 182 mm. (14) 336 bricks.
(15) 77/200. (16) 1200. (17) 5.7%. (18) 22.5%.
(19) £11 787.50. (20) £33.12. (21) 14%. (22) £170.50.

EXERCISE 5.2

(1) 1 : 0.5. (2) 1 : 8.4. (3) 1 : 0.2. (4) 1: 25. (5) 1 : 7.
(6) 1 : 48. (7) £29.25. (8) 0.9 km. (9) £5 402. (10) £5.07.
(11) 105. (12) 20 hours.

EXERCISE 5.3

(1) (a) 25 m/s, (b) 2.7 km/h. (2) (a) 3.2 km, (b) 15 mm. (3) 288 mm by 180 mm. (4) 227 mm. (5) 1 : 750. (6) (a) $85.50. (b) £94.59, (c) 5861.11 escudos.

EXERCISE 5.4

(1) (a) £11.70, (b) £17.64. (2) 8%. (3) £22.80; 10. (4) 75p. (5) £20.79. (6) (a) £36, (b) £101.50, (c) £41. (7) £2000. (8) 13.5%, 3 years.

EXERCISE 5.5

(1) £31.25, £25, £18.75. (2) 29.75 kg, 5.25 kg. (3) £5400, £2700, £1800, £900. (4) £3799.06. (5) 6000, 4000, 2000 tonnes.

EXERCISE 6.1

(1) 5. (2) 22. (3) 11. (4) 73. (5) 23. (6) 34. (7) 1001_2. (8) 1111_2. (9) 100011_2. (10) 101_5. (11) 157_9. (12) 533_6. (13) 734_8. (14) 10010. (15) 223. (16) 7. (17) 110110110. (18) 4. (19) (c). (20) $2113_4 > 141_6 > 28_9 > 1011_2$.

EXERCISE 6.2

(1) 1101_2, 56_7, 304_5. (2) 1010_2. (3) 1100_2. (4) 112_5. (5) 120_3. (6) 10_2. (7) 100_2. (8) 21_8. (9) 1222_7. (10) 1032_4. (11) 1010_2, (12) 21300_4. (13) 112_3. (14) 1111_2. (15) 442_6. (16) 354_8. (17) 1067_8. (18) $101 \times 11 \times 10$. (19) $11 \times 3 \times 2 \times 2$. (20) $2 \times 2 \times 3 \times 3$.

PROGRESS TEST 1

A

(1) (a) 1303, (b) 1265, (c) 281, (d) 58, (e) 3. (2) (a) $9\frac{7}{24}$, (b) $1\frac{7}{10}$, (c) $2\frac{3}{4}$, (d) $2\frac{5}{6}$, (e) 3, (f) $\frac{1}{3}$, (g) $\frac{1}{2}$. (3) (a) 5.2, (b) 0.584, (c) $6.\dot{6}$, (d) -1.35, (e) 66, (f) $1\frac{1}{3}$, (g) 2.52×10^{-3}, (h) $40\frac{1}{2}$ pence, (i) £11.44. (4) (a) $\overline{1}.5770$, (b) $\overline{1}.4642$, (c) 1.8141. (5) (a) (i) 0.01, (ii) 0.0060, (b) (i) 30.48, (ii) 30, (c) (i) 0.53, (ii) 0.53, (d) (i) 4.83, (ii) 4.8. (6) £24.50 : £61.25 : £98.00. (7) 7.5%. (8) 1272. (9) 73; 5 mm. (10) (a) £40, (b) £60, (c) £38. (11) (a) 10000, (b) 11011, (c) 1100, (d) 10.

B

(1) (a) 104, (b) 19.2, (c) $1\frac{41}{64}$, (d) $2\frac{25}{42}$, (e) 8, (f) 1.715 kg.
(2) (a) -2.84×10^{-3}, (b) 3, (c) $2\frac{1}{4}$, (d) 6.25, (e) 0.64, (f) 24, (g) 0.8, (h) 0.0001, (i) 77.5, (j) 1 000 010. (3) 1.8 km/h. (4) £250.
(5) 10. (6) 43 : 28. (7) 80. (8) 2.87×10^{-3}. (9) (a) \$9.69, (b) £7.50. (10) £6.75. (11) 22.5 km. (12) (a) 101 011 010, (b) 532, (c) 2341.

EXERCISE 7.1

(1) (a) $100x$, (b) $y/1000$, (c) $1000k$, (d) $1000(d+3)$.
(2) (a) $x+z$ years, (b) $2x+3$ years, (c) $x-2$ years, (d) $x-y$ years.
(3) (a) $2x$, (b) $x-4$, (c) $x/10$, (d) $23x/25$. (4) $X-Y+Z$.
(5) £$[y-(wx/100)]$, $100z/x$. (6) (a) $(ax+ty)$p, (b) $23a/20$ or $1.15a$.
(7) s/t hours; $(xt+yp)$ km.

EXERCISE 7.2

(1) a^4. (2) $8t^5$. (3) y. (4) b^2. (5) x. (6) s. (7) a^3b^3.
(8) $15z^3y^4x$. (9) $3y^3$. (10) $5a^3b$. (11) m^6. (12) $16x^6$.
(13) $y^6/27$. (14) $x/2w$. (15) $z^2/4xy$. (16) $12a^3b^2$ mm^2.
(17) $2x^4$ mm^3.

EXERCISE 7.3

(1) 1. (2) -4. (3) 0. (4) 12. (5) -36. (6) 16. (7) 4.
(8) 6. (9) 1/2. (10) 0. (11) 11. (12) 0. (13) -104.
(14) -324. (15) $3\frac{3}{4}$. (16) $1\frac{1}{2}$. (17) 100. (18) 76. (19) -4.
(20) (a) 56, (b) 0.56.

EXERCISE 7.4

(1) $3x^2$. (2) $5x+y$. (3) $3b^2-3a$. (4) t^2+2t-5.
(5) $-3x^2z - 2zx^2$. (6) $3m+2mn+2n$. (7) $9x-2y$.
(8) $4x^2+3y^2-10x+4$. (9) $2m^2+4n+8$. (10) $-x^2y-11x^2+12$.
(11) $5s^2+8t$. (12) $23x-8y$. (13) $5a^3+8$. (14) $18u$.
(15) $3x^2-2x^3$. (16) $p^2-3qp+2q^2$. (17) $2x^2+9x-18$.
(18) $8y^2-34y+21$. (19) $3x^2y^2+10x^2y-8x^2$. (20) $4a^2-b^2$.
(21) $9x^2+12x+4$. (22) (a) $22x-4$ m, (b) $8x+4$ m.

EXERCISE 7.5

(1) x/y^2. (2) $12a/x^2$. (3) $a^2x^2y^2$. (4) $4y^2/x$. (5) $4ab/3z$.
(6) $2x^2/[3(a-b)]$. (7) $11x/12$. (8) $x/6$. (9) $(x^2+1)/x$.
(10) $7/2x$. (11) $4/5b$. (12) $13/6y$. (13) $(17-7x)/12$.
(14) $(a-3)/10$. (15) 1. (16) $(6x-6)/[(x+1)(2x-1)]$.
(17) $(1-v)/2v$. (18) $(x+5z+x^2-z^2)/(x+z)^2$

EXERCISE 7.6

(1) $x^2(2-3x)$. (2) $pq(3q-pq+2x)$. (3) $(x-y)(4+x)$.
(4) $3(2t+1)(t-2x)$. (5) $(3z^2-4)(2x+3)$. (6) $(2z+3)(z+y)$.
(7) $(3a+2b)(4a^2-b)$. (8) $(3x+2y)(x-2)$. (9) $2(n+m)(m-p)$.
(10) $5(a-3b)(a+c)$.

EXERCISE 7.7

(1) $(x+6)(x+3)$. (2) $(x+7)(x-2)$. (3) $(y-5)(y-4)$.
(4) $(x-5)(x+3)$. (5) $(2t-1)(t+5)$. (6) $(3x+5)(x-4)$.
(7) $(5y-1)(y-4)$. (8) $2(3-x)(1+2x)$. (9) $(2x+3y)(2x-3y)$.
(10) $(3b+1)(3b-1)$. (11) $3x(y+3)(y-3)$. (12) $(2x+3)^2$.
(13) 45.994. (14) $t=7$, $(x-7)(x+1)$. (15) (a) 2, (b) -7.

EXERCISE 8.1

(1) -3. (2) $4\frac{1}{2}$. (3) 2. (4) -3. (5) 1. (6) 1/3. (7) 2.6.
(8) -3. (9) $-3\frac{1}{2}$. (10) 4. (11) -15. (12) 0.05. (13) -0.5.
(14) 20. (15) -1. (16) 5/36.

EXERCISE 8.2

(1) $x<5$. (2) $x\geqslant 11$. (3) $x<2$. (4) $x>-1$. (5) $x\leqslant 2\frac{1}{2}$.
(6) $x<6$. (7) $x>-1/4$. (8) $x\geqslant 4\frac{1}{2}$. (9) $x<-2$.
(10) $47\leqslant x\leqslant 60$.

EXERCISE 8.3

(1) 4. (2) £100, £50. (3) 6.5, 16.25, 22.75 cm. (4) £1.50.
(5) 27. (6) 9. (7) 2400 km.

EXERCISE 8.4

(1) 2, −1. (2) −3, 2. (3) 1/2, 2. (4) $1\frac{1}{2}$, $-1\frac{1}{3}$. (5) −0.2, 4. (6) 2. (7) 20 p, 70 p. (8) 18, 33. (9) $71\frac{2}{3}$, $-19\frac{1}{3}$; $-34\frac{2}{3}$. (10) 24, 12.

EXERCISE 8.5

(1) 5 or −5. (2) $\sqrt{(3/8)}$ or $-\sqrt{(3/8)}$. (3) 0 or 9. (4) −7 or 5. (5) $-1\frac{1}{2}$ or 1. (6) 1/5 or −7. (7) $1\frac{2}{3}$. (8) $1\frac{1}{3}$ or $1\frac{1}{2}$. (9) 1/4 or −2. (10) −2/3 or $1\frac{1}{2}$. (11) 2 or 1/3. (12) $x \geqslant 1\frac{2}{3}$, $x \leqslant -1\frac{2}{3}$. (13) $4 \leqslant x \leqslant 6$. (14) $-4 \leqslant y \leqslant 1\frac{1}{2}$ (15) $t \leqslant -2\frac{1}{2}$, $t \geqslant 2/3$.

EXERCISE 8.6

(1) 4.24 or −4.24. (2) 7.61 or 0.394. (3) −0.232 or −1.43. (4) 3.44 or −0.436. (5) 1.33 or −0.500. (6) 3.00 or 0.500. (7) 1.18 or −0.847. (8) 3.30 or −0.303. (9) 0.0873 or −19.1. (10) −0.923 or −1.93. (11) 0.550 or −1.36. (12) 13 or −1. (13) 2.5 or 0.4. (14) 11, 13 or −13, −11.

EXERCISE 8.7

(1) 3, −1. (2) 14, −2 or −4, 7. (3) 5, 2 or $-3\frac{10}{13}$, $-3\frac{11}{13}$. (4) 2/5, 2/5 or 1/2, 1/3. (5) −1, 4 or $2\frac{9}{11}$, $-1\frac{8}{11}$. (6) $4\frac{1}{2}$, $3\frac{1}{2}$.

EXERCISE 8.8

(1) $t = (v-u)/a$. (2) $k = (u-m)/t$. (3) $E = (V^2M)/2$. (4) $y = \pm\sqrt{[(x-b)/a]}$. (5) $v = uf/(u-f)$. (6) $x = (yb-a)/(1+y)$. (7) $t = \pm\sqrt{[uz/(x+z)]}$. (8) $R = (E-Ir)/I$. (9) $3\frac{1}{4}$. (10) 2.5×10^{-19}.

EXERCISE 9.1

(1) (a) 10, (b) 5. (2) 15. (3) 26. (4) (a) 15, (b) 1, (c) 14; $A \cap B \{4, 8, 12, 16, 20, 24, 28\}$; $A \cap C \{2\}$. (5) (a) 1, 2, 3, 6, 9, 11, 12, (b) 1, 2, 4, 5, 7, 8, 10, 11, (c) 2, 6, (d) 4, 8, 10, 12, (e) 3, 5, 7, 9, (f) 6, 12, (g) 1, 2, 6, 11, 12, (h) 3, 5, 7, 9. (6) (a) 3, (b) 22, (c) 9, (d) 23, (e) 2, (f) 10, (g) 19. (7) (a), (c). (8) (a) a, (b) m, n, s, a, d, r, (c) d, e, f, r, s, (d) m, n, (e) e, f, (f) a, d, m, n, r, s, (g) a, d, r, s, m, n.

EXERCISE 9.2

(1) (a) 37%, (b) 29%. (2) 16. (3) 6. (4) 7. (5) (a) 7, (b) 4.

EXERCISE 9.3

(1) (a) 12, (b) 32, (c) 50. (2) (a) 4, (b) (i) 90, (ii) 14, (iii) 5 or -5.
(3) (a) 0, (b) 4, (c) 7. (4) 2/3. (5) (a) 18, (b) 8 or -8.
(6) -4; 1 or 2.

EXERCISE 9.4

(1) A (i) yes, (ii) 3, (iii) 2, 3 self inverse; 2 inverse of 4, B (i) no, (ii) no, (iii) no, C (i) yes, (ii) d, (iii) a, d self inverse; c inverse of b, a inverse of c, D (i) yes, (ii) no, (iii) no.
(2) (ii) a. (3) (a) 5, (b) 4. (4) (a) 3, 5, 7, 9, (b) 7, (c) 3 or 9.

EXERCISE 10.1

(1) (a) $\begin{pmatrix} 1 & -3 \\ 5 & 4 \end{pmatrix}$, (b) $\begin{pmatrix} 3 & 3 \\ 1 & 6 \end{pmatrix}$, (c) $\begin{pmatrix} -1 & 0 \\ 9 & 3 \end{pmatrix}$. 2 (a) $\begin{pmatrix} 12 & 3 \\ -6 & 15 \end{pmatrix}$,

(b) $\begin{pmatrix} -4 & 12 \\ 6 & -2 \end{pmatrix}$, (c) $\begin{pmatrix} -2 & 19 \\ 7 & 2 \end{pmatrix}$, (d) $\begin{pmatrix} -8 & 11 \\ 8 & -7 \end{pmatrix}$, (e) $\begin{pmatrix} 2 & \frac{1}{2} \\ -1 & 2\frac{1}{2} \end{pmatrix}$.

(3) $\begin{pmatrix} 0 & 8 \\ 4 & -2 \end{pmatrix}$. (4) (a) $\begin{pmatrix} 1 & -1\frac{1}{2} \\ -\frac{1}{2} & 2 \\ 1\frac{1}{2} & 1 \end{pmatrix}$, (b) $\begin{pmatrix} 2 & -3 \\ -1 & 4 \\ 3 & 2 \end{pmatrix}$.

(5) (a) $x = 1$, $y = -4$, (b) $x = 1$, $y = -3$.

EXERCISE 10.2

(1) (a) 2, (b) 3, (c) no value. (2) (a) $\begin{pmatrix} 4 \\ 11 \end{pmatrix}$, (b) $\begin{pmatrix} -4 & 11 \\ -11 & 29 \end{pmatrix}$,

(c) $\begin{pmatrix} -7 \\ 12 \end{pmatrix}$, (d) $\begin{pmatrix} -2 & 13 \\ 4 & -26 \end{pmatrix}$, (e) -28, (g) $\begin{pmatrix} 1 & 2 \\ -6 & 13 \end{pmatrix}$.

(3) (a) -16, (b) $\begin{pmatrix} 6 & 1 & 5 \\ 5 & -2 & 2 \\ 3 & 13 & -3 \end{pmatrix}$. (4) (a) $\begin{pmatrix} x - 2y \\ 3x + 4y \end{pmatrix}$, (b) $\begin{pmatrix} -6 - 4x \\ 9 - 4y \end{pmatrix}$.

(5) (a) $\begin{pmatrix} 2 \\ -3 \end{pmatrix}$, (b) $\begin{pmatrix} -2 \\ 7 \end{pmatrix}$, (c) $\begin{pmatrix} -1 \\ 3 \end{pmatrix}$.

EXERCISE 10.3

(1) (a) 5, (b) 10, (c) 1, (d) $10ab$. (2) (a) 8/3, (b) 1/2, (c) $3b$.
(3) (a) $\frac{1}{25}\begin{pmatrix} 7 & -3 \\ -1 & 4 \end{pmatrix}$, (b) $-\frac{1}{5}\begin{pmatrix} -4 & 1 \\ -3 & 2 \end{pmatrix}$, (c) $\frac{1}{17}\begin{pmatrix} 5 & 6 \\ -2 & 1 \end{pmatrix}$,
(d) $\frac{1}{14}\begin{pmatrix} -14 & -12 \\ 0 & -1 \end{pmatrix}$. (4) $-\frac{1}{2}\begin{pmatrix} 4 & 3 \\ -6 & -5 \end{pmatrix}$, $-\frac{1}{2}\begin{pmatrix} -1 & 19 \\ 3 & -29 \end{pmatrix}$.
(5) $\begin{pmatrix} 1 & -2 \\ -1 & 3 \end{pmatrix}$. (6) 5. (7) $2x + 4y = 10$, $x - 3y = -5$.
(8) (a) -3, 2, (b) 1/2, 4.

EXERCISE 10.4

(1) (a) £18.87, £20.08, £20.29, (b) number of cups of each sold in 3 days.
(2) (a) 122, (b) £4075, £4340, £3330. (3) (a) NM, (b) total revenue from the four shops = £12.98.

PROGRESS TEST 2

A

(1) $-6x^3y^7$, (b) $x^2y + 3xy^2$, (c) $10x + 11y$, (d) $5x + 3y$,
(e) $-2m^2 + 8mn - 2n^2$. (2) (a) $10x/3y$, (b) $2t^2/3$, (c) $(5x^2 + 13y)/12$,
(d) $(18x - 15)/20y$. (3) (a) -2, (b) 9, (c) $-1/6$, (d) 1. (4) (a) 5,
(b) $2\frac{1}{6}$, (c) 1, (d) 1/2 or $-1/2$, (e) 3 or -4, (f) 0.908 or -2.57.
(5) (a) 2, -1, (b) $1\frac{1}{2}$, -2, (c) 4, $-1/2$ or -1, 2. (6) (a) $(x - 2b)(y - 4)$,
(b) $2(3 - b)(3 + b)$, (c) $(x - 4)(x + 2)$, (d) $2(y - 3)^2$.
(7) (a) $x \leqslant 3\frac{2}{3}$, (b) $x \leqslant 48$, (c) $-2/3 \leqslant x \leqslant 5$. (8) $x = v/(y - u)$.
(9) (a) $\{2, 4, 6\}$, (b) $\{1, 3, 5, 7, 8\}$, (c) $\{1, 2, 3, 4, 5, 6, 7\}$, (10) 69.
(11) (a) 1, (b) -11, (c) -3. (12) (a) $\begin{pmatrix} 2 & 3 \\ 1 & 2 \end{pmatrix}$, (b) $\begin{pmatrix} -14 & 11 \\ 8 & -7 \end{pmatrix}$,
(c) $\begin{pmatrix} 17 \\ -11 \end{pmatrix}$, (d) $\begin{pmatrix} -5 & -3 \\ 1 & 5 \end{pmatrix}$, (e) $\begin{pmatrix} 11/10 \\ 4/10 \end{pmatrix}$.

B

(1) (a) $-x-6y$, (b) $(2x+8)/[(x-1)(x+1)]$, (c) x^2-3x
(2) (a) 2, (b) $-2/3$ or 6, (c) 19, (d) -1.2. (3) (a) $(z^2-18b^2x)/9b^2$, (b) $(b+2c^2)/(2c-1)$. (4) (a) $2(x+2y)(x-2y)$, (b) $3(x-1)(3x-2)$, (c) $(2a-b)(2y-3x)$, (d) $x(3x-4)$, (e) $(3a+4b)^2$. (5) (a) $0.6\dot{3}$ or $-2.6\dot{3}$, (b) $2\frac{1}{14}$ or 3/7, (c) 1, -2 or -2, 1. (6) 4. (7) 25.44%.
(8) $\begin{pmatrix} 2 & 3 \\ -1 & 2 \end{pmatrix}$. (9) 18, 0. (10) (a) $\sqrt{2}$, (b) -5, (c) -3. (11) $\{2, 3\}$.
(12) (a) $x \leqslant 7\frac{1}{2}$, (b) $x \leqslant 2, x \geqslant 3$.

EXERCISE 11.1

(1) 134°, (2) 72°. (3) 302°. (4) $x = 35°, y = 145°$. (5) 162°. (6) $x = 135°$, $y = 135°, z = 45°$. (7) $x = 60°, y = 120°$, $z = 60°$ (8) 120°. (9) $x = 88°$, $y = 92°$, $z = 88°$. (10) $x = 125°$, $y = 55°$. (11) $x = 145°$, $y = 40°$, $z = 140°$. (12) $a = 120°$, $b = 120°$, $c = 60°$, $d = 60°$, $x = 60°$, $y = 120°$, $z = 120°$. (13) 70°. (14) $x = 80°$, $y = 30°, z = 80°$. (15) $x = 24°$. (16) $x = 27\frac{1}{2}°$. (17) (a) 36°, (b) 216°.

EXERCISE 11.2

(1) $x = 80$, $y = 70$, $z = 100$. (2) 80. (3) $x = 80$, $y = 50$. (4) $x = 140$, $y = 50$. (5) $x = 14.5$, $y = 110$. (6) $x = \sqrt{3}a$, $y = 60$, $z = a$. (7) $x = 6$, $y = 3$. (8) 10. (9) (a) 120°, (b) 30°. (10) $\sqrt{14}$ m. (11) 5.48 cm, 7.75 cm.

EXERCISE 11.3

(1) (i) $a = 3.75$ cm, $b = 3.45$ cm, (ii) $a = 15$ cm, $b = 2\frac{2}{3}$ cm, (iii) $a = 3\frac{1}{5}$ cm, $b = 5\frac{1}{2}$ cm. (2) $s = 35$ mm, $t = 30$ mm. (3) (a) $6\frac{2}{3}$ cm, (b) 5.4 cm, (c) 8.1 cm. (4) yes, S-A-S, (b) no, (c) no, (d) yes, A-A-S, (e) no, (f) yes, A-A-S. (5) 80 mm, (a) $160/\sqrt{3}$ mm, (b) $80/\sqrt{3}$ mm. (6) 3 cm, 3.75 cm.

EXERCISE 11.4

(1) 67°. (2) (a) 3.9 cm, (b) 43.0 mm. (3) (a) 63.6 mm, (b) 7.07 m.
(4) (a) 60, parallelogram, (b) 60, kite.
(5) (a) 110°, (b) 110°, (c) 30°. (6) (b), (d).

EXERCISE 12.1

(1) 56.5 mm. (2) 49 mm, 49°. (3) 53 mm, 75.5 mm. (4) 90°.

EXERCISE 12.2

(1) 101 mm. (2) 130 mm. (3) 70 mm.

EXERCISE 12.3

(1) 54°. (2) 39.5 mm, 32.5 mm. (3) 50 mm. (4) 51.5 mm, 66 mm. (5) 42°, 42°. (6) 65 mm. (7) 18 mm.

EXERCISE 12.4

(1) 112 mm, 49.5 mm. (2) 48 mm. (3) 42.5 mm, 15.5°. (4) 61.5 mm, 61.5 mm, 26.5°, 47.5 mm.

EXERCISE 13.1

(1) 8.7 cm, 5.8 cm. (2) 249.4 cm^2. (3) 10 cm. (4) 36 mm, 44%. (5) 546. (6) (a) 1176 mm^2, (b) 2850 cm^2, (c) 113 cm^2, (d) 1130 mm^2, (e) 576 cm^2, (7) 60 cm^2. (8) 0.64 ha. (9) 40.8 cm, 133 cm^2, 14.9 cm. (10) 30.1 cm^2, (a) 120°, (b) 17.3 cm. (11) 4630.

EXERCISE 13.2

(1) 57.0 m^2. (2) 198 cm^2. (3) 94.5 cm^2. (4) 4640 mm^2. (5) 2065 mm^2. (6) (a) 208 cm^2, (b) 312 cm^2. (7) 1 : 4, (a) 100 mm, (b) 45 mm, (c) 30 mm. (8) (a) 3 : 26, (b) 10 : 3. (9) 0.675 ha. (10) (a) 2.38 m, (b) 4.046 m^2.

EXERCISE 13.3

(1) 4 m^3. (2) 50 mm. (3) (a) 8.18 m^3, 19.6 m^2, (b) 42.8 cm^3, 76.5 cm^2, (c) 4.0×10^4 mm^3, 6.95×10^3 mm^2. (4) 855 litre, 1.11 m. (5) 302 m. (6) (a) 425 cm^3, (b) 373 cm^2. (7) (a) 810 cm^3, (b) 2490 cm^3.

EXERCISE 13.4

(1) (a) 9 : 25, (b) 27 : 125. (2) 4 : 9, 3.70 cm^3. (3) (a) 560 mm, (b) 64 : 1. (4) 26.4 litres. (5) 54 mm^2. (6) (a) 6.74 cm, (b) 763 cm^3. (7) 5.23 m^3.

EXERCISE 14.1

(1) (i) 1, (ii) 2, (iii) 2. (2) (a) $4x$, (b) $x/4$, (c) $2x/3$. (3) $\{0, 1, 8, 27, 64\}$. (4) (a) 3, (b) 1, (c) 7. (5) (a) 5, (b) 11, (c) 3. (6) (a) 1, 0, 15, (b) $1\frac{2}{7}$, $1\frac{1}{2}$, -1. (7) (a) 2, (b) 3, (c) 9. (8) $a = 3$, $b = -1/4$, $27\frac{3}{4}$. (9) -14, -4, 24; $3\frac{1}{7}$.

EXERCISE 14.2

(1) (a) $6x - 2$, (b) $(12/x) - 1$, (c) $4/(3x - 1)$, (d) $8/x$. (2) (a) $fg(x) = ax + 1$, (b) $gf(x) = a(x + 2) - 1$. (3) (a) $x/2$, (b) $x - 1$, (c) $5x/3$, (d) $(2x + 4)/3$, (e) $\sqrt[3]{[(x - 4)/2]}$. (4) (a) $f^{-1} : x \to (2/x) - (3/2)$, yes, $x < 4/9$, (b) $f^{-1} : x \to (2/x) - 2$; yes; $x < 2/5$, (c) $f^{-1} : x \to x^2$; yes; $x > \sqrt{3}$, (d) $f^{-1} : x \to (\sqrt{x} + 1)/3$; yes; $x > 64$, (e) $f^{-1} : x \to 1/\sqrt{x}$; yes; $x < 1/9$. (5) (a) $x \to (x + 1)/2$, (b) $x \to (x + 1)/4$, (c) $x \to 3(x + 1)/2$, (d) $x \to x$, (e) $x \to x + 1$. (6) $f^{-1} : x \to (4x - 1)/3$.

EXERCISE 14.3

(1) (a) $(x - 4)$, (b) $(x + 3)$. (2) $(2x - 3)$, 1, -1, $1\frac{1}{2}$. (3) 3, $(2x - 1)$.

EXERCISE 14.4

(1) (a) 150, (b) 2 or -2. (2) 12.5. (3) (a) 8, (b) 8/9. (4) 64. (5) 159. (6) $7\frac{2}{3}$. (7) (a) $31\frac{1}{4}$, (b) 0.244. (8) 4 s.

PROGRESS TEST 3

A

(1) (a) 32 cm, (b) 36.9°. (2) (a) 10.4 cm, (b) 10.6 cm. (3) 16. (4) (a) 3850 mm^2, (b) 132 cm^2, (c) 176 cm^2. (5) 449 mm^2. (6) (a) $179\frac{2}{3}$ cm^3, (b) 4224 cm^3, (c) 138.6 cm^3. (7) 14 mm. (8) (a) 63 mm, (b) 1403 mm^2. (9) 3. (10) 8000 : 1. (11) $(2x - 1)$. (12) (a) (i) 7, (ii) 1, (iii) 6, (b) 1.

B

(1) 36°, 144°. (2) 78.1 mm. (3) (a) 5 cm, (b) 90 cm^2. (4) 119 cm^2. (5) (b) and (c). (6) 9.08 cm^2. (7) 6.6 m^3. (8) (a) 97 cm^2, (b) 51.3 cm^3. (9) 4.24 or -4.24. (10) 36 000 m^2. (11) (a) 32 mm, (b) 60 mm, (c) $6.25P$ mm^2. (12) 15 cm.

EXERCISE 15.1

(1) (a) 7, (b) 2.5, (c) 3, (d) 2. (2) (a) (i) 2, (ii) 5, (iii) $(-2\frac{1}{2}, 0)$, (b) (i) -4, (ii) 2, (iii) $(\frac{1}{2}, 0)$, (c) (i) 2, (ii) $-2\frac{1}{2}$, (iii) $(1\frac{1}{4}, 0)$, (d) (i) $-1\frac{1}{2}$, (ii) 2, (iii) $(1\frac{1}{3}, 0)$, (e) (i) 5/7, (ii) $-1\frac{1}{7}$, (iii) $(1\frac{3}{5}, 0)$, (f) (i) 2, (ii) $-1/2$, (iii) $(\frac{1}{4}, 0)$. (3) (a), (c), (f). (4) (a) yes, (b) yes, (c) no, (d) yes. (5) (a) (i) $-1/3$, (ii) $3y + x - 11 = 0$, (b) (i) -2, (ii) $y + 2x = 0$, (c) (i) 1, (ii) $y = x - 5$, (d) (i) $-2\frac{2}{3}$, (ii) $8x + 3y = 28$. (6) $3x + 4y = 5$; $3x + 4y = 0$.

EXERCISE 15.2

(1) (a) $-2, 3$, (b) $1.2, -0.7$. (2) (3, 3), (4, 3). (3) 2. (4) (a) $x + y \geqslant 2$, $x \leqslant 2$, $y \leqslant 2$, (b) $3y - 2x \leqslant 6$, $x \leqslant 0$, $y \geqslant 0$, (c) $y \leqslant 6 - 3x$, $y \leqslant x + 2$, $y \geqslant 0$, (d) $s \leqslant t$, $t \leqslant 3$, $s \geqslant 0$, (e) $-2 \leqslant n \leqslant 3$, $-3 \leqslant m \leqslant 4$. (5) $\{(2, 3)\}$. (6) (1/2, 1/2), (1, 0), (1, 1). (7) $\{(3, 1)\}$. (8) (1, 3).

EXERCISE 15.3

(1) (a) 1 or -3, (b) 3.55 or -0.55, (c) 0.25 or -0.5, (d) -3.55 or 0.55. (2) (a) -8.1, 13, (b) -4, 6. (3) (a) 2 or -0.5, (b) $-0.5 < x < 2$, (c) 0.5 or 1, (d) 1 (equal roots). (4) (a) 8.1, (b) -1, 1.7, (c) $-0.8 < x < 0.8$. (5) $\{(-1, 1), (4, 16)\}$.

EXERCISE 15.4

(1) (i) (a), (ii) (c), (iii) (g). (2) (a) 89.2, (b) 3.7. (3) (a) -1.73, 1.73, (b) 0.1, (c) -6, -2.25. (4) (a) $x^3 + x^2 - 3 = 0$; 1.175, (b) 1.62, (c) 4, $-1\frac{3}{4}$.

EXERCISE 16.1

(1) (a) 4, (b) -4, (c) 7. (2) (a) (3, 3), (1, 1), (b) (1, 4). (3) (3, 3), (4, 3), (5, 2), (5, 3), (6, 2); 3 Alphas, 3 Betas. (4) 14. (5) $X = 2$, $Y = 4$ or 5; $X = 3$, $Y = 3$ or 4.

EXERCISE 16.2

(1) (a) 6.2, (b) 48.5. (2) 380. (3) $s = 0.25r^2$. (4) (a) $s = 30t - 5t^2$, (b) 6 seconds. (5) (a) 57 cm^2, (b) $8x^2 + 5x - 87 = 0$. (6) 98 cm^2; -1.6. (7) (a) 30.6 cm^2, (b) $x = 3$ or 0.73.

EXERCISE 16.3

(1) (a) 37.5 km/h, (b) 48 minutes. (2) (a) 4 hours, (b) 75 km/h. (3) (a) 200 cm, (b) 850 cm, (c) 1.5 cm/s^2, (d) – 3 cm/s^2. (4) (a) 8 to 10 s, (b) 1 cm/s^2, (c) 115 cm. (5) (a) 3 cm/s^2, (b) 862.5 cm, (c) 14.38 cm/s. (6) (a) 1062.5 m, (b) 19.32 m/s, (c) 2.5 m/s^2, (d) $1\frac{2}{3}$ m/s^2.

EXERCISE 17.1

(1) 3. (2) 4. (3) $2x - 3$. (4) $12x^2 + 2x$. (5) 0. (6) $2x + 4$. (7) $-6/x^2$. (8) $160x^4$. (9) $9x^2 - 8x$. (10) $2x - (2/x^3)$. (11) $-(12/x^5) + 12x$. (12) $9x^2 + (8/x^3)$. (13) $3x^2 + 8x - 3$. (14) $nx^{n-1} + m$. (15) $3(n - 2)x^{n-3}$.

EXERCISE 17.2

(1) (a) 4, (b) 8, (c) 19, (d) 12, (e) –7, (f) –3, (g) –1. (2) (a) (3, 15), (b) (–1/2, –6), (c) (–1/4, $-6\frac{1}{8}$). (3) (2, 4); $y = 9x - 14$. (4) 4. (5) (a) $y + 3x + 10 = 0$, (b) $y - 5x + 10 = 0$.

EXERCISE 17.3

(1) (a) (0, –1), min., (b) ($-1\frac{1}{3}$, $10\frac{1}{3}$), max., (c) (0, –1), min., (–1, 0), max., (d) (0, 7), max., ($1\frac{1}{3}$, $5\frac{22}{27}$), min., (e) (1/3, $12\frac{14}{27}$), max., (–3, –6), min. (2) (3, –54), (–3, 54); (2, –46), (–2, 46). (3) $\sqrt{3}$ or $-\sqrt{3}$. (4) $R = 2t^2 - 4t$; –2. (5) (a) $(15 - x)$ m, (b) $x(15 - x)$ m^2, (c) $(15 - 2x)$ m^2/m, (d) 56.25 m^2. (6) 22 or 7π cm^3/cm. (7) (a) 2π cm^3/cm, (b) $9\pi/2$ cm^3/cm. (8) $2\frac{1}{2}$ s, 5 cm^3/s.

EXERCISE 17.4

(1) 10 m/s, 12 m/s^2. (2) (a) (i) 34 m/s, (ii) 8 m/s^2, (b) (i) 20 m/s, (ii) 28 m/s^2. (3) (a) 16 cm/s, (b) 93 cm/s^2, (c) 71 cm/s^2. (4) (a) 57 cm/s, 0 cm/s^2, (b) 7.36 s, (c) 3 s.

EXERCISE 18.1

(1) (a) $(x^5/5) + C$, (b) $(x^6/2) + C$, (c) $(5x^2/2) - (4x^7/7) + C$, (d) $-(3/x) + C$, (e) $-(2/x^2) + (2x^3/3) + C$, (f) $(3x^4/4) - (2x^3/3) + (x^2/2) + C$. (2) $x^3 - (5x^2/2) + 4x - 2$. (3) $(2t^3/3) - (3t^2/2) + 4t$. (4) (a) 192, (b) $57\frac{1}{3}$, (c) $6\frac{2}{3}$, (d) $84\frac{3}{4}$. (5) $y = 3x^3 - x + 4$; 1/3, –1/3. (6) $y = 4x^2 - 4x + 1$.

EXERCISE 18.2

(1) 60 unit^2. (2) $1\frac{1}{3}$ unit^2. (3) (a) 36 unit^2, (b) $4\frac{1}{2}$ unit^2, (c) $55\frac{11}{24}$ unit^2, (d) 1/2 unit^2. (4) (a) $52\frac{2}{3}\pi$ unit^3, (b) $8\frac{2}{3}\pi$ unit^3, (c) $48\frac{2}{5}\pi$ unit^3. (5) 259.2π unit^3. (6) 2/3 unit^2; $26\pi/81$ unit^3. (7) 1/6 unit^2; $7\pi/10$ unit^3.

EXERCISE 18.3

(1) (a) $s = 4t + 3t^2$; $a = 6$, (b) $v = t^2 - t^3 + 6$; $s = (t^3/3) - (t^4/4) + 6t$, (c) $v = 14t - 15t^2$; $a = 14 - 30t$. (2) (a) $s = 9t^2 + 12t - 4$, (b) 18 m/s^2. (3) (a) 24 m/s, (b) 75 m. (4) (a) 36 m, (b) $13\frac{1}{8}$ m. (5) (a) $21\frac{1}{3}$ mm, (b) 152.5 mm, (c) 49.3 mm/s, (d) 12 mm/s^2.

EXERCISE 19.1

(1) (a) £1 125 000, £315 000, (b) 94 mm, 26 mm. (2) (a) 162°, 144°, 54°, (b) 90°, 72°, 108°, 90°. (3) 1.49 kg; 1.21 kg. (4) (c) 72.2%.

EXERCISE 19.2

(1) (a) $3\frac{1}{2}$, 16, (b) 19, 7. (2) 6. (3) 11. (4) (a) 14.1, 13.5, 16.5, (b) 12.8, 10, 15. (5) $1571\frac{1}{2}$ h; 1435 h; 28%. (6) 42.2%. (7) 2.15 g, 2.25 g, 2.26 g, 2.22 g. (8) 7.4 s, 7.1 s, 6.25 s. (9) (a) 2.536 g, (b) 0.036 g. (10) (a) £78, (b) £26, (c) 53%. (11) (a) 44%, (b) 785, (c) 47%, (d) 20%.

EXERCISE 20.1

(1) 1/2. (2) (a) 1/4, (b) 1/13, (c) 1/52, (d) 1/26. (3) (a) 4.3, (b) (i) 1/2, (ii) 7/15. (4) 1/8; 1/64. (5) 1/10. (6) x/14. (7) 16. (8) 1/625. (9) (a) 1/6, (b) 23/36. (10) (a) 1/36, (b) 1/36, (c) 1/18. (11) (a) 8/343, (b) 600/2401. (12) (a) 4%, (b) 20%, (c) 40.32%.

EXERCISE 20.2

(1) (a) 1/13, (b) 1/13, (c) 1/221. (2) (a) 1/27, (b) 2/91. (3) (a) 3/7, (b) 3/8, (c) 3/7, (d) 1/56, (e) $5x = 2y$. (4) (a) 1/2, (b) 1/6, (c) 1/3, (d) 1/3. (5) (a) 15/64, (b) 29/64, (c) 35/64. (6) 13.

PROGRESS TEST 4

A

(1) (a) (i) $y = 8 - 4x$, (ii) $2y + 8x + 17 = 0$, (b) $y + 2x + 3 = 0$, (c) (i) $y + 3x - 12 = 0$, (ii) $y + 3x + 1 = 0$, (d) $y + 3x = 0$. (2) (a) 22°, (b) 116.5°. (3) (1, −1/2). (4) $y = (1/6)(x^2 + x - 12)$. (5) (a) $y = 0$, (b) $y = -3$, (c) $y + x + 16 = 0$. (6) (a) $x^3 - 2x - 1 = 0$, (b) $2x^5 - x^3 - 4x^2 + 2 = 0$. (7) 18 cm², 8 cm². (8) 5; −1. (9) (a) 72 km/h, (b) 90 km/h. (10) $(2y + 5)/4$ h. (11) 10; (a) 12.5 cm/s, (b) $37\frac{1}{3}$ cm. (12) 17. (13) 28; 8. (14) 12.5; 13. (15) (a) 13, (b) 4, (c) 14, (d) 0.55. (16) (a) (i) 1/4, (ii) 1/6, (b) (i) 1/5, (ii) 1/5.

B

(1) (a) $2x + 4$, (b) $4x^3 - (12/x^3)$, (c) $(3/2)\sqrt{(6x)}$. (2) (a) 1, (b) 5. (3) $y = -4$. (4) (a) −9 m/s; −12 m/s², (b) 4 m. (5) (a) $4\sqrt{x} + C$, (b) $(3/4)\sqrt[3]{x^4} + C$, (c) $3x^3 - 6x^2 + 4x + C$, (d) 16. (6) (a) $10\frac{2}{3}$ unit², (b) $34\frac{2}{15}\pi$ unit³. (7) $2\pi r$. (8) (a) 67 cm/s², (b) $194\frac{2}{3}$ cm.

EXERCISE 21.1

(1) (a) 3/5, (b) 3/4, (c) 3/5, (d) 4/3. (2) (a) (i) 24/25, (ii) 7/24, (b) (i) 8/17, (ii) 15/17. (3) (a) 1/2, (b) $1/\sqrt{3}$, (c) $1/\sqrt{2}$, (d) 1, (e) 1/2. (4) (a) 0.8192, (b) 0.8878, (c) 0.8536, (d) 7.9158, (e) 0.5340. (5) (a) 26.4°, (b) 47.2°, (c) 65.7°, (d) 54.3°, (e) 61.3°. (6) 39.8. (7) (i) 12.9 mm, 15.3 mm, (ii) 37.9°, 5.53 cm, 4.30 cm, (iii) 15.3 units. (8) 79.3 mm. (9) 23.9°, 66.1°. (10) (a) 76.0°, (b) 56.3°, (c) 53.1°.

EXERCISE 21.2

(1) 35.6 m. (2) 125 m. (3) 129.8°. (4) (a) 291.8°, (b) 20 km. (5) (a) 29.1 km, (b) 344.4°. (6) (a) 15.1 km, (b) 9.54 km, (c) 29.0 km.

EXERCISE 21.3

(1) (a) 0.8480, (b) 0.7254, (c) −0.2588, (d) −0.9848, (e) −5.2422, (f) −0.4040. (2) (i) 116 mm, (ii) 10.3 cm, 60.4 cm², (iii) 33.2°, 44.8°. (3) (a) 0.5672, (b) −0.6886, (c) 0.2970. (4) (a) 110.7°, (b) 89.0°, (c) 114.6°. (5) 60.6°, 12.6 cm. (6) 6.87 km. (7) 5864 mm²; 113.3 mm, 146.2 mm. (8) 31.2 cm. (9) 22.4 km/h. (10) (a) 12 m, (b) 15.6°. (11) (a) 1944 cm², (b) 36.9°, (c) 49.4°.

EXERCISE 21.4

(1) (a) 6120 km, (b) 3890 km. (2) (a) 25 700 km, (b) 32 800 km, (c) 35 400 km. (3) 75.1° S. (4) 22.1° S, 12.1° W. (5) (a) 32° S, 163° E, (b) 28° N, 135° W. (6) (a) 1083 km/h, (b) 1373 km/h.

EXERCISE 22.1

(1) (a) $-\boldsymbol{c}$, (b) $-\boldsymbol{b}$, (c) $\boldsymbol{c}$, (d) $\mathbf{0}$. (2) $\boldsymbol{a}-\boldsymbol{b}$, $\boldsymbol{a}=k\boldsymbol{b}$. (3) (a) $2\boldsymbol{a}+\boldsymbol{b}$, (b) $3\boldsymbol{a}-\boldsymbol{b}$, (c) $\boldsymbol{a}-\boldsymbol{b}$; $\frac{1}{8}\boldsymbol{a}+\boldsymbol{b}$. (4) (a) $\boldsymbol{v}-\boldsymbol{w}+\boldsymbol{x}+\boldsymbol{y}$, (b) $\boldsymbol{w}-\boldsymbol{x}-\boldsymbol{v}$, (c) $\boldsymbol{x}+\boldsymbol{y}$, (d) $\boldsymbol{w}-\boldsymbol{v}$. (5) (a) $\boldsymbol{b}-\boldsymbol{a}$, (b) $\frac{1}{2}(\boldsymbol{b}-\boldsymbol{a})$; PQ ∥ AB, PQ = $\frac{1}{2}$AB. (6) (a) $\boldsymbol{q}-\boldsymbol{p}$, (b) $\frac{1}{4}\boldsymbol{q}$, (c) $\boldsymbol{q}-\frac{3}{4}\boldsymbol{p}$. (7) $\overrightarrow{QP}$. (8) $a=1/4$, $b=4/5$.

EXERCISE 22.2

(1) (a) $\begin{pmatrix}10\\-12\end{pmatrix}$, (b) $\begin{pmatrix}8\\4\end{pmatrix}$, (c) $\begin{pmatrix}8\\-8\end{pmatrix}$. (2) (a) (10, −20), (b) (0, 5), (c) (10, −22). (3) (a) 13, (b) 5, (c) $\sqrt{32}$ or $4\sqrt{2}$. (4) (a) (−2, 4), (b) (5, −2). (5) (a) $\sqrt{109}$, (b) $2\sqrt{13}$, (c) $\sqrt{13}$. (6) (a) 53.1°, (b) 28.3°. (7) (c) and (e). (8) (−1, 3), (−2, 0), (0, 6); $\sqrt{10}$ units. (10) (3, 10); $\begin{pmatrix}-4\\-5\end{pmatrix}$.

EXERCISE 23.1

(1) (2, 4). (2) (4, 8). (3) (a) (−3, 5), (b) (−4, 2), (c) (−2, 8), (d) (4, 9). (4) $a=1$, $b=2$. (5) (−2, −1), (0, −3), (−1, 4). (6) (7, 1); (14/11, 9/11). (7) $\begin{pmatrix}1&0\\-3&1\end{pmatrix}$; (−1, 1). (8) (a) (9, −63), (b) (22, −44), (c) (−11, −44).

EXERCISE 23.2

(1) (a) (2.5, −7.5), (b) (−1.7, 5.1). (2) 7. (3) (a) 14 mm, (b) 6 mm. (4) (a) (−8, 8), (8, 8), (8, −4), (−8, −4), (b) (−5, 14), (11, 14), (11, 2), (−5, 2). (5) Enlargement with centre A, scale factor $\frac{1}{2}$; $\begin{pmatrix}\frac{1}{2}&0\\0&\frac{1}{2}\end{pmatrix}$. (6) (7, 0); (−3, 0).

EXERCISE 23.3

(1) (a) (2, −1), (3, −1), (3, −5), (2, −4), (b) (−2, 1), (−3, 1), (−3, 5), (−2, 4), (c) (1, 2), (1, 3), (5, 3), (4, 2), (d) (1, −2), (1, −3), (5, −3), (4, −2). (2) (−3, −4), (−3, −5), (−7, −5), (−6, −4). (3) (a) (−2, 4), (b) (3, 1), (c) (−3, 2), (d) (−2, 1), (e) (12, 4),

(f) $(-7, 2)$, (g) $(-4, 5)$, (h) $(0, 3)$. (4) (a) $x = 3$, (b) $y = 3$, (c) $y = x + 2$.

EXERCISE 23.4

(1) (a) $(4, -3)$, $(-4, 3)$, (b) $(2, 1)$, $(0, 5)$, (c) $(1, -2)$, $(-3, 6)$.
(2) (a) $(-3, -4)$, (b) $(1, 3)$. (3) (a) $(0, 0)$, $(2, 0)$, $(2, -2)$, $(0, -2)$, (b) $(0, 0)$, $(1, 0)$, $(1, -1)$, $(0, -1)$, (c) $(0, 0)$, $(1, 0)$, $(1, 1)$, $(0, 1)$, (d) $(0, 0)$, $(1, 2)$, $(1, 3)$, $(0, 1)$, (e) $(0, 0)$, $(0, 1)$, $(-1, 1)$, $(-1, 0)$.
(4) (a) $(-1, 1)$, $(-3, 3)$, (b) $(-1, 1)$, $(-3, 3)$, (c) $(4, 4)$, $(12, 12)$, (d) $(0, \sqrt{2})$, $(0, 3\sqrt{2})$, (e) $(0, 8)$, $(0, 24)$, (f) $(-1, -1)$, $(-3, -3)$.
(5) (i) (b), (ii) (a), (iii) (d), (iv) (f), (v) (c), (vi) (e).
(6) (a) $\begin{pmatrix} -3 \\ 1 \end{pmatrix}$, (b) $\begin{pmatrix} -1 \\ -13 \end{pmatrix}$, (c) $\begin{pmatrix} -7 \\ -11 \end{pmatrix}$. (7) $(10, 19)$, $(4, 15)$, $(2, 23)$.

EXERCISE 24.1

(1) (a) 37.5 mm, (b) 9 mm, (c) 18 mm. (2) (a) 30 mm, (b) 54 mm, (c) 135 mm, (d) 54 mm. (3) (a) $6\sqrt{2}$ or 8.48 cm, (b) 12 cm.
(4) (a) 6 cm, (b) 79.6°.

EXERCISE 24.2

(1) (a) 140°, (b) 45°, 85°, (c) 124°, (d) 36°, 34°, 60°, (e) 60°, 60°, 40°, (f) $x = 115°$, $y = 25°$, $z = 65°$. (2) 130°. (3) (a) 60°, (b) $8\sqrt{3}$ cm, 12 cm. (4) (a) 55°, (b) 55°. (5) (a) 26.5 cm, (b) 25.2 cm.

EXERCISE 24.3

(1) 30 mm. (2) 11.25 cm. (3) (a) 4 cm, (b) $\sqrt{52}$ cm. (4) $\sqrt{27}$ cm.
(5) 42.5 mm. (6) (a) 30°, (b) 60°. (7) (a) 33.85 mm, (b) 686 mm^2.
(8) (a) 12 cm, (b) 10 cm; 10 cm. (9) 12.5 mm.

EXERCISE 25.1

(1) (a) 45 mm, (b) 43.5 mm. (2) (a) 200 mm, (b) 187 mm.
(3) $65\frac{1}{2}°$, 19 cm^2. (4) (a) 108 mm, (b) 65 mm. (5) 46 mm.
(6) (a) 18 mm, (b) 39.5 mm. (7) 32 mm. (8) (a) 60 mm, (b) 85 mm.
(9) (a) 33 mm, (b) 60 mm. (10) (a) 49 mm, (b) 58 mm.
(11) 52 mm. (12) 59 mm; 63 mm.

PROGRESS TEST 5

A

(1) (a) 9.71 cm, (b) 7.05 cm, (c) 15.8 cm. (2) (a) $\pm(2\sqrt{22})/13$, (b) $\pm 9/(2\sqrt{22})$. (3) (a) 0.5736, (b) −0.7431, (c) −6.314, (d) −0.2679. (4) (a) 367 m, (b) 164 m, (c) 147 m. (5) (a) 13.9 cm, 25.9°, 34.1°; 27.3 cm^2, (b) 43.9°, 60.1°, 12.5 cm; 60.7 cm^2. (6) (a) $\boldsymbol{p} + \boldsymbol{q} - \boldsymbol{r}$, (b) $\boldsymbol{p} + \frac{3}{4}\boldsymbol{q}$, (c) $\frac{1}{3}\boldsymbol{p} + \frac{3}{4}\boldsymbol{q}$, (d) $\frac{1}{3}\boldsymbol{p} + \boldsymbol{q} - \boldsymbol{r}$.. (7) (a) (i) $\sqrt{65}$, (ii) $(1/\sqrt{65})\mathbf{i} + (8/\sqrt{65})\mathbf{j}$, (b) (i) $5\sqrt{2}$, (ii) $(7/\sqrt{50})\mathbf{i} + (1/\sqrt{50})\mathbf{j}$.
(8) $\begin{pmatrix} 1 \\ 2 \end{pmatrix}$, (3, 4). (9) (a) 17°, (b) 28°, (c) 28°.

B

(1) (a) 0.7333, (b) 13.21, (c) 0.3652. (2) 2.54 km, 248°. (3) (a) 136 cm^2, (b) 13.1 cm, 21.5 cm. (4) (a) 5560 km, (b) (i) 4470 km, (ii) 2120 km. (5) (a) (i) $\begin{pmatrix} 5 \\ 5 \end{pmatrix}$, (ii) $5\sqrt{2}$, (b) (i) $\begin{pmatrix} 6 \\ -2 \end{pmatrix}$, (ii) $2\sqrt{10}$, (c) (i) $\begin{pmatrix} 1 \\ 3 \end{pmatrix}$, (ii) $\sqrt{10}$.

(6) Enlargement, centre (0, 0) SF 2 units; $\begin{pmatrix} 2 & 0 \\ 0 & 2 \end{pmatrix}$; $\begin{pmatrix} -1 \\ 5 \end{pmatrix}$.

(7) (a) (3, 0), (b) 90° anticlockwise. (8) (a) 45.3 mm, (b) 61.0°.
(9) 12 cm.

INDEX